# AIR QUALITY MANAGEMENT
## IN THE UNITED STATES

Committee on Air Quality Management in the United States

Board on Environmental Studies and Toxicology

Board on Atmospheric Sciences and Climate

Division on Earth and Life Studies

NATIONAL RESEARCH COUNCIL
*OF THE NATIONAL ACADEMIES*

THE NATIONAL ACADEMIES PRESS
Washington, D.C.
**www.nap.edu**

THE NATIONAL ACADEMIES PRESS    500 Fifth Street, NW    Washington, DC 20001

NOTICE: The project that is the subject of this report was approved by the Governing Board of the National Research Council, whose members are drawn from the councils of the National Academy of Sciences, the National Academy of Engineering, and the Institute of Medicine. The members of the committee responsible for the report were chosen for their special competences and with regard for appropriate balance.

This project was supported by Grant No. X-82822101-0 between the National Academy of Sciences and the U.S. Environmental Protection Agency. Any opinions, findings, conclusions, or recommendations expressed in this publication are those of the author(s) and do not necessarily reflect the view of the organizations or agencies that provided support for this project.

**Library of Congress Cataloging-in-Publication Data**

Air quality management in the United States / Committee on Air Quality Management in the United States, Board on Environmental Studies and Toxicology, Board on Atmospheric Sciences and Climate, Division on Earth and Life Studies.
    p. cm.
  Includes bibliographical references and index.
  ISBN 0-309-08932-8 (hardback)—ISBN 0-309-51142-9 (pdf)
  1. Air quality management—United States. I. National Research Council (U.S.).
Committee on Air Quality Management in the United States.
  TD883.2.A64325 2004
  363.739′25′0973—dc22
                                        2004014594

Additional copies of this report are available from:

The National Academies Press
500 Fifth Street, NW
Box 285
Washington, DC 20055

800-624-6242
202-334-3313 (in the Washington metropolitan area)
http://www.nap.edu

Printed in the United States of America

# THE NATIONAL ACADEMIES
*Advisers to the Nation on Science, Engineering, and Medicine*

The **National Academy of Sciences** is a private, nonprofit, self-perpetuating society of distinguished scholars engaged in scientific and engineering research, dedicated to the furtherance of science and technology and to their use for the general welfare. Upon the authority of the charter granted to it by the Congress in 1863, the Academy has a mandate that requires it to advise the federal government on scientific and technical matters. Dr. Bruce M. Alberts is president of the National Academy of Sciences.

The **National Academy of Engineering** was established in 1964, under the charter of the National Academy of Sciences, as a parallel organization of outstanding engineers. It is autonomous in its administration and in the selection of its members, sharing with the National Academy of Sciences the responsibility for advising the federal government. The National Academy of Engineering also sponsors engineering programs aimed at meeting national needs, encourages education and research, and recognizes the superior achievements of engineers. Dr. Wm. A. Wulf is president of the National Academy of Engineering.

The **Institute of Medicine** was established in 1970 by the National Academy of Sciences to secure the services of eminent members of appropriate professions in the examination of policy matters pertaining to the health of the public. The Institute acts under the responsibility given to the National Academy of Sciences by its congressional charter to be an adviser to the federal government and, upon its own initiative, to identify issues of medical care, research, and education. Dr. Harvey V. Fineberg is president of the Institute of Medicine.

The **National Research Council** was organized by the National Academy of Sciences in 1916 to associate the broad community of science and technology with the Academy's purposes of furthering knowledge and advising the federal government. Functioning in accordance with general policies determined by the Academy, the Council has become the principal operating agency of both the National Academy of Sciences and the National Academy of Engineering in providing services to the government, the public, and the scientific and engineering communities. The Council is administered jointly by both Academies and the Institute of Medicine. Dr. Bruce M. Alberts and Dr. Wm. A. Wulf are chair and vice chair, respectively, of the National Research Council.

**www.national-academies.org**

Pesticides in the Diets of Infants and Children (1993)
Dolphins and the Tuna Industry (1992)
Science and the National Parks (1992)
Human Exposure Assessment for Airborne Pollutants (1991)
Rethinking the Ozone Problem in Urban and Regional Air Pollution
   (1991)
Decline of the Sea Turtles (1990)

*Copies of these reports may be ordered from The National Academies Press*
*(800) 624-6242 or (202) 334-3313*
*www.nap.edu*

# Preface

Recognizing the central role that science and engineering plays in air quality management and anticipating the next congressional reauthorization of the Clean Air Act and its amendments, the United States Congress directed the U.S. Environmental Protection Agency (EPA) to arrange for a study by the National Academy of Sciences (1) to evaluate from a scientific and technical perspective the effectiveness of the major air quality provisions of the Clean Air Act and their implementation by federal, state, tribal, and local government agencies; and (2) to develop scientific and technical recommendations for strengthening the nation's air quality management system with respect to the way it identifies and incorporates important sources of exposure to humans and ecosystems and integrates new understandings of human and ecosystem risks. In response, the National Research Council established the Committee on Air Quality Management in the United States, which prepared this report. Biosketches of the committee members are presented in Appendix A.

In the course of preparing this report, the committee met in public sessions in Washington, D.C.; Denver, Colorado; Los Angeles, California; and Atlanta, Georgia, where local, state, and federal officials and representatives from the private sector and nongovernmental organizations, including regulated industries and advocacy groups, were invited to meet with the committee and present their views on air quality management. Interested members of the public at large were also given an opportunity to speak on these occasions. The committee received oral and written presentations from the following individuals:

Daniel Albritton, National Oceanic and Atmospheric Administration Aeronomy Laboratory; William Becker, Association of Local Air Pollution Control Officials; Robert Brenner, EPA; Cynthia Burbank, Federal Highway Administration; Tim Carmichael, Coalition for Clean Air; Michael Chang and Rodney Weber, Georgia Institute of Technology; Patrick Cummins, Western Governors' Association; Gregory Dana, Alliance of Automobile Manufacturers; Frank Danchetz, Georgia Department of Transportation; Joan Denton, California Office of Environmental Health Hazard Assessment; Howard Feldman, American Petroleum Institute; John Froines, University of California at Los Angeles; Mike Kenny, California Air Resources Board; Eric Fujita, Desert Research Institute; Norma Glover and Barry Wallerstein, South Coast Air Quality Management District; Charles Goodman, Southern Company; Richard Jackson, Centers for Disease Control and Prevention; Kip Lipper, California State Senator Sher's office; Patricia Mariella, Gila River Indian Community; Barry McNutt, U.S. Department of Energy; Christopher Miller, Environment and Public Works Committee, U.S. Senate; Frank O'Donnell, Clean Air Trust; Harold Reheis, Georgia Department of Natural Resources; Catherine Ross, Georgia Regional Transportation Authority; Chet Tisdale, King and Spaulding; Paige Tolbert, Emory University; Cindy Tuck, California Council for Environmental and Economic Balance; Andrew Wheeler, Clean Air, Wetlands, Private Property, and Nuclear Safety Subcommittee, U.S. Senate; and Robert Yuhnke, Robert Yuhnke and Associates.

In addition to the information from those presentations, the committee made use of the peer-reviewed scientific literature, government agency reports, and unpublished databases, as well as related statistics and data directly obtained from EPA.

This report consists of seven chapters. The first chapter provides an overview of the committee's charge, the issues related to this charge, and the approach the committee took in completing its task. Chapters 2–6 review the current air quality management system in the United States and assess how well this system is operating. Chapter 7 looks to the future; it identifies the major air quality challenges the nation is likely to face in the coming decade and advances a set of five interrelated recommendations for enhancing the nation's air quality management system to meet these challenges. The Executive Summary provides a brief overview of the committee's findings and recommendations. The more-detailed Summary is presented immediately after the Executive Summary. Readers who are well versed in the current operation of air quality management in the United States or who do not need to become well versed may wish to move directly from the Executive Summary or Summary to Chapter 7. The recommendations in

Chapter 7 relate to Chapters 2–6, where the detailed background information and justification for the recommendations are provided.

We wish to thank James Mahoney for his valuable service as a member of the committee during the early stages of this study. He resigned appropriately from the committee upon becoming assistant secretary of commerce for oceans and atmosphere and deputy administrator of the National Oceanic and Atmospheric Administration. The committee's work was assisted by staff of the NRC's Board on Environmental Studies and Toxicology (BEST) and its Board on Atmospheric Science and Climate. We wish to thank Raymond Wassel, project director, and James Reisa, director of BEST. Scientific and technical information was provided by Laurie Geller, K. John Holmes, Karl Gustavson, Amanda Staudt, Chad Tolman, Jhumoor Biswas, Ramya Chari, Mirsada Karalic-Loncarevic, and Rachel Hoffman. Craig Hicks assisted with science writing. Invaluable logistical support was provided by Emily Brady and Dominic Brose. The report was ably edited by Ruth Crossgrove.

William L. Chameides, *Chair*
Daniel S. Greenbaum, *Vice Chair*

# Acknowledgment of Review Participants

This report has been reviewed in draft form by individuals chosen for their diverse perspectives and technical expertise, in accordance with procedures approved by NRC's Report Review Committee. The purpose of this independent review is to provide candid and critical comments that will assist the institution in making its published report as sound as possible and to ensure that the report meets institutional standards for objectivity, evidence, and responsiveness to the study charge. The review comments and draft manuscript remain confidential to protect the integrity of the deliberative process. We wish to thank the following individuals for their review of this report:

William Agnew, General Motors (retired); Thomas Burke, Johns Hopkins University; Paul Crutzen, Max Planck Institute for Chemistry; Gregory Dana, Alliance of Automobile Manufacturers; E. Donald Elliott, Willkie, Farr & Gallagher, LLP; David Hawkins, Natural Resources Defense Council; Walter Heck, North Carolina State University; Timothy Larson, University of Washington; Leonard Levin, Electric Power Research Institute; Arthur Marin, Northeast States for Coordinated Air Use Management; Michael Myer, Georgia Institute of Technology; Joseph Norbeck, University of California, Riverside; John Seitz, Sonnenschein, Nath & Rosenthal, LLP; Thomas Tietenberg, Colby College; John Watson, Desert Research Institute; Catherine Witherspoon, California Air Resources Board; Terry Yosie, American Chemistry Council.

Although the reviewers listed above have provided many constructive comments and suggestions, they were not asked to endorse the conclusions or recommendations, nor did they see the final draft of the report before its release. The review of this report was overseen by Robert Frosch, Harvard University, and Edwin Clark II, Clean Sites. Appointed by the NRC, they were responsible for making certain that an independent examination of this report was carried out in accordance with institutional procedures and that all review comments were carefully considered. Responsibility for the final content of this report rests entirely with the authoring committee and the institution.

# Contents

Principles for Enhancing the AQM System, 278
Recommendations for an Enhanced AQM System, 283
Conclusion, 313

# Figures and Tables

**FIGURES**

## TABLES

# AIR QUALITY MANAGEMENT
## IN THE UNITED STATES

# Executive Summary[1]

The Clean Air Act (CAA) provides a legal framework for promoting public health and public welfare[2] by pursuing five major air quality goals (see Box 1). For the first goal, the CAA authorizes the U.S. Environmental Protection Agency (EPA) to set maximum allowable atmospheric concentrations of six major "criteria" pollutants by establishing National Ambient Air Quality Standards (NAAQS). Individual states then develop state implementation plans (SIPs) that show how, with the assistance of national control programs, they will meet these standards. Such efforts, as well as those in pursuit of the other CAA goals, seek to regulate emissions from a variety of stationary and mobile sources through the nation's air quality management (AQM) system (see Figure ES-1). Since passage of the CAA Amendments of 1970, the nation has devoted significant efforts and resources to AQM, and substantial progress has been made.

The Committee on Air Quality Management was formed by the National Research Council to examine the role of science and technology in the implementation of the CAA and to recommend ways in which the scientific and technical foundations for AQM in the United States can be enhanced. Over a 2-year period, the committee heard briefings from experts

---

[1]This Executive Summary provides a brief overview of the committee's findings and recommendations. The detailed Summary is presented after the Executive Summary.

[2]Within the framework of the CAA, "welfare" refers to the viability of agriculture and ecosystems (such as forests and wildlands), the protection of materials (such as monuments and buildings), and the maintenance of visibility.

## BOX 1  Goals of the Clean Air Act

• Mitigate potentially harmful ambient concentrations of six "criteria" pollutants: carbon monoxide (CO), nitrogen dioxide ($NO_2$), sulfur dioxide ($SO_2$), ozone ($O_3$), particulate matter (PM), and lead (Pb).
   • Limit sources of exposure to hazardous air pollutants (HAPs).
   • Protect and improve visibility in wilderness areas and national parks.
   • Reduce emissions of substances that cause acid deposition, specifically sulfur dioxide and nitrogen oxides ($NO_x$).
   • Curb use of chemicals that have the potential to deplete the stratospheric ozone layer.

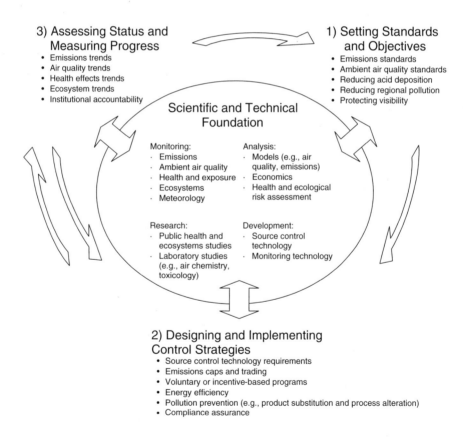

**3) Assessing Status and Measuring Progress**
• Emissions trends
• Air quality trends
• Health effects trends
• Ecosystem trends
• Institutional accountability

**1) Setting Standards and Objectives**
• Emissions standards
• Ambient air quality standards
• Reducing acid deposition
• Reducing regional pollution
• Protecting visibility

**Scientific and Technical Foundation**

Monitoring:
· Emissions
· Ambient air quality
· Health and exposure
· Ecosystems
· Meteorology

Analysis:
· Models (e.g., air quality, emissions)
· Economics
· Health and ecological risk assessment

Research:
· Public health and ecosystems studies
· Laboratory studies (e.g., air chemistry, toxicology)

Development:
· Source control technology
· Monitoring technology

**2) Designing and Implementing Control Strategies**
• Source control technology requirements
• Emissions caps and trading
• Voluntary or incentive-based programs
• Energy efficiency
• Pollution prevention (e.g., product substitution and process alteration)
• Compliance assurance

FIGURE ES-1   Idealized schematic showing the iterative nature of air quality management.  Bullets under each heading provide examples.

and stakeholders and examined the operation, successes, and limitations of the many components of the nation's AQM system.

## PROGRESS

The committee concluded that implementation of the CAA has contributed to substantial decreases in emissions of several pollutants. Regulations for light-duty vehicles, light-duty trucks, and fuel properties have greatly reduced emissions per mile traveled. Programs for stationary sources, such as power plants and large factories, have also achieved substantial reductions of pollutant emissions. However, most of the reductions have been accomplished through regulations on new facilities, while many older, often higher-emitting facilities can be a substantial source of emissions. Emission "cap and trade" has also provided a mechanism for achieving emission reductions at reduced costs. Air quality monitoring networks have confirmed that ambient pollutant concentrations, especially in urban areas, have decreased over the past three decades, and monitoring has documented a reduction in sulfate deposition in the eastern United States. Economic assessments of the overall costs and benefits of AQM in the United States indicate, despite uncertainties, that implementation of the CAA has had and will probably continue to have substantial net economic benefits.

## CHALLENGES AHEAD

Despite the progress, the committee identified scientific and technical limitations in the current AQM system that will hinder future progress, especially as the nation attempts to meet the following key challenges in the coming decade:

- Meeting new standards for ozone, particulate matter, and regional haze.
- Understanding and addressing the human health risks from exposure to air toxics.
- Responding to the evidence that, for some pollutants, there may be no identifiable threshold exposure below which harmful effects cease to occur.
- Mitigating pollution effects that might disproportionately occur in minority and low-income communities in densely populated urban areas.
- Enhancing understanding and protection of ecosystems affected by air pollution.
- Understanding and addressing multistate and international transport of pollutants.
- Adapting the AQM system to a changing (and most likely warmer) climate.

## MEETING THE CHALLENGES:
## THE COMMITTEE RECOMMENDATIONS

To meet these challenges and remedy current limitations, the committee identified a set of overarching long-term objectives that should guide future improvement of the AQM system. In the committee's view, AQM should

• Strive to identify and assess more clearly the most significant exposures, risks, and uncertainties.
• Strive to take an integrated multipollutant approach to controlling emissions of pollutants posing the most significant risks.
• Strive to take an airshed[3]-based approach by assessing and controlling emissions of important pollutants arising from local, multistate, national, and international sources.
• Strive to emphasize results over process, create accountability for the results, and dynamically adjust and correct the system as data on progress are assessed.

Immediate attainment of these objectives is unrealistic. It would require a level of scientific understanding that has yet to be developed, a commitment of new resources that would be difficult to obtain in the short term, and a rapid transformation of the AQM system that is undesirable in light of the system's past successes. The committee proposes, therefore, that the AQM system be enhanced so that it steadily evolves toward meeting these objectives. In that spirit, the committee makes five interrelated recommendations to be implemented through specific actions:

1. Strengthen the scientific and technical capacity of the AQM system to assess risk and track progress. *Recommended actions* include enhancing assessments of air quality and health, ecosystem monitoring, emissions tracking, exposure assessment (both outdoors and indoors), and other components of the scientific and technical foundation of AQM.

2. Expand national and multistate performance-oriented control strategies to support local, state, and tribal efforts. *Recommended actions* include controlling currently unregulated and underregulated sources; expanding use of performance-oriented, market-based (where appropriate) multipollutant control strategies; and enhancing authority to identify and address multistate and international air pollutant transport.

3. Transform the SIP process into a more dynamic and collaborative performance-oriented, multipollutant air quality management plan (AQMP)

---

[3]Airshed is used here to denote the broader geographic extent of the emissions that contribute to the deleterious effects of a pollutant in a given location.

process. *Recommended actions* include enhancing the effectiveness and innovation of state and local air quality planning, while maintaining federal oversight and retaining requirements for conformity with regional transportation planning.

4. Develop an integrated program for criteria pollutants and hazardous air pollutants (HAPs). *Recommended actions* include establishing a more unified assessment of criteria and hazardous air pollutants, setting priorities for those pollutants, establishing a more dynamic process for considering new pollutants, and considering multiple pollutants in forming the scientific basis for NAAQS.

5. Enhance protection of ecosystems and other aspects of public welfare. *Recommended actions* include better tracking of ecosystem effects and building an improved basis for implementing secondary or alternative standards to protect ecosystems.

Implementation of these recommendations will still require substantial resources, but they should not be overwhelming, especially when compared with current expenditures for CAA compliance and costs resulting from harmful effects of air pollution on human health and welfare. Implementing these recommendations will also require a commitment by all parties to adjust and change; it may also require new legislation from Congress. As the transition occurs, however, it is imperative that ongoing programs to reduce emissions continue so that progress toward cleaner air is maintained.

# Summary

Over the past three decades, the nation has devoted substantial efforts and resources to protect and improve air quality through implementation of the Clean Air Act (CAA). The U.S. Environmental Protection Agency (EPA) estimates that the direct costs of this implementation have been as high as \$20–30 billion per year. There is little doubt that these expenditures have helped reduce pollutant emissions despite the substantial increases in activities that produce these emissions (see Figure S-1). Although it is not possible to know what the exact concentrations of pollutant emissions might be in the absence of the CAA, it is reasonable to conclude that implementation of the act played an important role in lowering these emissions. Cost-benefit analyses have generally concluded that the economic value of the benefits to public health and welfare[1] have equaled or exceeded the costs of implementation.

Despite substantial progress in improving air quality, the problems posed by pollutant emissions in the United States are by no means solved. Future economic and population expansions and the concomitant increased needs, for example, for electricity and transportation, will undoubtedly increase the potential for emissions. Consequently, additional effort will almost certainly be needed to maintain current air quality; even more effort will be needed to make further improvements. The CAA prescribes a com-

---

[1]Within the framework of the CAA, "welfare" refers to the viability of agriculture and ecosystems (such as forests and wildlands), the protection of materials (such as monuments and buildings), and the maintenance of visibility.

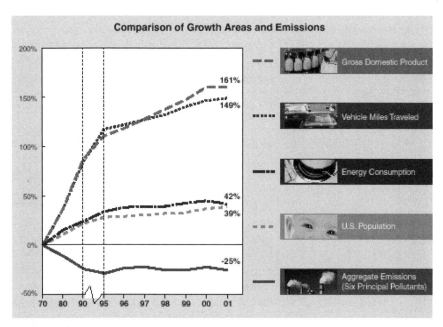

FIGURE S-1   Comparison of growth areas and emission trends. Note that the trends in the graph (except for aggregate emissions) did not change substantially in 1995; only the scale of the graph changed. SOURCE: EPA 2002a.

plex set of responsibilities and relationships among federal, state, tribal,[2] and local agencies for implementing the CAA. This is essentially the nation's air quality management (AQM) system.

## CHARGE TO COMMITTEE

The Committee on Air Quality Management in the United States was formed by the National Research Council in response to a congressional request for an independent evaluation of the overall effectiveness of the CAA and its implementation by federal, state, and local government agencies. The committee was asked to develop scientific and technical recommendations for strengthening the nation's AQM system. In response to its charge,[3] the committee examined in detail the operation, successes, and limitations of the many components of the nation's AQM system and developed a set of unanimous findings and recommendations, as discussed below and outlined in Figure S-2.

---

[2]Hereafter, "state" will be used to denote both state and tribal authorities.

[3]See Chapter 1 for a discussion of the committee's approach to carrying out its charge.

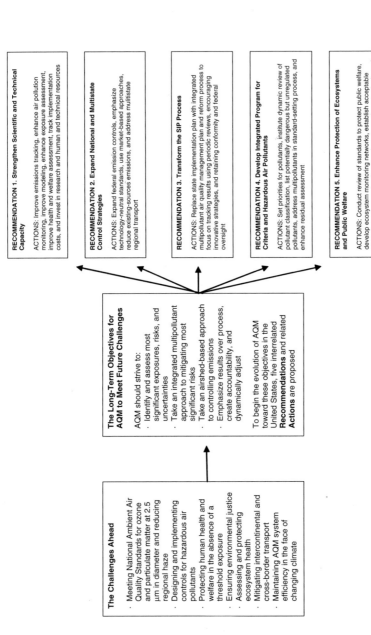

FIGURE S-2 To meet the major challenges that will face air quality management (AQM) in the coming decade, the committee identified a set of overarching long-term objectives. Because immediate attainment of these objectives is unrealistic, the committee made five interrelated recommendations to be implemented through specific actions.

## THE CURRENT AQM SYSTEM

Two landmark events in 1970 helped to establish the basic framework for managing air quality in the United States: the enactment of the CAA Amendments and the creation of EPA. The CAA and its subsequent amendments (such as those in 1977 and 1990) endeavor to protect and promote public health and public welfare by pursuing the following goals:

- Mitigate potentially harmful ambient concentrations of six so-called criteria pollutants: carbon monoxide (CO), nitrogen dioxide ($NO_2$), sulfur dioxide ($SO_2$), ozone ($O_3$), particulate matter (PM), and lead (Pb).
- Limit the sources of exposure to hazardous air pollutants (HAPs), also called "air toxics."
- Protect and improve visibility in wilderness areas and national parks.
- Reduce emissions of substances that cause acid deposition, specifically sulfur dioxide and nitrogen oxides ($NO_x$).
- Curb use of chemicals that have the potential to deplete stratospheric ozone.[4]

The nation's AQM system operates through three broad kinds of activities (Figure ES-1): (1) setting standards and objectives, (2) designing and implementing control strategies, and (3) assessing status and measuring progress. The committee's detailed assessments of the strengths and limitations of these activities are presented in Chapter 2 (Setting Standards and Objectives), Chapter 3 (Implementation Planning), Chapter 4 (Mobile-Source Controls), Chapter 5 (Stationary-Source Controls), and Chapter 6 (Measuring Progress). Overall, the committee found that the AQM system has made substantial progress, especially in the following ways:

- Setting National Ambient Air Quality Standards (NAAQS) for criteria pollutants, designing and implementing state implementation plans (SIPs) to comply with the NAAQS, and implementing other CAA programs to address hazardous air pollutants, acid rain, and other issues have all promoted enhanced technologies for pollution control and have contributed to substantial decreases in pollutant emissions.
- Air quality monitoring networks have documented decreases in ambient concentrations of the criteria pollutants, especially in urban areas, and despite growth in power production and transportation uses. The NAAQS for sulfur dioxide, nitrogen dioxide, and carbon monoxide have

---

[4]The NRC charged the committee only to address air quality in the troposphere (lower atmosphere). The NRC has elsewhere provided treatment of issues related to stratospheric ozone depletion and global climate change, see NRC (1998a, 2001a,b, 2003a) and NAE/NRC (2003).

been largely attained. Monitoring networks have also documented a reduction in sulfate deposition in the eastern United States.

• Economic assessments of the overall costs and benefits of AQM in the United States conclude that, even recognizing the considerable uncertainties, implementation of the CAA has had net economic benefits.

With regard to the three broad activities in AQM (Figure ES-1), the committee found the following:

*Standard Setting*

• Standard setting, planning and control strategies for criteria pollutants and hazardous air pollutants have largely focused on single pollutants instead of potentially more protective and more cost-effective multipollutant strategies. Integrated assessments that consider multiple pollutants (ozone, particulate matter, and hazardous air pollutants) and multiple effects (health, ecosystem, visibility, and global climate change) in a single approach are needed.

• Current risk assessment and standard-setting programs do not account sufficiently for all the hazardous air pollutants that may pose a significant risk to human health and ecosystems or for the complete range of human exposures both outdoors and indoors.

• EPA's current practice for setting secondary standards[5] for most criteria pollutants does not appear to be sufficiently protective of sensitive crops and ecosystems.

*Designing and Implementing Controls*

• Although pollutant concentrations have decreased, the federal, regional, and state emission-control programs implemented under the SIP process have not resulted in NAAQS attainment for ozone and particulate matter in many areas. In addition, the SIP process has become overly bureaucratic, places too much emphasis on uncertain emission-based modeling simulations of future air pollution episodes, and has become a barrier to technological and programmatic innovation.

• Air quality models have often played a major role in designing air pollution control strategies. Much effort has gone into the development and improvement of these models; as a result, they are highly sophisticated. Limitations remain, however, in large part due to a lack of data to adequately evaluate their performance in specific applications for specific locations and an inability to rigorously quantify their uncertainty.

---

[5]Secondary standards are intended to protect against adverse public welfare effects (such as deleterious effects on ecosystems and materials).

• Progress has been made in recognizing and addressing multistate transport of air pollution, especially for ozone and atmospheric haze and their precursors, in some parts of the nation. However, transport issues need to be identified and addressed more proactively and the scope broadened to include international transport.

• For mobile sources, regulations for light-duty vehicles, light-duty trucks, and fuel properties have greatly reduced emissions per mile traveled. Gaps remain, however, in the ability to monitor, predict, and control vehicular emissions, especially from nonroad vehicles, heavy-duty diesel trucks, and malfunctioning automobiles.

• Emission reductions from stationary sources (for example, power plants and large factories) have also been substantial. However, most of the reductions have been accomplished through regulations on new facilities, while many older higher-emitting facilities continue to be a substantial source of emissions.

• In recent years, emissions cap and trade has provided an effective mechanism for achieving stationary-source emission reductions at reduced costs. However, cap-and-trade programs have been limited to relatively few pollutants, and the process of revising caps and targets in response to new technical and scientific knowledge has been cumbersome.

*Assessing Status and Measuring Progress*

• With the exception of continuous emissions monitoring at some large stationary sources, the nation's AQM system lacks a comprehensive and quantitative program to confirm the emission reductions claimed to have occurred as a result of AQM.

• The air quality network in the United States is a national resource but is nevertheless inadequate to meet important objectives, especially that of tracking regional patterns of pollutant concentrations, transport, and trends (see Figure S-3).

• The AQM system has not developed a program to track health and ecosystem exposures and effects and to document improvements in health and ecosystem outcomes achieved from improvements in air quality. Ecosystem effects have not been reliably and consistently accounted for in cost-benefit analyses.

## THE CHALLENGES AHEAD

Although the nation's AQM system has been effective in addressing some of the most serious air quality problems, it has a number of limitations, as outlined above. In addressing how those limitations can best be remedied,

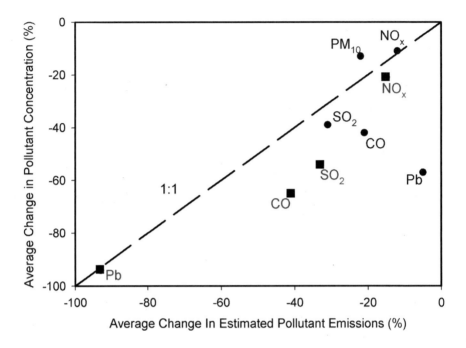

FIGURE S-3    Plot of the estimated relative trends in emissions versus ambient concentrations of various primary pollutants ($PM_{10}$, $NO_x$, $SO_2$, Pb, and CO). Emission trends, which were derived from emission inventories, are shown along the x-axis, and the trends in average concentrations, which were derived from air quality monitoring networks, are shown along the y-axis. The squares are the relative trends in emissions and ambient concentrations for the 20-year period spanning 1983–2002 (except for $PM_{10}$ emissions, which are for the trend period 1985–2002), and the circles are the relative trends for the 10-year period of 1993–2002. If the emission inventory trends were accurate and the nation's air quality monitoring networks were able to accurately measure the average concentration of primary pollutants in the air overlying the United States, all the points on the graph would fall on the 1:1 (diagonal) line. However, the fact that most of the points on the graph do not fall on the 1:1 line indicates that the emission inventory trends are inaccurate and/or that the nation's air quality network, which was initially designed to monitor urban pollution and compliance with NAAQS, has not been able to track trends in pollutant concentrations quantitatively across urban, suburban, and rural settings. Despite such uncertainty, it is important to note that the downward trend in ambient pollutant concentrations provides qualitative confirmation that pollutant emissions have been decreasing.    SOURCE: Data from EPA 2003.

it is important to consider the major air quality challenges facing the nation in the coming decades. Seven major challenges are outlined below.

- **New Standards.** Additional reductions in pollutant emissions will be required to meet the EPA 1997 standards for ozone and particulate matter and the 1999 regulations for regional haze. Improvements to the AQM system will be needed to best identify what emissions to reduce and to monitor the progress toward meeting new standards.
- **Toxic Air Pollutants.** The human health risks from exposure to toxic pollutants remain significant and poorly quantified. A greater research effort that focuses on the sources, atmospheric distribution, and effects of most toxic air pollutants will be needed to address health risks and ensure adequate protection to the public.
- **Health Effects at Low Pollutant Concentrations.** There is increasing evidence that there might not be an identifiable exposure concentration (threshold) for some criteria pollutants below which human health effects would cease to occur. A better understanding of the reducible (human-induced) and irreducible components of pollution, as well as the health and ecosystem impacts at low levels of exposure, is needed. Once improved scientific understanding is developed, it might be necessary to reconsider how to set standards to protect public health from pollutants for which thresholds can not be identified.
- **Environmental Justice.** The CAA does not have any programs explicitly aimed at mitigating pollution effects that might be borne disproportionately by minority and low-income communities in densely populated urban areas. Addressing this need will require enhancing the science base for determining exposures of selected communities to air pollution and incorporating environmental equity concepts in the earliest stages of air quality planning. Native American tribes should be given help to develop and implement AQM programs for reasons of environmental justice and tribal self-determination.
- **Protecting Ecosystem Health.** Although mandated in the CAA, the protection of ecosystems affected by air pollution has not received appropriate attention in the implementation of the act. A research and monitoring program is needed that can quantify the effects of air pollution on the structure and functions of ecosystems. That information can be used to establish realistic and protective goals, standards, and implementation strategies for ecosystem protection.
- **Multistate, Cross-Border, and Intercontinental Transport.** Evidence is accumulating that shows that air quality in a specific area can be influenced by pollutant transport across multistate regions, national boundaries, and continents. To address multistate pollutant transport, the AQM system must improve the techniques for tracking and documenting pollutant

transport and develop more effective mechanisms for coordinating multi-state regional air pollution control strategies. In addition, the nation should continue to pursue collaborative projects and enter into agreements and treaties with other nations to help minimize pollution transport to and from the United States.

• **AQM and Climate Change.** The earth's climate is warming. Although uncertainties remain, the general consensus within the scientific community is that this warming trend will continue or even accelerate in the coming decades. The AQM system will need to ensure that pollution reduction strategies remain effective as the climate changes, because some forms of air pollution, such as ground-level ozone, might be exacerbated. In addition, emissions that contribute to air pollution and climate change are fostered by similar anthropogenic activities, that is, fossil fuel burning. Multipollutant approaches that include reducing emissions contributing to climate warming as well as air pollution may prove to be desirable.

## RECOMMENDATIONS

To meet the challenges of the coming decades and remedy current limitations, the committee identified a set of long-term, overarching objectives to guide future improvement of the AQM system. In the committee's view, AQM should

• Strive to identify and assess more clearly the *most significant exposures, risks, and uncertainties.*
• Strive to take an *integrated multipollutant approach* to controlling emissions of pollutants posing the most significant risks.
• Strive to take an *airshed[6]-based approach* by assessing and controlling emissions of important pollutants arising from local, multistate, national, and international sources.
• Strive to emphasize *results over process*, create accountability for the results, and dynamically adjust and correct the system as data on progress are assessed.

Immediate attainment of these objectives is unrealistic. It would require a level of scientific understanding that has yet to be developed, a commitment of new resources that would be difficult to obtain in the short term, and a rapid transformation of the AQM system that is uncalled for in light of the system's past successes. The committee proposes, therefore, that the AQM system be enhanced so that it steadily evolves towards meeting these

---

[6]Airshed is used here to denote the broader geographic extent of the emissions that contribute to the deleterious effects of a pollutant in a given location.

objectives. In that spirit, the committee makes five interrelated recommendations, to be implemented in concert through some 30 specific actions described in this report.

### Recommendation One

**Strengthen scientific and technical capacity to assess risk and track progress.**

Improving the nation's AQM system will depend heavily on reassessing and investing in relevant scientific and technical capacity to help evolve the AQM system to one that can focus on risk in priority setting and on performance in measuring progress. Without the enhancement of the nation's scientific and technical capacity, implementation of the other four recommendations will be more difficult. The most critical actions are

- *Improve emissions tracking,* including new emissions monitoring techniques and regularly updated and field-evaluated inventories.
- *Enhance air pollution monitoring,* including new monitoring methods, expanded geographic coverage, improved trend analysis, and enhanced data accessibility.
- *Improve modeling,* including enhanced emission and air measurement programs to provide data for model inputs and model evaluation and continued development of shared modeling resources.
- *Enhance exposure assessment,* including improved techniques for measuring personal and ecosystem exposure and designing strategies to control the most significant sources of ambient, hot-spot,[7] and indoor exposures.
- *Develop and implement a system to assess and monitor human health and welfare effects* through the identification of indicators capable of characterizing and tracking the effects of criteria pollutants and hazardous air pollutants and the benefits of pollution control measures and their sustained use in assessments, such as the 2003 EPA *Draft Report on the Environment.*
- *Continue to track implementation costs* by supporting the Pollution Abatement Cost and Expenditure (PACE) survey and conducting detailed

---

[7]Hot spots are locales where pollutant concentrations are substantially higher than concentrations indicated by ambient outdoor monitors located in adjacent or surrounding areas. Hot spots can occur in indoor areas (for example, public buildings, schools, homes, and factories), inside vehicles (for example, cars, buses, and airplanes), and outdoor microenvironments (for example, a busy intersection, a tunnel, a depressed roadway canyon, toll plazas, truck terminals, airport aprons, or nearby one or many stationary sources). The pollutant concentrations within hot spots can vary over time depending on various factors including the emission rates, activity levels of contributing sources, and meteorological conditions.

and periodic examinations of actual costs incurred in a subset of past regulatory programs, including comparisons of actual costs to the costs predicted by various parties prior to adoption of regulations.

• *Invest in research to facilitate multipollutant approaches that targets the most significant risks,* including enhanced research into the full range of ambient, hot-spot, and indoor exposures and their potential risks.

• *Invest in human and technical resources* through programs and incentives to attract and train a diverse corps of scientists and engineers contributing to AQM and the development of an environmental extension service.

## Recommendation Two

**Expand national and multistate performance-oriented control strategies to support local, state, and tribal efforts.**

The role of EPA in establishing and implementing national and multistate emission-control measures should be expanded so that states can focus their efforts on local emission concerns. The most critical actions are

• *Expand federal emission-control measures* especially for nonroad mobile sources (for example, aircraft, ships, locomotives, and construction equipment), area sources (relatively small dispersion), and building and consumer products. Development of these measures should actively involve states, local agencies, and stakeholders and allow for continued control-measure innovation at the state and local level.

• *Emphasize technology-neutral standards for emission control.* Whenever practical, control measures should cap the total emissions from a given source or group of sources, as opposed to limiting the rate of emissions per unit of resource input or product output. In cases where a cap is not practical, standards should be set that promote improved technologies rather than being tied to a single technology and that are stringent enough to offset projected emission increases caused by future growth in economic activity.

• *Use market-based approaches whenever practical and effective* through the expanded use of approaches, such as the acid rain $SO_2$ emissions cap-and-trade program, that have the potential to be highly effective and realize substantial cost savings. Such programs must incorporate continuous emissions monitoring to ensure that emission goals are met and be designed to identify and minimize geographic and temporal disparities in results. Expansion to new industrial sectors will require enhanced continuous emission-monitoring systems, technologies to ensure that required reductions are achieved, and ambient monitoring to establish that the program does not inadvertently result in geographic and temporal disparities in results.

* *Reduce emissions from existing facilities and vehicles,* to the extent practical by promulgating standards for sources regardless of their age, status, or fuel. Older stationary sources and mobile nonroad sources are of particular concern.

* *Address multistate transport problems* by providing EPA with greater statutory responsibility to assess multistate air quality issues on an ongoing basis and the regulatory authority to deal with them in a regional context. Constitutionally, interstate environmental rules and regulations must be based on federal authority, but EPA has not been given sufficient tools under the CAA to address the multistate aspects of most air quality problems.

### Recommendation Three

**Transform the SIP process.**

Implementation planning at the state and local levels should be changed to place greater emphasis on performance and results and to facilitate development of multipollutant strategies. Critical actions include

* *Transform the SIP into an AQM plan.* Each state should be required to prepare an air quality management plan (AQMP) that integrates the relevant air quality measures and activities into a single, internally consistent plan. An evolution of the SIP process to an AQMP approach should involve the following:

—Given the similarity of sources, precursors, and control strategies, the AQMP should encompass *all* criteria pollutants in an integrated multipollutant plan.

—EPA should identify key hazardous air pollutants that have diverse sources or substantial public health impacts. These pollutants should be included in an integrated multipollutant control strategy and addressed in each state's AQMP.

—The scope of the AQMP should explicitly identify and propose control strategies for air pollution hot spots and situations where disadvantaged groups may be disproportionately exposed and should provide incentives to implement the strategies.

—Given the current statutory requirements and rules associated with the SIP, it might be necessary to implement this recommendation in stages and provide incentives to facilitate the transition to an AQMP approach.

• *Reform the planning and implementation process* by

—Encouraging regulatory agencies to concentrate their resources on tracking and assessing the performance of the strategies that have been implemented rather than on preparing detailed documents to justify the effectiveness of strategies in advance of their implementation.

—Carrying out a formal and periodic process of review and re-analysis of the AQMP to identify and implement revisions to the plan when progress toward attainment of standards falls below expectations or when conditions change sufficiently to invalidate the underlying assumptions of the plan. Given the large contributions of federal and multistate measures to the success of any plan, it is essential that this review process be collaborative and include all relevant federal and state agencies.

—Encouraging the development and testing of innovative strategies and technologies by not requiring predetermined and agreed-upon benefits for every strategy but periodically evaluating their effectiveness.

—Retaining the federal requirement for conformity between air quality planning and transportation planning. Conformity could be improved by mandating greater consistency between the data, models, and time frames used in air quality and transportation plans.

—Continuing to require that states implement agreed upon strategies, ensure private-sector compliance, and are held accountable for failure to meet the AQMP commitments through federally mandated sanctions.

## Recommendation Four

**Develop an integrated program for criteria pollutants and hazardous air pollutants.**

The time has come for the nation's AQM system to begin the transition toward an integrated, multipollutant approach that targets the most significant exposures and risks. The critical actions include

• *Develop a system to set priorities for hazardous air pollutants* by expanding the approach embodied in EPA's urban air toxics program. As proposed in Recommendation Three, a few hazardous air pollutants, because of their diverse sources, ubiquitous presence in the atmosphere, or exceptionally high risk to human health and welfare, might warrant treatment similar to criteria pollutants and be included in AQMPs.

• *Institute a dynamic review of pollutant classification,* and reclassify and revise priorities for criteria pollutants and hazardous air pollutants accordingly.

• *List potentially dangerous but unregulated air pollutants for regulatory attention.* Determine whether there are sufficient data on adverse impacts, chemical structure, and potential for population exposure and identify whether some level of regulatory response would be prudent and appropriate.

• *Address multiple pollutants in the NAAQS review and standard-setting process* by beginning to review and develop NAAQS for related pollutants simultaneously.

• *Enhance assessment of residual risk* by performing an increased number of assessments in the years to come and by attempting to include in the assessments other major sources of the same chemicals.

## Recommendation Five

**Enhance protection of ecosystems and other aspects of public welfare.**

Many of the programs and actions undertaken in response to the CAA have focused almost entirely on the protection of human health. Further efforts are needed to protect ecosystems and other aspects of public welfare. The critical actions include

• *Completion of a comprehensive review of standards to protect public welfare.*

• *Develop and implement networks for comprehensive ecosystem monitoring* to quantify the exposure of natural and managed resources to air pollution and the effects of air pollutants on ecosystems.

• *Establish acceptable exposure levels for natural and managed ecosystems* by evaluating data on the effects of air pollutants on ecosystems at least every 10 years.

• *Promulgate secondary standards* where needed that take the appropriate form. For example, in some cases a standard based on the amount of a pollutant that is deposited on the earth's surface over a particular area may be more appropriate than a standard based on the atmospheric concentration of that pollutant. Allow for consideration of regionally distinct standards.

• *Track progress toward attainment of secondary standards* by using the aforementioned monitoring of ecosystem exposure and response.

## MOVING FORWARD

Because the nation's AQM system has been effective in many aspects over the past three decades, much of the system is good and warrants retaining. Thus, the recommendations proposed here are intended to evolve the AQM system incrementally rather than to transform it radically. The recommendations are also not intended to deter current ongoing AQM activities aimed at improving air quality. Indeed, even as these recommendations are implemented, there can be little doubt that important decisions to safeguard public health and welfare should continue to be made, often in the face of scientific uncertainty. Moreover, new opportunities and approaches for managing air quality will appear. These include addressing pollution problems in multiple environmental media, such as air and water, taking advantage of new technologies, and undertaking pollution prevention activities rather than controlling air pollutants after they have been produced.

Implementation of the recommendations will require the development of a detailed plan and schedule of steps. *The committee urges EPA to convene an implementation task force from the key parties to prepare a plan of action and an analysis of legislative actions, if any are needed.*

Implementation of the recommendations will also require additional resources. Although these resources are not insignificant, they should not be overwhelming. For example, consider the costs associated with air quality research and monitoring. Even a doubling of the approximate $200 million in EPA funds currently dedicated to air quality monitoring and research would represent about 1% of annual expenditures nationwide for complying with the CAA. Such resources are even smaller when compared with the costs imposed by the deleterious effects of air pollution on human health and welfare.

Implementation of the recommendations will require a commitment by all parties to stages of implementation over several years. As that transition occurs, it is important that action on individual programs to reduce emissions continues to maintain progress toward cleaner air.

The full complement of scientific and engineering disciplines will need to be prepared to take up the substantial challenges embodied in these recommendations. Given the opportunity, the committee believes that the scientific and engineering communities can provide the human resources and technologies needed to underpin an enhanced AQM system and to achieve clean air in the most expeditious and effective way possible.

# 1

# Introduction

The goal of protecting and enhancing air quality to protect and promote human health and public welfare[1] has been consistently set forward in the United States during the latter part of the twentieth century. To accomplish this goal, numerous regulations and standards, a broad suite of management tools, and several monitoring networks to track progress have been established. All of these components depend on robust and up-to-date scientific and technical input, which includes an understanding of relationships between air pollutant levels and impacts on human health, ecosystems, atmospheric visibility, and materials. The National Research Council Committee on Air Quality Management in the United States was asked to evaluate the effectiveness of the nation's air quality management (AQM) system and the extent to which it is informed by the most advanced science and technology. This chapter begins with a brief summary of the current scientific and technical understanding of air pollution and its impacts, as well as an overview of the AQM system in the United States and the federal legislation that has motivated and driven much of its development in the latter part of the twentieth century. This overview discussion is intended to provide an introduction to aspects of the AQM system that are described in more detail and critiqued in later chapters of this report. The report is not intended to provide a comprehensive description of AQM activities in the

---

[1]Within the framework of the Clean Air Act, "welfare" refers to the viability of agriculture and ecosystems (such as forests and wildlands), the protection of materials (such as monuments and buildings), and the maintenance of visibility.

United States; pertinent references are provided for those who desire more background information on general concepts of scientific and technical understanding of air pollution and its impacts.

## AIR POLLUTION SCIENCE

The atmosphere is composed of a mixture of gases and particles. An air pollutant is generally defined as any substance in air that, in high enough concentrations, harms humans, ecosystems (other animals and vegetation), or materials (such as buildings and monuments) and reduces visibility. In this report, the committee uses the term air pollutant to denote the subset of harmful atmospheric substances that are present, at least in part, because of human activities rather than natural production and whose principal deleterious effects occur as a result of exposure at ground level. Greenhouse gases, as well as pollutants that cause depletion of ozone ($O_3$) in the stratosphere, the layer of atmosphere extending from about 10 to 16 kilometers (km) up to 50 km altitude, are addressed only in the context of managing ground-level air quality.[2]

The science of air pollution is primarily concerned with quantitatively understanding the so-called "source-receptor relationships" that link specific pollutant emissions to the pollutant concentrations and deposition observed in the environment as a function of space and time. This quantitative understanding is developed through extensive field and laboratory measurements and analysis and is then tested and documented in air quality models that use mathematical and numerical techniques to simulate the physical and chemical processes that affect air pollutants as they disperse and react in the atmosphere. As illustrated in Figure 1-1, the pollutants at a particular time and place depend on the proximity to sources that emit pollutants or their precursors; the chemical reactions that pollutants or their precursors undergo once in the atmosphere; and the impact of mixing, dilution, transport, and removal or deposition processes (Seinfeld and Pandis 1998). The areas of air quality science and air quality management are closely coupled, because the tools developed by scientists and engineers to carry out the tasks described above are also widely used by the agencies tasked with controlling air pollution. For example, the instrumentation used by scientists in field experiments is also used by regulatory agencies to monitor air pollution exposures, trends, and compliance. Similarly, the models developed by scientists to simulate and better understand air pollution are used in AQM to help design effective strategies for air pollution mitigation.

---

[2]The NRC charged the committee to address only air quality in the troposphere (lower atmosphere). For treatment of issues related to stratospheric ozone depletion and global climate change, see NRC (1998a, 2001a,b, 2003a) and NAE/NRC (2003).

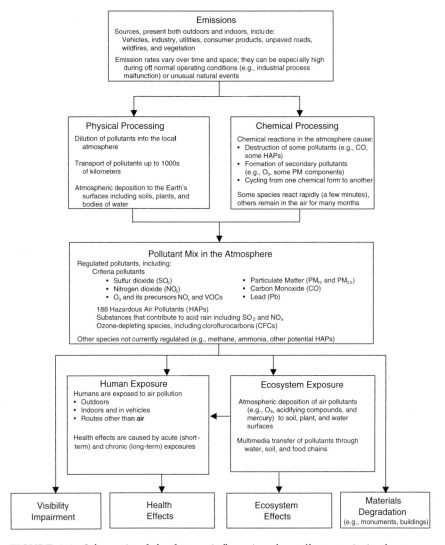

FIGURE 1-1 Schematic of the factors influencing the pollutant mix in the atmosphere and the resultant impacts of pollution. Greenhouse gases and climate change impacts are not included because they fall outside the committee's charge.


Air pollutants are often characterized by how they originate: pollutants emitted directly into the atmosphere are called primary pollutants; those formed as a result of chemical reactions within the atmosphere are called secondary pollutants. Control of secondary pollutants is generally more problematic than that of primary pollutants, because mitigation of secondary pollutants requires identification of the precursor compounds and their sources as well as an understanding of the specific chemical reactions that result in the formation of the secondary pollutants. Control can be further complicated when the chemical reactions resulting in secondary-pollutant formation involve complex, nonlinear interactions among the precursors. Under those conditions, a 1:1 relationship might not exist between a reduction in precursor emissions and reductions in secondary-pollutant concentrations. Ground-level $O_3$ is an example of such a secondary pollutant; it is formed by reactions of nitrogen oxides ($NO_x$) and volatile organic compound (VOC) species[3] in the presence of sunlight. In some circumstances, $O_3$ concentrations are most effectively controlled by lowering both VOC and $NO_x$ emissions. For other circumstances, lowering VOC or $NO_x$ emissions may be most effective (NRC 1991).

Similar complications arise in the mitigation of suspended particulate matter (PM), which refers to a heterogeneous collection of solid and liquid particles that include ultrafine particles (diameters of less than 0.1 micrometers [µm]); fine particles (diameters of 0.1 to a few micrometers), which are commonly dominated by sulfate, nitrate, organic, and metal components; and relatively coarse particles (diameters of a few micrometers or more), which are often dominated by dust and sea salt. PM can be a primary or secondary pollutant. As a primary pollutant, PM is emitted directly to the atmosphere, for instance, as a result of fossil fuel combustion. As a secondary pollutant, PM is formed in the atmosphere as a result of such processes as oxidation of sulfur dioxide ($SO_2$) gas to form sulfate particles. Because the reactions that result in the formation of secondary PM often depend on the concentration and composition of preexisting airborne PM, control strategies that lower the emissions of one chemical constituent of airborne PM might not affect or might in some cases increase the concentrations of other components of PM. Even though pollutants have been typically treated independently in many of the air quality regulations in the United States, pollutants are often closely coupled. For example, most pollutants are emitted into the atmosphere by the same source types (see Figure 1-2). They also often share similar precursors and similar chemical interactions once in the atmosphere. For example, many of the VOCs that react to form $O_3$ are also identified as hazardous air pollutants

---

[3]Nitric oxide (NO) and nitrogen dioxide ($NO_2$) are referred to together as $NO_x$. VOCs are organic compounds present in the gas phase at ambient conditions. Several other terms are used operationally to refer to and classify organic compounds. For example, reactive VOCs are sometimes designated as reactive organic gases (ROG); however, because hydrocarbons make up most of the organic gas emissions, this category is also called reactive hydrocarbons (RHC). Moreover, because methane dominates the unreactive category, the term nonmethane hydrocarbons (or NMHC) is often used. Unless noted otherwise, VOCs will be used in this report to represent the general class of gaseous organic compounds.

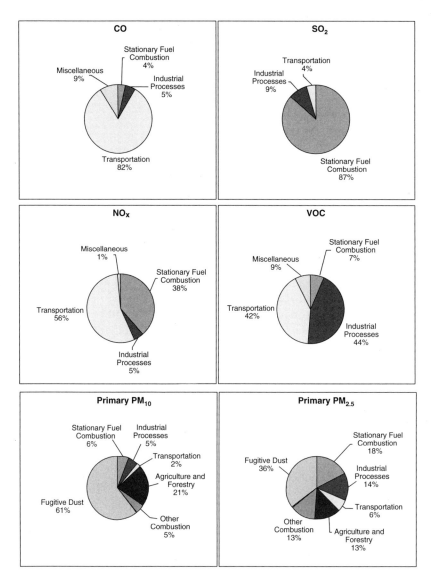

FIGURE 1-2 National average emission categories for carbon monoxide (CO), sulfur dioxide ($SO_2$), nitric oxide and nitrogen dioxide ($NO_x$), volatile organic compounds (VOCs), particulate matter less than 10 micrometers in diameter ($PM_{10}$), and particulate matter less than 2.5 micrometers in diameter ($PM_{2.5}$) for 2001. For primary $PM_{10}$ and $PM_{2.5}$, a significant fraction of the PM fugitive dust emissions return to the ground after a few minutes. Biogenic emissions (such as VOC emissions from vegetation and $NO_x$ emissions from soil microbial processes) are not included. SOURCE: Adapted from EPA 2003a.

(HAPs). VOCs and HAPs can also be precursors to or components of PM. In a similar manner, $NO_x$ has several significant environmental effects that warrant its control. In addition to being an $O_3$ precursor, it has a direct impact on human health, is a precursor for acid rain and the formation of inhalable fine particles, decreases atmospheric visibility, and contributes to the eutrophication of water bodies.

## AIR POLLUTION IMPACTS

The primary objective of most air quality standards in the United States is the protection of human health. Humans are exposed to air pollution outdoors and indoors, including during transit in vehicles. Indoor air pollution comprises a mixture of contaminants penetrating from outdoors and those generated indoors. Especially high exposure to air pollution can also occur in various microenvironments, referred to here as "hot spots," which include highway toll plazas; truck stops; airport aprons; and areas adjacent to industrial facilities, busy roadways, and idling vehicles. Some studies have suggested that disproportionate exposures may be found in low-income and predominantly minority communities and have raised concerns about environmental justice (see discussion on environmental justice in Chapter 2 and references cited therein).

Many types of health effects have been attributed to air pollution, including pulmonary, cardiac, vascular, and neurological impairments—all of which can lead directly to mortality. In addition, a number of regulated air pollutants are known or probable carcinogens. Some health effects, such as an increase in asthma attacks, have been observed in conjunction with episodes of high pollution concentrations lasting 1 or 2 days. Such effects are considered acute, because they are associated with short-term exposures to a pollutant. Other health effects, particularly increased risk of cancer, are associated with long-term exposure (EPA 2002b).

The scientific techniques for assessing health impacts of air pollution include air pollutant monitoring, exposure assessment, dosimetry, toxicology, and epidemiology (NRC 1998b, 1999a). Because most of the health effects attributable to air pollutants can also be attributable to a wide variety of other risk factors, the impact of air pollution on human health can be difficult to distinguish and quantify. Determining the impact of air pollution on human health is further complicated by human exposure to a mixture of substances at various concentrations present in the air. Also, a number of subgroups within the human population at large are considered more susceptible to the effects of air pollution. They include people who have coronary disease, asthma, or chronic pulmonary diseases; the elderly; and infants. Fetuses are also possibly susceptible (Wilhelm and Ritz 2003).

In addition to air pollution effects on human health, impacts on ecosystem form and function are also a serious concern. Moreover, because ecosystems often supply society with valuable services (such as cleaning and purifying water), damage to ecosystems from air pollution can exact a significant economic as well as an environmental cost (Daily 1997). Terrestrial, aquatic, and coastal ecosystems are exposed to air pollution via atmospheric substances (such as $O_3$) or by deposition of substances (such as acids, nutrients, and metals). In terrestrial ecosystems, air pollution deposition affects plant physiology; microbial processes; biogeochemical cycles of substances, such as nitrogen; and plant community dynamics. In aquatic ecosystems, acidic deposition results in acidification of waterways, the mobilization of trace metals in surface waters, and ultimately, the loss of aquatic biodiversity (Driscoll et al. 2001a). Atmospheric deposition is also a major source of mercury to some aquatic ecosystems in North America. When mercury is present as methylmercury in sufficient quantities in the food chain, this contaminant is toxic to humans and animals (NRC 2000a). In addition, atmospheric deposition of nitrate and ammonium might be an important source of nitrogen in coastal regions, contributing to eutrophication, increased or harmful algal blooms, hypoxic and anoxic bottom waters, loss of sea grasses, and reduced fish stocks (Fisher and Oppenheimer 1991; D'Elia et al. 1992; Boynton et al. 1995; Paerl 1997; Castro and Driscoll 2002).

Protection of visibility in national parks and wilderness areas has traditionally played a smaller but nonetheless important role in driving air quality regulation. Scenic vistas in most U.S. parklands are often diminished by haze that reduces contrast, washes out colors, and renders distant landscape features indistinct or invisible. Haze degrades visibility primarily through scattering or absorption of light by fine atmospheric particles (NRC 1993a; Watson 2002).

Air pollution can discolor or damage commonly used building materials and works of art. In addition, such pollutants as sulfate can accelerate the natural weathering process of materials, including metals, painted surfaces, stone, and concrete.

## AIR QUALITY MANAGEMENT IN THE UNITED STATES

As the health, ecological, and economic impacts of air pollution in the United States have become increasingly evident through more sophisticated scientific approaches, the nation has endeavored to protect air quality through increasingly complex and ambitious legislation (Table 1-1). The federal government's first major efforts in this regard began in 1955 with the Air Pollution Control Act. These efforts were enhanced over the next 15 years through a series of enactments, including the Clean Air Act (CAA)

TABLE 1-1   Federal Air Quality Management Legislation

| Date | Legislation | Authorization |
|------|-------------|---------------|
| 1955 | Air Pollution Control Act | Provided funds to local and state agencies for research and training |
| 1959 | Air Pollution Control Act Extension | Extended the 1955 act |
| 1960 | Motor Vehicle Exhaust Study | Authorized the Public Health Service (PHS) to study automotive emissions and health |
| 1962 | Air Pollution Control Act Extension | Extended 1955 act and required PHS to include auto emissions in their program |
| 1963 | Clean Air Act | Research at the federal level<br>Aid to states for training<br>Federal authority to abate interstate pollution<br>Matching grants to local/state agencies for air pollution control |
| 1965 | Motor Vehicle Air Pollution Control Act | National standards for auto emissions<br>Coordinated pollution control between United States, Canada, and Mexico<br>Research into $SO_2$ and auto emissions |
| 1967 | Air Quality Act | Air quality control regions (AQCRs)<br>Air quality criteria<br>Control technology documents<br>State implementation plans (SIPs)<br>Separate automotive emission standards for California |

President Nixon (1970) created the Environmental Protection Agency (EPA) by Executive Order

| Date | Legislation | Authorization |
|------|-------------|---------------|
| 1970 | Clean Air Act Amendments | National ambient air quality standards (NAAQS)<br>SIPs to achieve NAAQS by 1975<br>New source performance standards (NSPS)<br>National emission standards for hazardous air pollutants (NESHAP)<br>Aircraft emission standards to be developed by EPA<br>Automotive emission standards for hydrocarbons and CO for 1975 models and for $NO_x$ for 1976 models<br>States allowed to adopt air quality standards more stringent than federal standards<br>Motor vehicle emissions inspection and maintenance (I/M) program<br>Citizens allowed to sue for air pollution violations |
| 1977 | Clean Air Act Amendments | Geographic regions (Classes I, II, III) to preserve air quality<br>EPA-sanctioned emission offsets and emission banking within nonattainment regions<br>State permits that require prevention of significant deterioration (PSD) studies |

TABLE 1-1   continued

| Date | Legislation | Authorization |
|------|-------------|---------------|
| | | Lowest-achievable emission rate (LAER) in nonattainment regions |
| | | Delayed auto emission standards set in 1970 Clean Air Act Amendments |
| | | Section 169A declared a national goal of preventing and remedying visibility impairment due to anthropogenic pollution in mandatory Class I areas |
| 1990 | Clean Air Act Amendments | *Title I: Nonattainment regions*<br>Nonattainment regions for ozone are ranked in terms of pollution severity; each has a deadline to achieve NAAQS. New and amended NAAQS must be attained in 5 years with a possible extension for another 5 years. |

| Classification (applicable only to ozone nonattainment areas) | Years to achieve NAAQS |
|---|---|
| Marginal | 3 |
| Moderate | 6 |
| Serious | 9 |
| Severe | 15 (17 with a 1988 design value between 0.190 and 0.280 ppm) |
| Extreme (Los Angeles) | 20 |

*Title II: Mobile sources*
Gasoline reformulation toward lower toxic and VOC generation by 1997
Reduction in 1990 $NO_x$ emissions standards for light-duty vehicles (LDVs) by 60% beginning in 1994
Reduction in 1990 hydrocarbons emissions standards for LDVs by 40% beginning in 1994
Introduction of cold temperature (20°F) CO emissions standards set at 10 g/mile beginning in 1994
"Clean car" (ZEV, electric car) pilot program in California
150,000 vehicles by model year 1996
300,000 vehicles by model year 1999

*Title III: Toxics*
Emissions of 189 hazardous air pollutants (HAPs) controlled[a]
Mass $\geq$ 10 tons/yr for a specific HAP
Mass $\geq$ 25 tons/yr for a combination of HAPs
EPA-approved maximum achievable control technology (MACT) mandated
After 8 years, EPA must promulgate more stringent standards to address residual risks where necessary.

*(continued on next page)*

TABLE 1-1 continued

| Date | Legislation | Authorization |
|------|-------------|---------------|
| | | *Title IV: Acid rain (Electricity Generation Facilities)* |
| | | $NO_x$: cut emissions by $2.0 \times 10^6$ tons/yr |
| | | $SO_2$: by 2000, reduce to $9.2 \times 10^6$ tons/yr (U.S. total); by 2010, reduce to $8.9 \times 10^6$ tons/yr (U.S. total) |
| | | Phase I (beginning 1995): 110 large power plants |
| | | Phase II (beginning 2000): remaining units |
| | | Policy: market-based "cap and trade" rather than "command and control" |
| | | If a utility reduces $SO_2$ emissions below its emissions "allowance," the utility can sell its extra "allowance" to another utility |
| | | *Title V: Permits* |
| | | New and existing major sources must secure permits, duration ≤ 5 yr |
| | | Fees to sustain state air pollution control agencies |
| | | *Title VI: Stratospheric ozone* |
| | | Phase out chlorofluorocarbons, halons, and carbon tetrachloride by 2000 |
| | | Phase out methylchloroform ($CH_3CCl_3$) by 2002 |
| | | Phase out hydrochlorofluorocarbons by 2030 |
| | | *Title VII: Enforcement* |
| | | Larger penalties |

[a]Since passage of the 1990 CAA Amendments, one compound (caprolactam) has been deleted from the list of 189 pollutants.
SOURCES: Heinsohn and Kabel 1998; Wark et al. 1998.

of 1963. In 1970, two landmark events took place that helped to establish the basic framework by which air quality is managed in the United States. These events were the creation of the U.S. Environmental Protection Agency (EPA) and the passage of the CAA Amendments of 1970. This framework was further developed and refined with the passage of the CAA Amendments of 1977 and 1990.

Five major goals for protecting and promoting human health and public welfare are identified in the CAA as amended:

- Mitigating potentially harmful human and ecosystem exposure to six criteria pollutants: CO, $NO_2$, $SO_2$, $O_3$, PM, and lead (Pb).[4]
- Limiting the sources of and risks from exposure to HAPs, which are also called air toxics.

---

[4]The term "criteria pollutants" derives from the requirement that EPA must describe the characteristics and potential health and welfare effects of these pollutants (see Chapter 2).

- Protecting and improving visibility impairment in wilderness areas and national parks.
- Reducing the emissions of species that cause acid rain, specifically $SO_2$ and $NO_x$.
- Curbing the use of chemicals that have the potential to deplete the stratospheric $O_3$ layer.

The CAA prescribes a complicated set of responsibilities and relationships among federal, state, tribal,[5,6] and local agencies. This matrix is referred to in this report as the nation's air quality management (AQM) system. The federal government's role is coordinated by EPA and is intended in part to provide a degree of national uniformity in air quality standards and approaches to pollution mitigation so that all individuals in America are assured a basic level of environmental protection. State and local governments are given much of the responsibility for implementing and enforcing the federally mandated rules and regulations within their jurisdictional domains, including developing and implementing specific strategies and control measures to meet national air quality standards and goals. Although many aspects of the AQM system assume a collaborative relationship between the federal, state, and local agencies, the CAA empowers EPA to oversee the activities carried out by those agencies. This oversight includes the power to impose federal sanctions and federally devised pollution-control plans on delinquent areas in some cases.

The federal courts also have a role in AQM. Final agency rules promulgated under the CAA are subject to judicial review, usually in the Court of Appeals for the District of Columbia Circuit, and the reviewing court may set aside any portion of a regulation found to be "arbitrary and capricious." Under this standard, the courts take a "hard look" at the agency's reasoning and the support in the rule-making record for critical factual conclusions, but the court is not supposed to substitute its judgment for that of the agency (Motor Vehicle Manufacturer's Association v. State Farm Mutual Automobile Insurance Co., 463 U.S. 29, 43, 1983). A court will set aside an agency rule only if it finds that the decision was not based on a consideration of the relevant factors or that the agency committed a clear error of judgment. In addition, any citizen may file a civil action in district court against EPA that challenges the agency's failure to perform any nondiscretionary act or duty, and the courts have the authority to order EPA to perform that act or duty and to compel agency action that is

---

[5]Hereafter, "state" will be used as shorthand to denote both state and tribal authorities.

[6]EPA's adoption of 40 CFR Part 49, the Tribal CAA Authority rule, generally authorizes eligible tribes to exercise the same rights and have the same responsibilities as states have under the CAA.

"unreasonably delayed" (CAA § 304(a)(2), 42 U.S.C. § 7604(a)(2)).  The courts have vigorously implemented this authority in the past to mandate, for example, a schedule for EPA action (Melnick 1983).

The nation's AQM system may be conceptualized to operate in a variety of ways to meet the goals and requirements set forth in the CAA.  The committee chose a model that describes the system's operation in terms of four broadly defined sequential activities (Figure 1-3).  The first three are the following:

• Setting air quality standards and objectives (either in the CAA or by EPA).

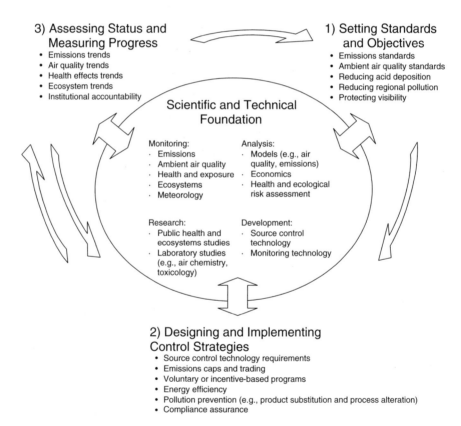

FIGURE 1-3  Idealized schematic showing three of the four sequential activities carried out by the nation's air quality management system.  The fourth iterative activity involves a return to activities 1 and/or 2 to account for new information to correct deficiencies identified in step 3. Bullets under each heading provide examples; listings are not necessarily comprehensive.

- Designing and implementing control strategies to meet those standards and objectives.
- Assessing status and progress.

Because the AQM system typically functions with substantial scientific, technological, and societal uncertainties, there is a need for a fourth iterative activity: to revisit the first and second activities, taking advantage of new information and any deficiencies identified in the third. Examples of the four activities are provided in Figure 1-3. A more-detailed discussion of how each of these activities is carried out in the United States is provided in subsequent chapters.

## THE ROLE OF SCIENCE

Although an understanding of the causes and remedies of air pollution is not yet complete, it is now well-established that the vast majority of the air pollutants addressed by the CAA arise from the burning of fossil fuels and the emission of the myriad of materials and chemicals produced and used in the commerce of this country. However, for a number of broader societal and technical reasons, a total termination of the nation's dependence on fossil fuels and the products and industrial processes that result in pollutant emissions is not a viable option. Indeed, the substantial disruption likely to result could conceivably cause greater damage to human health and welfare than that caused by air pollution in the United States. A more viable option, and the one our society uses, is to control air pollutants at concentrations that pose a minimal or acceptable level of risk to human health and welfare without unduly disrupting the technological infrastructure and economic engine that underpins the nation's economy. To accomplish such control, science and technology are required. Their roles include the following:

1. Quantifying risks to human health and public welfare (such as ecosystems) associated with varying concentrations, mixtures, and rates of deposition of air pollutants to establish air quality standards and goals.
2. Quantifying the source-receptor relationships that relate pollutant emission rates to ambient pollutant concentrations and deposition rates in order to develop air pollution mitigation strategies to maximize benefits and minimize costs.
3. Quantifying the expected demographic and economic trends with and without air pollution control strategies to better account for growth in activity that might offset pollution control measures and to better design control strategies that are compatible with the economic incentives of those who must implement them.

4. Designing and implementing air quality monitoring technologies and methods for documenting pollutant exposures to identify risks and set priorities.

5. Designing, testing, and implementing technologies and systems for efficiently preventing or reducing air pollutant emissions.

6. Designing and implementing methods and technologies for tracking changes in pollutant emissions, pollutant concentrations, and human health and welfare outcomes to document and ultimately improve the effectiveness of air pollution mitigation activities.

As indicated in Figure 1-3, the aforementioned contributions of science and technology are made through monitoring, analysis, research, and development. Monitoring provides the data necessary to determine trends in emissions, air quality, and various health and ecosystem outcomes. Such observational data are essential for determining the effectiveness of regulations and assuring compliance, providing valuable input to air quality models, and supporting long-term health and ecosystem assessments. In addition, the data are used by the scientific research community. Analysis activities also provide critical information to air quality managers who use model results, risk assessments, and economic and other analyses to better characterize their air quality problems and the impacts of various control strategies. Finally, research and development efforts furnish advances in the fundamental understanding of the science and impacts of air pollution, the instruments needed for monitoring, and the technology available for controlling emissions. Thus, at each stage of CAA implementation, science and technology provide a fundamental basis for sound decisions; at the same time, the requirements of the CAA to continually improve air quality and the understanding of it serves as an important incentive to promote scientific and technological advances.

Although the inputs of science and technology are important, they are not the sole determinants of the success of an AQM system. Effective AQM decisions are made and implemented by elected and appointed leaders in the context of diverse social, economic, and political considerations. Successful AQM requires that the input from the scientific and technological communities is utilized by those leaders to produce adequate and cost-effective pollutant emission reductions for which a variety of societal considerations, including environmental justice, are taken into account. The U.S. AQM system entails the promulgation of rules and regulations on specific types of emissions, the institution of programs that provide incentives for the creative development of new technologies, and the use of emission control technologies and systems that reduce air pollution to a sufficient degree to protect public health. However, the effectiveness of AQM can be undermined by a breakdown in any of these components.

This report's recommendations are aimed at improving the nation's AQM system to better integrate the tools and methods of the scientific and technological communities, and to provide an improved mechanism for assuring that the system and its components are achieving the intended public benefits.

## ESTIMATING THE COSTS AND BENEFITS OF THE FEDERALLY MANDATED AIR QUALITY MANAGEMENT SYSTEM

The management of the nation's air quality is a major and complex undertaking. AQM in the United States involves the work of tens of thousands of people who monitor the concentrations of various air pollutants at over a thousand sites, regulate thousands of different emission sources, and maintain a multimillion dollar research and development program to better understand the sources, fate, and effects of air pollutants. The CAA requires regulatory control of air pollutants that have widely varying properties. Some pollutants are rapidly removed from the atmosphere so that effects are largely limited to the immediate source area. Other pollutants (such as $O_3$ and PM) can be transported in significant amounts for hundreds to thousands of miles; therefore, their control requires cooperation between states and, in some instances, nations.

Implementation of the CAA has clearly contributed to the reduction in pollutant emissions in the United States. For example, over the past 30 years, the nation's gross domestic product and total vehicle miles traveled increased by more than 2-fold, and its energy consumption increased by a factor of about 1.5. However, over the same period, EPA (2002a) reported that the total aggregate of emissions that directly affect the ambient concentrations of six criteria pollutants has decreased by 25% (see Figure 1-4). This trend suggests that the nation has been able to decouple, to some extent, pollutant emission rates from economic activity. EPA argues that the CAA played a major role in bringing about this decoupling. In the absence of the CAA, EPA estimated that emissions of CO, $SO_2$, $NO_x$, and PM in 1990 would have been larger by factors of about 2, 1.6, 1.4, and 3, respectively (EPA 1997). Others argued that factors other than environmental regulation (for example, increased income and technological advances) might be the main causes of the decrease in pollutant emissions (for example, Lomborg 2001; Pacala et al. 2003). However, Pacala et al. (2003) concluded that although a variety of factors contribute to observed benefits, regulation plays a prominent role. Although it is not possible to know precisely what the levels of pollutant emissions in the United States would be in the absence of a federally mandated AQM system, it is reasonable to conclude that this system has had an important role in controlling and lowering these emissions over the past 30 years.

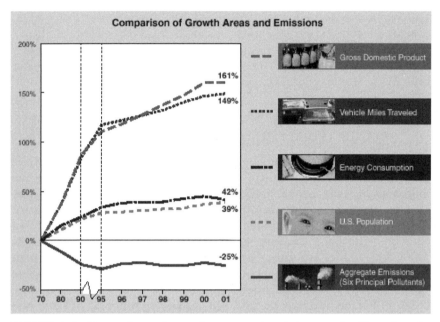

FIGURE 1-4  Comparison of growth areas and emission trends. Note that the trends in the graph (except for aggregate emissions) did not change substantially in 1995; only the scale of the graph changed. SOURCE: EPA 2002a.

EPA (1997) estimated that the benefits that have accrued in human health and welfare as a result of the aforementioned decreases in pollutant emissions have been substantial. The estimated benefits include about 100,000 to 300,000 fewer premature deaths per year and 30,000 to 60,000 fewer children each year with intelligence quotients below 70. However, these estimates are highly uncertain (OMB 1997). They require quantification of the air quality responses to pollutant emission changes and quantification of the human health and welfare responses to those air quality changes. Because it is difficult to isolate the effects of air pollution exposures from those of other risk factors that humans face daily, little direct empirical evidence is available to carry out the latter quantification. As a result, assessments of the benefits of pollution control often rely on complex models instead of direct empirical evidence.[7] These models, in turn, tend to depend on a variety of estimated input parameters and assumptions.

Even more uncertainty is added when the benefits are compared with the costs of regulatory compliance by using a cost-benefit analysis, which

---

[7]A notable exception is lead, for which blood lead concentrations are used as a relatively straightforward indication of human health impacts (Mendelsohn et al. 1998).

requires a monetization of the benefits—a typically controversial and difficult process (Croote 1999). Despite such difficulties, cost-benefit analysis is often used in government to evaluate the merit of environmental regulations. In fact, the 1990 CAA Amendments specifically require EPA to carry out periodic evaluations of the costs and benefits of the implementation of the act. Despite arguments by others identified above, most comprehensive cost-benefit analyses of the nation's AQM system have suggested that the benefits of air pollution control have been equivalent to or exceeded the costs, albeit with significant uncertainties. For example, EPA estimated that the benefits of implementation of the CAA between 1970 and 1990 were $5–50 trillion greater than the costs (EPA 1997). Although others (Lutter and Belzer 2000; Brown et al. 2001) argue over whether EPA's analysis overstates likely benefits and understates costs (also see Chapter 6), the White House Office of Management and Budget (OMB) recently reported that benefits of environmental regulations far outweigh the costs. OMB found monetary benefits over the years of regulation from 1992 to 2002 by EPA to range from roughly $121 billion to $193 billion and costs to range from $23 billion to $27 billion (OMB 2003a). A large fraction of aggregate benefits found by OMB pertain to rules limiting PM, $NO_x$, and $SO_2$. The $SO_2$ provisions of the 1990 CAA Amendments alone account for $80 billion of the aggregate benefit estimate.

## THE FUTURE

Despite the nation's significant progress in improving air quality, the problems posed by pollutant emissions in the United States are by no means solved. For example, it is estimated that the demand for electrical power in the United States will increase by 40% over the next 20 years (DOE 2003), and a substantial amount of the increased demand will be met by the burning of fossil fuels (see Figure 1-5). Increases of over 50% in total vehicle miles traveled by light-duty vehicles on the nation's roads and highways, as well as increases of off-road vehicular use, are also projected (DOE 2003). Thus, substantial improvements in cleaner power-generating and automotive technologies will be needed if the nation is to maintain the current level of air quality. Some of these improvements are already under way (for example, response to Tier II emission standards for automobiles [see Chapter 4]), and others are being considered (for example, multipollutant emission caps for power plants [see Chapter 5]).

However, even as additional emission reductions and new technologies are needed in the coming years just to maintain the current level of air quality, additional effort is likely to be deemed necessary to adequately protect human health and welfare. A number of major goals and requirements of the CAA Amendments of 1990 have yet to be met (Figure 1-6);

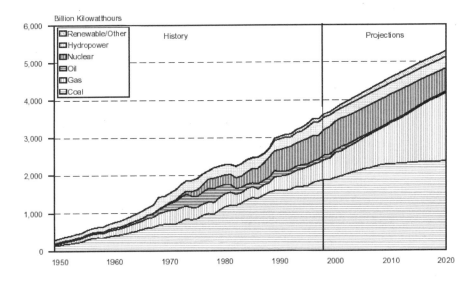

FIGURE 1-5   Electricity generation by fuel in billion kilowatt hours, 1949–1999, and projections for the Reference Case, 2000–2020.  Projections: National Energy Modeling System, run M2BASE.D060801A. SOURCE: EIA 2000.

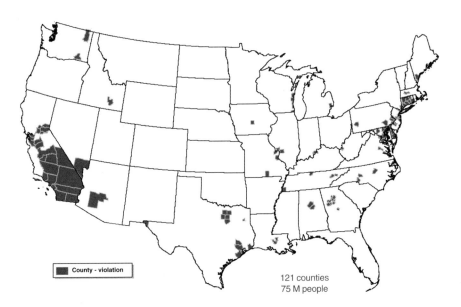

FIGURE 1-6   Counties in the continental United States where any NAAQS were violated in 1999. SOURCE: EPA 2002c.

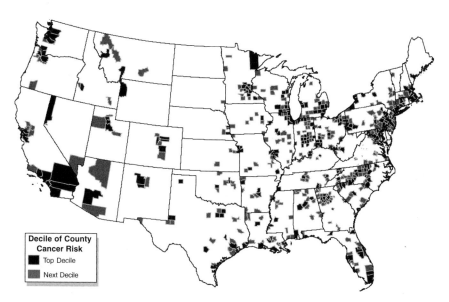

FIGURE 1-7  High cancer risk counties for urban air toxics in 1996. SOURCE: EPA 2002c.

those include total compliance with the NAAQS for $O_3$ and PM and the establishment of a technology-based regulatory program to reduce the emissions of all 188 HAPs (see Figure 1-7). Perhaps even more important, new data on the health effects of $O_3$ and PM led to the promulgation in 1997 of stricter NAAQS for $O_3$ and new NAAQS for $PM_{2.5}$ that will require even greater reductions in pollutant emissions than had been envisioned at the time the CAA Amendments of 1990 were enacted (see Figure 1-8). EPA (1999a) estimated that complete implementation of the 1990 CAA Amendments costs the nation about $27 billion annually. Most of the costs are directly borne by industry and consumers. Approximately $600 million is expended annually in federal funds, and a large amount is expended by states, tribes, and localities. Of the federal funds, approximately $200 million is dedicated to air quality research and monitoring (OMB 2003b).

## CHARGE TO THE COMMITTEE ON AIR QUALITY MANAGEMENT IN THE UNITED STATES

Given the sizable investment in air quality management envisioned for the nation over the next decade and the role science and technology can have in optimizing the effectiveness of this investment, the following ques-

FIGURE 1-8   Potential violations of the $PM_{2.5}$ (1999–2000 data) and 8-hr $O_3$ (1997–1999 data) NAAQS by county. SOURCE: EPA 2002c.

tions arise: (1) how well are scientific and technological advances incorporated into the current AQM system? and (2) to what extent could the AQM system be improved by changing the way scientific understanding and approaches and new technologies are used for air quality management in the United States? These questions are the basis for this report.

The committee has been charged to develop scientific and technical recommendations for strengthening the nation's AQM system with respect to the way it identifies and incorporates important sources of exposure to humans and ecosystems and integrates new understanding of human and ecosystem risks.[8]

In carrying out its charge, the committee has evaluated the effectiveness of the major air quality provisions of the CAA and their implementation by federal, state, tribal, and local government agencies. It also reviewed scientific and technical aspects of the policies and programs intended to manage important air pollutants, including but not limited to criteria pollutants and HAPs. In addition, the committee evaluated scientific and technical aspects of current approaches for health and environmental problem identification, regulatory standards development, AQM plan development, plan imple-

---

[8]The committee's full Statement of Task is included in Appendix B.

mentation, compliance assurance, and progress evaluation. Stratospheric $O_3$ protection and greenhouse gas emission control were not included in the scope of the study except in regard to strategies in tropospheric air quality control programs to control emissions.

A wide range of external factors beyond the scientific and technical aspects of air quality can drive the character and effectiveness of an AQM system and are relevant to our review. Governmental policies on economic growth, energy production and use, transportation, and land use, for example, affect pollutant emissions and can therefore reinforce or frustrate AQM policies. Decisions and practices of consumers concerning the technologies and products they purchase and use also affect pollutant emissions.

Legal and institutional factors also affect the implementation of AQM. Most important, the nation's federal system of government, with specific authorities assigned to the federal government and others to the states, limits the kinds of regulatory structures that can be used to administer the AQM system in the United States. For example, although air pollution issues often demand regional controls, all such controls can only be enforced at the state or federal levels.

It is beyond the scope of this report to comprehensively analyze these external factors and assess how they could be changed to enhance the effectiveness of the nation's AQM system. However, at each stage in considering implementation of the CAA, the committee attempted to take into account the degree to which these larger factors could reduce the effectiveness of specific control measures (for example, the growth in travel and its relation to automobile emission standards). The committee found that considerable progress in air quality improvement has been accomplished in the United States over the past 2-3 decades even in the face of these confounding external factors. Thus, the recommendations advanced here tend to be evolutionary in nature and do not involve a major overhaul of the AQM system.

## REPORT STRUCTURE

To provide a basic foundation for conclusions and recommendations, the committee reviewed and critiqued the key elements of the CAA and the concomitant methods and approaches to manage air quality in the United States. The discussion in Chapter 2 focuses on how standards and goals are set. Chapter 3, 4, and 5 describe the design and implementation of control strategies adopted by federal, regional, state, and local governments. Chapter 6 discusses how progress in meeting the AQM goals is measured, particularly with respect to health and ecosystem outcomes. After consideration of the major air quality challenges facing the nation in the coming decades, Chapter 7 provides a series of recommendations. In formulating these recommendations, the committee endeavored to look beyond the statutorily mandated constraints, methods, and approaches currently imposed on the nation's AQM system by the CAA and other relevant acts. Thus, pursuit of many of the recommendations will require broad acceptance within the policy-making communities and perhaps legislative action.

# 2

# Setting Goals and Standards

## INTRODUCTION

The introduction to the Clean Air Act (CAA) (42 USC § 7401) lists four overarching goals or purposes for the legislation:

(1) to protect and enhance the quality of the Nation's air resources so as to promote the public health and welfare and the productive capacity of its population;

(2) to initiate and accelerate a national research and development program to achieve the prevention and control of air pollution;

(3) to provide technical and financial assistance to State and local governments in connection with the development and execution of their air pollution prevention and control programs; and

(4) to encourage and assist the development and operation of regional air pollution prevention and control programs.

In the subsequent sections or titles to the act and its amendments, these overarching goals are further delineated into a variety of more tangible air quality goals and standards, such as those listed in Chapter 1, that are to be achieved through the implementation of rules, regulations, and practices designed to control and limit pollutant emissions. In this chapter, the committee focuses on these air quality goals and standards. We begin with a descriptive discussion of the various standards and the combined responsibilities of Congress and the administrator of the U.S. Environmental Protection Agency (EPA) in setting them. That discussion is followed by a critical analysis of specific aspects of the standard-setting procedure, especially those aspects relating to the scientific basis for the standards and the procedures used to set them.

## OVERVIEW OF AIR QUALITY STANDARDS

The CAA sets standards in a number of ways:

- The setting of National Ambient Air Quality Standards (NAAQS) for six principal pollutants (known as criteria pollutants).
- The setting of emission standards for a variety of stationary and mobile sources for substances that are the criteria pollutants, their precursors, or hazardous air pollutants (HAPs).
- Promulgate additional emission standards for HAPs that continue to pose a significant residual risk following the implementation of the first round of emission standards.
- The setting of fuel and product reformulation standards (for example, reformulated gasoline and requirements for chlorinated fluorocarbons).
- The setting of reduced caps for emissions of certain pollutants from certain industries (for example, the sulfur dioxide [$SO_2$] cap-and-trade program).

The CAA also contains many provisions for attaining and maintaining these standards. In this chapter, the committee focuses on, and critiques, the process by which many of these standards are set. Subsequent chapters discuss how they are implemented.

The CAA begins by addressing two major categories of pollutants for which standards are set differently: criteria pollutants and HAPs. The principal difference between the two arises from the specification in the CAA that the presence of criteria pollutants "in the ambient air results from numerous or diverse mobile or stationary sources." No such requirement is stated for HAPs.[1] Thus, presumably, criteria pollutants are more ubiquitous, pose a risk to a larger fraction of the general population, and have more widespread impacts on ecosystems and natural resources than HAPs. Criteria pollutants and HAPs are managed through fundamentally different regulatory frameworks. Criteria pollutants are regulated primarily through the setting of ambient-air-concentration and time standards, known as the NAAQS, and taking action to attain these standards. HAPs are regulated through the promulgation of standards that limit the release or emissions of such compounds (as opposed to their ambient concentrations), followed in

---

[1]The only specific limitation currently placed on HAPs in the CAA is that they cannot be criteria pollutants.

the cases of major stationary sources and area sources by assessment of residual risk. The responsibility for setting the standards for both types of pollutants is assigned to the EPA administrator.

In addition to the programs to control criteria pollutants and HAPs, the CAA includes provisions to control emissions from mobile sources, protect areas with good air quality, reduce the effects of acid deposition (or acid rain), safeguard stratospheric ozone ($O_3$), and reduce visibility impairment resulting from regional haze. However, the tenor of these provisions is substantially different from those relating to the programs for criteria pollutants and HAPs. Although the CAA directs the EPA administrator to set air quality and emission standards for criteria pollutants and HAPs, it is more explicit about setting standards for the other provisions, with Congress itself often setting the standards. For example, the CAA now includes specific standards for evaporative and exhaust emissions from light-duty and heavy-duty on-road vehicles and engines. Controls are also required for a wide range of nonroad engines (such as lawnmowers, construction equipment, and locomotives), and programs are mandated for clean fuel and inspection and maintenance of light-duty vehicles. Similarly, in the case of acid rain mitigation, the CAA Amendments of 1990 contain language that establishes a specific nationwide cap for $SO_2$ emissions and standards for nitrogen oxides ($NO_x$) emissions from electric utilities.

## THE STANDARD-SETTING PROCESSES

### Criteria Pollutants

Criteria pollutants were first defined in the 1970 Amendments to the CAA, which directed the administrator of EPA to identify those widespread ambient air pollutants that are reasonably expected to present a danger to public health or welfare.[2] On the basis of air quality criteria[3]—that is, the current state of scientific knowledge on the effects of these pollutants on health and welfare—the administrator is directed to develop and promulgate primary and secondary NAAQS for each criteria air pollutant. In addition to specifying a maximum ambient concentration for each pollutant, promulgation of a standard must also include descriptions of the moni-

---

[2]Within the framework of the CAA, "welfare" refers to the viability of agriculture and ecosystems (such as forests and wildlands), the protection of materials (such as monuments and buildings), and the maintenance of visibility (EPA 2002d).

[3]Air quality criteria are defined in Section 108 of the CAA as a summary of the "latest scientific knowledge useful in indicating the kind and extent of all identifiable effects on public health or welfare which may be expected from the presence of such pollutant in the ambient air."

toring and statistical methods that are to be used to determine whether an area is in compliance with the standard.[4] Primary standards are intended to protect public health "with an adequate margin of safety" for the most sensitive population subgroups. Secondary standards are intended to protect against adverse public welfare effects. Although the CAA specifies a date when a given primary standard is to be achieved and provides EPA with authority to enforce state and tribal compliance, no such timetable and enforcement authority are provided for secondary standards (see Chapter 3).

In 1971, NAAQS were established for the first time for six criteria pollutants: carbon monoxide (CO), nitrogen dioxide ($NO_2$), $SO_2$, total suspended particulate matter (TSP), hydrocarbons (HCs), and photochemical oxidants. Lead (Pb) was added to the list in 1976, photochemical oxidants were replaced by $O_3$ in 1979, and HCs were removed in 1983. The definition of suspended particles as a criteria pollutant has also changed. TSP was revised in 1987 to include only particles with an equivalent aerodynamic particle diameter of less than or equal to 10 micrometers (μm) ($PM_{10}$) and was further revised in 1997 to include a separate standard for particles with an equivalent aerodynamic particle diameter of less than or equal to 2.5 μm ($PM_{2.5}$). The current standards for criteria pollutants are provided in Table 2-1. Although efforts to meet the NAAQS have not always successfully resulted in attainment of the standards, they appear to have been responsible for sizable reductions in pollutant emissions across the nation (see Chapter 6 for further discussion).

### The Procedure for Setting NAAQS

The CAA instructs the EPA administrator to specify primary and secondary NAAQS and to conduct a review of the air quality criteria and NAAQS for each pollutant at least every 5 years. The review process is a complex one that includes input and comment from independent scientific bodies as well as the general public (see Figure 2-1).

EPA's Office of Research and Development (ORD) prepares a detailed summary, called a criteria document, for each criteria air pollutant. The criteria document is based on the existing body of scientific and technical information and typically includes chapters on emission sources, air concentrations, exposure, dosimetry, and health and welfare effects, as well as a concluding synthesis chapter. The research findings summarized include results from studies supported by EPA, other federal agencies, industry, and

---

[4]A discussion of the methods used to establish nonattainment of NAAQS and the subsequent actions a state or local authority must undertake to bring about attainment is presented in Chapter 3.

TABLE 2-1  NAAQS in Effect as of January 2003[a]

| Pollutant | Primary Standard (Health-Based) Type of Average | Primary Standard Level Concentration | Secondary Standard (Welfare-Based) Type of Average | Secondary Standard Level Concentration |
|---|---|---|---|---|
| $PM_{10}$ | Annual arithmetic mean | 50 $\mu g/m^3$ | | Same as primary standard |
| | 24-hr average not to be exceeded more than once per year on average over 3 yr | 150 $\mu g/m^3$ | | Same as primary standard |
| $PM_{2.5}$ | Spatial and annual arithmetic mean in area | 15 $\mu g/m^3$ | | Same as primary standard |
| | 98th percentile of the 24-hr average | 65 $\mu g/m^3$ | | Same as primary standard |
| $O_3$[b] | Maximum daily 1-hr average to be exceeded no more than once per year averaged over 3 consecutive years | 0.12 ppm | | Same as primary standard |
| | 3-yr average of the annual fourth highest daily 8-hr average | 0.08 ppm | | Same as primary standard |
| $NO_2$ | Annual arithmetic mean | 0.053 ppm | | Same as primary standard |
| $SO_2$ | Annual arithmetic mean | 0.03 ppm | 3 hr | 0.50 ppm |
| | 24-hr average | 0.14 ppm | | |
| CO | 8 hr (not to be exceeded more than once per year) | 9 ppm | | No secondary standard |
| | 1 hr (not to be exceeded more than once per year) | 35 ppm | | No secondary standard |
| Lead | Maximum quarterly average | 1.5 $\mu g/m^3$ | | Same as primary standard |

[a]A more detailed discussion of how an area is determined to be in attainment or nonattainment of a NAAQS is presented in Chapter 3.
[b]EPA is phasing out the 1-hr, 0.12-ppm standards (primary and secondary) and putting in place the 8-hr, 0.08-ppm standards. However, the 0.12-ppm standards will not be revoked in a given area until that area has achieved 3 consecutive years of air quality data meeting the 1-hr standard.
Abbreviations: $\mu g/m^3$, micrograms per cubic meter; ppm, parts per million (by volume); hr, hour; yr, year.
SOURCE: Adapted from EPA 2001a.

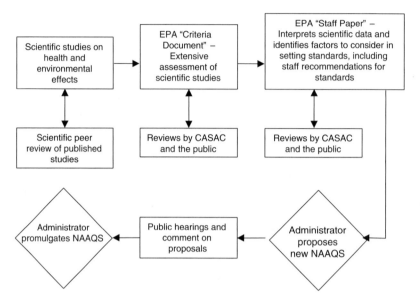

FIGURE 2-1    Flow diagram illustrating the process by which the EPA administrator reviews and sets a new NAAQS. CASAC refers to the Clean Air Scientific Advisory Committee. Diamonds are used to denote official actions by the administrator. SOURCE: Greenbaum et al. 2001. Reprinted with permission from the *American Journal of Epidemiology*; copyright 2001, Oxford Press.

private funding organizations. Investigators from both inside and outside EPA collaborate in writing the document.

Next, EPA's Office of Air Quality Planning and Standards (OAQPS) in the Office of Air and Radiation prepares a document, called the staff paper, which recommends and provides the justification for policy options presented to the EPA administrator, who makes a determination on whether to retain an existing standard or propose a new one. In preparing the staff paper, EPA uses the information included in the criteria document and also conducts analyses of population exposure to characterize risk and to recommend standards. Both the criteria document and the staff paper are made available for public comment. The two documents, along with the public comments, are reviewed by the Clean Air Scientific Advisory Committee (CASAC), which is a committee composed of independent experts from outside EPA and is organizationally situated within the EPA Science Advisory Board (SAB). CASAC makes recommendations to EPA staff on revisions to both the criteria document and the staff paper, resulting in one or more rounds of revision and review. When satisfied, CASAC informs the EPA administrator that the document fully and fairly represents the current state of the science. Because of the effort involved in their preparation and the rigorous review process involved, criteria documents and staff papers have traditionally served as comprehensive reviews of the current understanding of air pollution health effects at the time of their publication and, as such, have stimulated new focused research.

On the basis of the criteria document and staff paper, the EPA administrator publishes proposals for new or revised NAAQS in the *Federal Register*, and a new round of public comment ensues. The administrator then makes final decisions on the NAAQS, taking into account both public input and advice from CASAC. The primary and secondary standards, including the levels and the forms of the standards, together with their justification, are published in the *Federal Register* as part of the standard promulgation process. The Supreme Court has determined that the CAA requires that the setting of primary NAAQS is to be done without consideration of the economic consequences (Whitman v. American Trucking Association, 531 U.S. 457, 2001). However, there are two points in the process where costs are assessed: (1) under Executive Orders dating back to the Carter administration and the Small Business Regulatory Enforcement Fairness Act, major federal regulations are subject to a regulatory impact analysis by the Office of Management and Budget, and (2) the Congressional Review Act provides Congress up to 60 days following promulgation of any rule to conduct hearings and review the rule (although no congressional action is required for the NAAQS to take effect).

According to the CAA, each of the NAAQS for each of the criteria pollutants must be reviewed every 5 years. However, the complexity of the process and the sheer volume of new research results on some criteria pollutants have made it necessary to extend the periods between reviews (see Figure 2-2). As a result, EPA has been sued by some stakeholders and at times required by the courts to complete an overdue review on a court-ordered schedule.

## Protection of Ecosystems and the Establishment of Secondary NAAQS

Although the CAA empowers EPA to set independent primary and secondary standards for each criteria pollutant, and criteria documents prepared by EPA have included reviews of the available data on impacts to ecosystems, visibility, human-made structures, and other aspects of public welfare, $SO_2$ is the only criteria pollutant for which there is a unique secondary standard (Table 2-1). The promulgation of common standards for protecting public health and welfare could simply reflect a judgment on EPA's part that humans and ecosystems have similar sensitivity to air pollutants or that humans are more sensitive; thus, a single standard can adequately protect both. It is more likely, however, that the correspondence

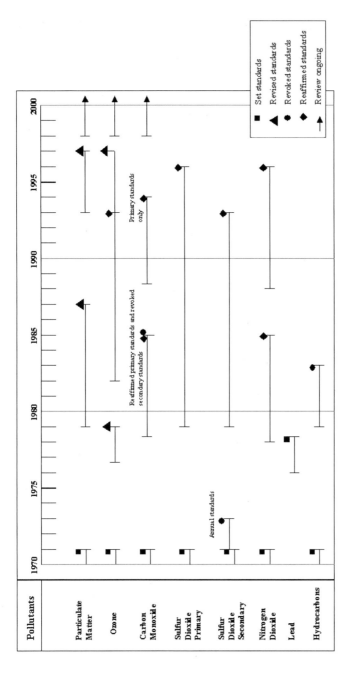

FIGURE 2-2 Timeline illustrating historical sequence of the periodic NAAQS reviews and final decisions carried out by EPA since the passage of the 1970 CAA Amendments. For particulate matter, standards were set in 1971 for total suspended particles (TSP). In 1987, the standards were revised to more stringent $PM_{10}$ standards. In 1997, those standards were revised by adding more stringent $PM_{2.5}$ standards. For ozone, photochemical oxidant standards were set in 1971. They were revised to less stringent 1-hr ozone standards in 1979. Then, in 1997, the standards were revised to more stringent 8-hr standards. For sulfur dioxide, the decision to reaffirm the primary standards was remanded in 1996. EPA's response to the remand is ongoing.

between the two standards reflects a historical and reasonable tendency in EPA to set priorities to protect human health over aspects of public welfare and to focus on urban rather than nonurban pollution. The inability to enforce standards to protect public welfare by the requirement for attainment by a specific date might also play a part.

Whatever the reason that led EPA to use identical primary and secondary NAAQS in the past, it is becoming increasingly evident that a new approach will be needed in the future. There is growing evidence that the current forms of the NAAQS are not providing adequate protection to sensitive ecosystems and crops (see Figure 2-3) (Driscoll et al. 2001a; Mauzerall and Wang 2001). Moreover, new research documenting the economic importance of the functions and services supplied to society by ecosystems (Daily 1997; Ecological Society of America [ESA] 1997a) suggests that air pollution damage to ecosystems exacts economic as well as environmental costs to the nation. At the same time, air quality management (AQM) and science are experiencing a shift in focus to problems related to multistate (and by extension rural) air quality problems.[5] Thus, the nation's AQM system may now be in a better position to tackle the problem of air pollution damage to sensitive ecosystems and crops. In the CAA Amendments of 1990, Congress instructed EPA to undertake a comprehensive review of the need for and use of standards to protect public welfare (42 USC § 7409 [1990]). However, such a study was never undertaken. In Chapter 7, the committee advances more specific recommendations for improving the scientific basis for setting secondary standards by strengthening the nation's ability to monitor ecosystems and their exposure and response to air pollution.

## National Emission Standards Mandated by Congress to Help Attain NAAQS

The general procedure for attainment of NAAQS is specified in the CAA. The process includes the monitoring of ambient air pollution concentrations, the designation of nonattainment areas on the basis of the data from this monitoring, and the development of state implementation plans (SIPs) to achieve the emission reductions necessary to bring areas into attainment.[6] However, Congress recognized that the states could not be expected to achieve NAAQS for criteria pollutants on their own without potential substantial economic disruptions. In particular, it was recognized that economies of scale could be realized through the promulgation of more

---

[5]Later in the report, a number of regional approaches to address these types of problems are discussed.

[6]A detailed discussion of the SIP process and its implementation is provided in Chapter 3.

FIGURE 2-3    Foliar injury to cotton induced by chronic exposure to ozone. Chronic exposures consist of relatively low concentrations (for example, less than 40 parts per billion), with periodic, random, intermittent episodes or relatively high ozone concentrations (for example, greater than 80 parts per billion) throughout the plant growth season. Symptoms of chronic injury include premature senescence and purple pigmentation in the interveinal areas. Photograph courtesy of P.J. Temple (retired, U.S. Department of Agriculture Forest Service). SOURCE: Krupa et al. 1998.

uniform national regulations for certain key sources of criteria pollutants. To that end, the CAA includes a number of programs to reduce criteria pollutant emissions from stationary and mobile sources. As noted earlier, these national controls have been implemented in large part by setting specific emission standards within the act itself or, barring that, by providing specific instructions to the EPA administrator. Detailed discussions of the emission-control programs mandated in the CAA for mobile and stationary sources are presented in Chapters 4 and 5, respectively.

## Hazardous Air Pollutants

The CAA Amendments of 1970 required EPA to identify and list all air pollutants (not already identified as criteria pollutants) that "may reasonably be anticipated to result in an increase in mortality or an increase in serious irreversible or incapacitating reversible illness." For each pollutant identified, EPA was to then promulgate national emission standards for hazardous air pollutants (NESHAPs) at levels that would ensure the protection of public health with "an ample margin of safety" and to prevent any significant and adverse environmental effects, which may reasonably be anticipated, on wildlife, aquatic life, or other natural resources. During the 1970s and 1980s, EPA began developing risk assessment methods necessary to establish the scientific basis for regulating HAPs (EPA 2000a). Despite advances in risk assessment methods gained through this work, the chemical-by-chemical regulatory approach based solely on risk proved difficult. Legal, scientific, and policy debates ensued over the risk assessment methods and assumptions, how much health risk data are needed to justify regulation, analyses of the costs to industry and benefits to human health and the environment, and decisions about "how safe is safe" (EPA 2000a). In the 20 years following enactment of the 1970 legislation, EPA identified only eight pollutants as HAPs and regulated sources of seven of them (asbestos, benzene, beryllium, inorganic arsenic, mercury, radionuclides, and vinyl chloride) (NRC 1994).

With the CAA Amendments of 1990, Congress mandated a new approach. To expedite control of HAPs without explicit consideration of their inherent toxicity and potential risk, Congress provided a list of 189 compounds to be controlled by EPA as HAPs (see Appendix C). The EPA administrator was given the responsibility to review and amend the list of regulated HAPs periodically as dictated by new scientific information. However, since passage of the CAA Amendments of 1990, one compound has been deleted from the list (caprolactam), the scope of chemicals covered by glycol ethers was reduced, and no compound has been added to the list.[7]

### Current Standard-Setting Procedure for HAPs

In contrast to criteria pollutants for which ambient concentration standards are used, the control of HAPs is based on an initial promulgation of emission standards and a subsequent assessment of risk that remains after implementation of these standards (the CAA defines this remaining risk as the "residual risk"). The standards are to be imposed on (1) all major

---

[7]On May 30, 2003, EPA proposed to remove the compound methyl ethyl ketone (MEK) from the HAPs list, and on November 21, 2003, EPA proposed to remove ethylene glycol monobutyl ether from the list.

sources, which are defined as "any stationary source or group of stationary sources located within a contiguous area and under common control that emits or has the potential to emit . . . 10 tons per year or more of any hazardous air pollutant or 25 tons per year or more of any combination of hazardous air pollutants"; and (2) a sufficient number of area sources "to ensure that area sources (excluding mobile sources) representing 90 percent of the area source emissions of the 30 (or more) hazardous air pollutants that present the greatest threat to public health in the largest number of urban areas are subject to regulation." In the case of major sources, 174 different types of sources were identified as emitters of HAPs and targeted for regulation by Congress. In a separate portion of the CAA Amendments of 1990 (Section 202), EPA was instructed to develop and promulgate emission standards for vehicles and fuels "on those categories of emissions that pose the greatest risk to human health . . . [and] at a minimum, apply to emissions of benzene and formaldehyde."

In the first regulatory phase of the program described in Section 112 of the 1990 CAA Amendments for major and area sources, EPA was directed to promulgate national technology-based emission standards for HAPs using available control technologies or work practices. In this regard, the CAA defined two types of emission standards for promulgation:

- Maximum achievable control technologies (MACTs) are emission standards that achieve "the maximum degree of reduction in emissions of the hazardous air pollutants . . . that the Administrator, taking into consideration the cost of achieving such emission reduction, and any non-air quality health and environmental impacts and energy requirements, determines is achievable."
- Generally available control technologies (GACTs) are less stringent emission standards based on the use of more standard technologies and work practices.

Despite the use of the word "technologies" in those definitions, Congress did not intend them to be technologically prescriptive. They were intended to be technology-neutral, performance-oriented standards that set a maximum allowable emission rate, based on the emissions obtained using MACT or GACT, and that allowed the affected industries and facilities to choose any combination of technologies and practices to achieve these performance levels (EPA 2000a).[8]

Congress mandated that MACT be applied to all major stationary sources of HAPs. For area sources, the EPA administrator was directed to

---

[8]A more detailed discussion of the implementation of MACT and GACT controls and their effectiveness as a promulgator of technology-neutral standards is presented in Chapter 5.

select between MACT and GACT as deemed appropriate. For HAPs with a health threshold, EPA could consider the threshold with an ample margin of safety in establishing the emission standard. Congress further mandated a 10-year schedule for the promulgation of MACT standards for major sources, with certain standards being promulgated in the first 2 years, 25% in the first 4 years, an additional 25% promulgated not later than the seventh year, and the remaining 50% not later than the tenth year. As of February 2003, EPA had promulgated 79 MACT standards affecting 123 source categories (T. Clemons, EPA, Washington, DC, personal commun., March 31, 2003). In addition, two standards regulating solid waste were promulgated under Section 129 of the CAA Amendments of 1990. In total, over 100 HAPs fell under these regulations (EPA 2000a). In May 2003, EPA indicated that it expects to finalize all MACT standards by the time that states will be required to set MACT limits on a facility-by-facility basis according to Section 112(j), as amended (EPA 2003b). Clearly, the promulgation of MACT has proved to be a complex task for EPA and has occurred more slowly than mandated by Congress. This delay has given rise to criticism of EPA by some environmental groups (for example, Williams 2003), as well as litigation (Sierra Club v. U.S. Environmental Protection Agency, No. 02-1135, 2002 [DC Circuit]).

In the second regulatory phase of the HAPs program, EPA was instructed to conduct an assessment of and report on the residual risk due to HAPs emitted from the regulated major and area sources discussed above. Then, in the absence of any specific action by Congress for a 2-year period following issuance of the report, the EPA administrator was to promulgate additional emission standards to "provide an ample margin of safety to protect public health or to prevent, taking into consideration costs, energy, safety, and other relevant factors, an adverse environmental effect." The promulgation of these additional emission standards was to occur no later than 8 years after EPA's initial promulgation of the technology-based standards. Although no formal standard for acceptable residual risk was mandated in the CAA Amendments of 1990, the act cited an example of such a standard: the reduction of excess cancer risk for the most exposed individuals to less than 1 in 1 million for a lifetime of exposure to a particular HAP.

Because of the difficulties in assessing residual risk, completion of the residual risk analysis and promulgation of additional emission controls mandated in the CAA are still some years off. The agency is proceeding to investigate the residual risk that is likely to remain after attainment of the MACT and GACT standards. In 1999, EPA reported to Congress on its progress in determining residual risk for cancer and non-cancer health effects. The report describes how EPA intends to calculate risk, what standard it will apply (the 1 in 1 million as applied in the 1989 benzene

NESHAP), and progress in furthering the understanding of the health effects of HAPs (EPA 1999b). The 1990 CAA Amendments also called for a number of special studies related to assessing risk resulting from exposure to HAPs, including studies on emissions from electric utility-steam generating units and publicly owned treatment works.

In the case of area sources, the EPA administrator was directed to first undertake a research program of monitoring and analysis to identify the area sources of HAPs in representative urban locations. On the basis of the research, the administrator was to then propose a "comprehensive strategy" to control the emissions of urban area sources that in the aggregate accounted for at least 90% of the emissions of the 30 or more HAPs that presented the greatest health risk in the largest number of urban areas. Congress further mandated that one specific goal of the strategy was to be a 75% reduction in the incidence of cancers attributable to all sources of HAPs. The comprehensive strategy was to be completed within 5 years of the passage of the CAA Amendments of 1990 and implemented "as expeditiously as practicable, assuring that all sources are in compliance with all requirements not later than 9 years after the . . . enactment of the" CAA Amendments of 1990. To meet the requirements, EPA developed the integrated urban air toxics strategy (EPA 2000b) and later issued a notice in the *Federal Register* (67 Fed. Reg. 43112 [2002]) that focused on 33 HAPs and listed 47 area sources that are or will be subject to standards. In developing the strategy, EPA attempted to address areas of high exposure by characterizing exposure and risk as a function of geography and demography as a means of selecting the highest priority HAPs from among the 188. Subsequently, EPA refined this approach by conducting the National Air Toxics Assessment (NATA), which summarizes the exposure and risk levels for each of the 33 highest priority HAPs (a subset of 32 of the list of 188 plus diesel particulate matter) (EPA 2002e). To better understand the sources and health risks associated with HAPs, the strategy includes activities to expand HAPs monitoring, improve emission inventories and national- and local-scale modeling, investigate health effects and exposures to HAPs in ambient and indoor air, and improve assessment tools.

In Title II of the CAA Amendments of 1990, several actions were included to reduce emissions of mobile-source HAPs. The act required that EPA implement, by January 1, 1995, a new reformulated gasoline (RFG) program for $O_3$ nonattainment areas that for the first time required caps on a number of toxic and volatile constituents of gasoline, especially benzene. In Section 202, the 1990 CAA Amendments also required the mobile-source HAPs program to conduct a motor-vehicle-source HAPs study and to promulgate regulations. Implementation of this requirement is discussed in Chapter 4.

## GOALS FOR MITIGATING VISIBILITY DEGRADATION

Most visibility impairment is caused by the presence of fine PM suspended in the atmosphere; these particles scatter and absorb light and, in so doing, reduce visibility (see Figure 2-4A and 2-4B). Haze is a common phenomenon in most parts of the United States. Although most severe in the East and California, hazy conditions are encountered episodically in virtually all areas of the country, even those remote from major centers of population and industrial development (see Figure 2-4C). The affected areas include many of the nation's most beautiful national parks and wilderness areas. Some visibility impairment is natural (for example, from wind-blown dust or wildfires); however, much of it is caused by pollutant emissions of particles and gases, such as $SO_2$, that are converted into particles in the atmosphere.

In response to growing concerns about deteriorating visibility in our nation's recreational areas, the CAA Amendments of 1977 (Section 169A) established a national goal of preventing and remedying visibility impairment due to anthropogenic pollution in Class I areas, which include most U.S. large parks and wilderness areas. The CAA Amendments of 1990 provided additional emphasis on regional haze issues. In 1999, EPA promulgated a Regional Haze Rule, which established a 65-year program to return 156 national parks and wilderness areas to their natural visibility conditions. To accomplish that, anthropogenic emissions in the United States would have to be reduced until visibility in all Class I areas is not noticeably poorer than that under natural conditions. The EPA rule establishes "reasonable-progress" goals that are based on a uniform rate of visibility improvement between baseline conditions (measured from 2000 to 2004) and natural visibility conditions to be achieved by 2064.[9] SIPs outlining plans for achieving visibility goals are due no later than 2008, and these plans are to be updated every 10 years thereafter to ensure that reasonable progress has been achieved.

## STANDARDS FOR MITIGATING EFFECTS OF ACID RAIN

"Acid rain" (also known as acid deposition) refers to the wet and dry deposition of acidic compounds to the earth's surface that can have deleterious effects on ecosystems and materials (see Figure 2-5). In the United States, most of the harmful acidity is in the form of sulfuric and nitric acids, which are produced in the atmosphere from the chemical conversion of $SO_2$ and $NO_x$. $SO_2$ enters the atmosphere primarily via the combustion of sulfur-containing fossil fuel largely in coal-fired power plants. $NO_x$ enters

---

[9]Implementation of the EPA residual haze rule is discussed in Chapter 3.

A

B

C

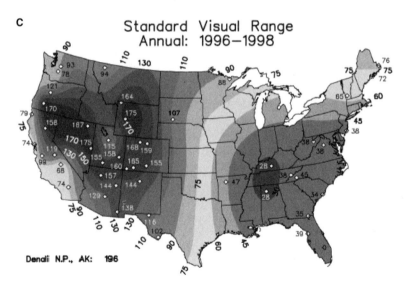

FIGURE 2-4  The impact of haze on visibility. (A) Good air quality contrasted with (B) poor air quality in Big Bend National Park. SOURCE: NPS 2002a,b,c. (C) Standard visual range for the period 1996–1998 in units of kilometers. SOURCE: VIEWS 2003.

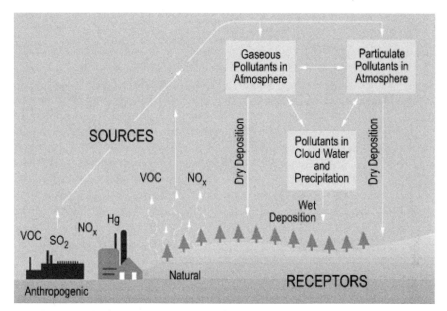

FIGURE 2-5   Anthropogenic sources and natural sources contribute emissions that result in the deposition of acidic compounds.  SOURCE: EPA 1999c.

the atmosphere primarily via high temperature combustion of coal, oil, and gas in power plants and of gasoline and diesel fuels in automobiles and trucks.  Because it typically takes days to weeks for atmospheric $SO_2$ and $NO_x$ to be converted to acids and deposited to the earth's surface, acid deposition occurs on a multistate scale hundreds (if not thousands) of miles away from its sources.

### Controls on Acid Rain Precursors before the CAA Amendments of 1990

The establishment of NAAQS for $SO_2$ and $NO_2$ as a result of the CAA Amendments of 1970 marked the beginning of a nationwide program to control their emissions.  (In addition to the NAAQS, EPA, pursuant to the 1970 CAA Amendments, imposed a new-source performance standard (NSPS) of a maximum of 1.2 pounds (lb) of $SO_2$ per million British thermal units (Btu) generated for all new power plants.)  However, the initial response to the establishment of NAAQS for $SO_2$ and $NO_2$ did not necessarily result in a reduction of their emissions from stationary sources.  It was generally found that the most cost effective method to limit the contribution of stationary-source emissions of $SO_2$ and $NO_2$ to local nonattainment events was to install tall stacks, which dispersed the emissions and thereby eliminated local areas of elevated concentrations in the immediate vicinity of the facility. As a result, 429 tall stacks, many over 500 feet tall, were constructed on coal-fired boilers in the electricity industry during the 1970s (Regens and Rycroft 1988), and, by 1980, the vast majority of urban areas in the United States were in attainment of the NAAQS for $SO_2$ and $NO_2$. However, the use of tall stacks had an unintended consequence: by facilitating the long-distance transport of $SO_2$ and $NO_2$ and their conversion to sulfuric and nitric acids before deposition, the installation of tall stacks exacerbated the acid rain problem.

The CAA Amendments of 1977 required EPA to address new coal utility plants (those built after 1978). EPA promulgated a standard allowing new plants to either (1) remove 90% of potential $SO_2$ emissions (as determined by the sulfur content of the fuel burned) and operate with an emission rate below 1.2 lb of $SO_2$ per million Btu, or (2) remove 70% of potential $SO_2$ emissions and operate with an emission rate less than 0.6 lb of $SO_2$ per 1 million Btu. (The 1977 Amendments also prohibited using tall stacks to comply with emission standards.) The percent-reduction requirement effectively forced all new coal plants to operate with flue gas desulfurization. It also substantially reduced the advantage of using low-sulfur coal as a means of compliance, because a facility using low-sulfur coal would still be required to remove at least 70% of the potential emissions. Because coal in the western United States has lower sulfur, the statute had the effect of imposing a more stringent emission cap on new sources in the West than in the East, perhaps to help prevent significant deterioration of existing air quality in the West.

As a result of the aforementioned emission standards, U.S. $SO_2$ emissions peaked in the early 1970s and declined steadily throughout the remainder of the 1970s and early 1980s (see Figure 2-6). By 1985, $SO_2$ emissions nationwide had declined by about 25% from the peak emissions of the 1970s. However, after 1985, the NSPS lost its ability to affect further $SO_2$ emission reductions. The change seems to have been caused by the focus of the statutes on new generating plants, which had two significant effects: (1) it created a significant gap between the $SO_2$ emissions of older plants and those permitted for new plants; and (2) it raised the costs of new plants and thus created an economic incentive for keeping old plants on line beyond their original design lifetimes. As a result, it became increasingly common to keep older plants on line through lifetime extension projects. By 1985, 83% of power-plant $SO_2$ emissions in the United States came from generating plants not meeting the 1971 NSPS, and $SO_2$ emissions in the United States were relatively stable (Ellerman et al. 2000).

In contrast to $SO_2$ emissions, $NO_x$ emissions in the United States remained fairly flat over the 1970s and 1980s (see Figure 2-6). The emissions peaked in the late 1970s and decreased somewhat over the next 5 years as

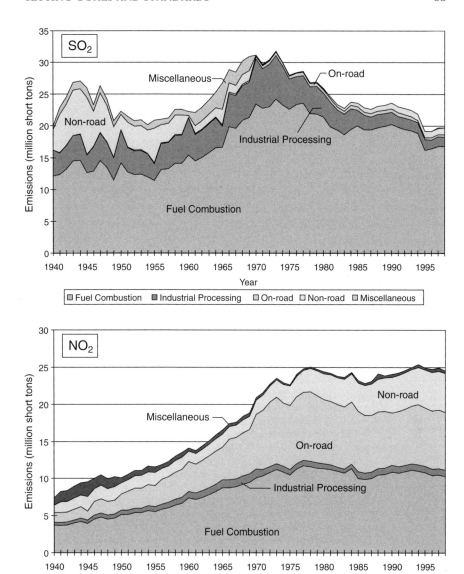

FIGURE 2-6    Trends in nationwide $SO_2$ and $NO_2$ emissions by year since 1940. In 2003, EPA revised some of its emission estimates for the years 1970 and later (see EPA [2003r] for the revised profiles).   SOURCE: EPA 2000c.

a result of the imposition of $NO_x$ emission standards on new motor vehicles (see Chapter 4) and then inched upward for the remainder of the 1980s.

### Acid Rain Goals Set by the CAA Amendments of 1990

Acid rain gradually emerged as a serious environmental concern in the late 1970s. A growing body of scientific evidence had accumulated that documented the deleterious impacts of acid rain on ecosystems, aquatic life, and property, particularly in regions where soils are acidic, such as eastern Canada and the northeastern United States. Pressure for remedial action began to build from environmental groups and officials from the northeastern states most affected by acid rain. The Canadian government also began to pressure the United States, claiming that its ecosystems were being damaged from the transport of acid rain precursors from the United States.

In response, Congress created the National Acid Precipitation Assessment Program (NAPAP) in 1980 to study the impacts of acid deposition, recommend if emission controls were needed to mitigate these impacts and, if so, the magnitude of the emission reductions needed (see Box 2-1). Ten years

---

**BOX 2-1  The Role of NAPAP in Shaping the Acid Rain Provision of the CAA Amendments of 1990**

The National Acid Precipitation Assessment Program (NAPAP) was established by Congress to study the impacts of acid deposition, recommend if emission controls were needed to mitigate these impacts, and, if so, the magnitude of the emissions reductions needed.

A major endeavor of NAPAP was the development of the regional acid deposition model (RADM) (Chang et al. 1987), a state-of-the-science 3-dimensional grid model capable of simulating the physical and chemical processes leading to the formation and deposition of acidic species[a] (NAPAP, 1991a). Model development started in 1983 and its application was completed in the early 1990s. The model (together with similar tools like the ADOM model developed for Canada) provided important insights into the source-receptor relationships of the acid deposition problem in the United States. However, these modeling exercises (together with the rest of the synthesis of the acid deposition research) came to fruition very close to the time of the completion of the CAA Amendments of 1990 and appear to have played a minor role in the development of the acid rain provisions in the 1990 Amendments. Instead it appears that a complex set of technological, legal, and political considerations played the most critical roles in shaping the emission targets for the acid rain program in the 1990 Amendments. These other considerations included the technological and economic feasibility of reducing emissions as well as the need to reach consensus on a regionally and politically divisive issue.

---

[a]A discussion of air quality models is presented in Chapter 3.

later, in Title IV of the CAA Amendments of 1990, Congress enacted specific legislation to mitigate the adverse effects of acid rain as well as to "encourage energy conservation, use of renewable and clean alternative technologies, and pollution prevention as a long-range strategy . . . for reducing air pollution and other adverse impacts of energy production and use."

The legislation in Title IV represented a significant departure from the regulatory approaches prescribed by Congress for criteria pollutants and HAPs. Instead of directing the EPA administrator to set standards to protect human health and welfare, Congress imposed their own standards—standards that, as noted in Box 2-1, were influenced to some extent by scientific understanding of the effects of acid rain but also by nontechnical economic and political considerations. The standards prescribed by Congress were aimed at the emissions of $SO_2$ and $NO_x$ from electric utilities and were designed to bring about significant reductions in these emissions nationwide. Electric utilities were targeted because they were estimated to contribute two-thirds of all $SO_2$ emissions and one-third of $NO_x$ emissions (EPA 1999c).

Because the scientific evidence suggested that $SO_2$ emissions made the largest contribution to acid rain (NAPAP 1991a), the most aggressive control program in Title IV was aimed at $SO_2$ emissions. Specifically, $SO_2$ emissions from electric utilities nationwide were capped at an amount that would require a decrease in total emissions by 2010 of 10 million tons (or about 50%) relative to emission levels of 1980. Instead of imposing a technological or emissions-based standard on power-generating facilities, Congress specified that the emission reductions were to be achieved through a market-based mechanism; specifically in this case, a cap-and-trade program. A smaller and more traditional program for reducing $NO_x$ emissions was also enacted. This program involved a two-phase strategy to reduce $NO_x$ emissions from coal-fired electric utility plants by over 400,000 tons per year between 1996 and 1999 (Phase I) and by approximately 1.17 million tons per year beginning in year 2000 (Phase II). To accomplish these reductions, Congress imposed emission standards on power plants; standards varied depending upon the type of facility. Language was also included to allow a state or group of states to petition the EPA administrator to use a cap-and-trade program instead of emission standards to meet Congress's $NO_x$ reductions goals. A more detailed discussion of the $SO_2$ and $NO_x$ acid-rain emission-control programs and their implementation can be found in Chapter 5.

## Environmental Justice as an Air Quality Goal

Environmental justice has been defined by EPA as the "fair treatment for people of all races, cultures, and incomes, regarding development of environmental laws, regulations and policies" (EPA 2003c). In practice, environmental justice issues have been concerned with the adverse health and economic effects of environmental hazards when disproportionately suffered by minority and low-income communities. Historically, Title VI of the Civil Rights Act of 1964 has been the primary instrument available to these communities to redress and ameliorate environmental injustices. Because this act specifically forbids funding recipients, such as state agencies, from using criteria or administrative methods that have the effect of subjecting individuals to discrimination on the grounds of race, color, or national origin, relief can be obtained in principle by alleging discriminatory environmental and health effects that have resulted from environmental permits issued by state agencies that receive federal funds.

The CAA and its amendments make no direct or specific reference to environmental justice. However, there are clearly environmental justice issues that can arise in the implementation of its many provisions. For example, trading programs, if not carefully designed (see Chapter 5), could result in hot spots[10] that might disproportionately affect minority and other low-income communities (Solomon and Lee 2000 and references therein). In addition, the highest ambient air pollution concentrations are most often found in densely populated urban centers, where the highest proportion of minority and other low-income populations are also found, resulting in a disproportionate burden of effects (NRC 2003b). Finally, recent health studies have suggested that people with lower socioeconomic status are more likely to suffer premature mortality from exposure to air pollution than higher-income populations (Krewski et al. 2000a,b). A number of studies have been conducted that show, with varying degrees of uncertainty, that a correlation exists between race, income level, and a disproportionate exposure to environmental toxicants (Brown 1995; Goldman 1994; Perlin et al. 2001; Sexton et al. 1993; Zimmerman 1993; Gwynn and Thurston 2001).

In response to the aforementioned concerns, EPA established the Office of Environmental Justice in 1992 to integrate environmental justice into EPA's policies, programs, and activities. In February 1994, the President of the United States issued Executive Order 12898 on environmental justice. That order was designed to focus federal attention on the environmental

---

[10]Hot spots are locales where pollutant concentrations are substantially higher than concentrations indicated by ambient outdoor monitors located in adjacent or surrounding areas. Hot spots can occur in indoor areas (for example, public buildings, schools, homes, and factories), inside vehicles (for example, cars, buses, and airplanes), and outdoor microenvironments (for example, a busy intersection, a tunnel, a depressed roadway canyon, toll plazas, truck terminals, airport aprons, or nearby one or many stationary sources). The pollutant concentrations within hot spots can vary over time depending on various factors including the emission rates, activity levels of contributing sources, and meteorological conditions.

and health conditions of minority communities and low-income communities. It also called on federal agencies to make environmental justice a part of their missions and to develop an environmental justice strategy. An Interagency Working Group on Environmental Justice (IWG) was established to implement the order, and the EPA administrator was designated to serve as the convener of the IWG. Thus, the assurance of environmental justice has become a general goal of the nation's AQM system. It remains to be seen how this goal will ultimately affect environmental policy and to what extent improved scientific tools and enhanced monitoring can be used to aid in this effort.

## THE SCIENTIFIC BASIS FOR SETTING STANDARDS

The CAA directs the EPA administrator to set primary NAAQS to protect human health with "an adequate margin of safety," and to set secondary NAAQS to "protect the public welfare from any known or anticipated adverse effects associated with" a criteria pollutant. It similarly directs the EPA administrator to set emission standards for HAPs on the basis of an assessment of the residual risk they pose following implementation of MACT and GACT. To do that, the administrator must have reliable and quantitative information on how human health and welfare outcomes are affected by varying concentrations of air pollutants. This information is typically expressed in terms of a dose-response relationship, which relates an undesirable health or welfare outcome to the concentration of a pollutant and the level of exposure to a pollutant over some specified period of time. Schematic illustrations of such dose-response relationships are presented in Figure 2-7. In this section, we discuss the scientific basis for establishing these relationships in the cases of human health and ecosystem effects.

### Health Effects Studies

The effects data used to set health-related standards and goals (such as the primary NAAQS) are generated typically from two types of studies: (1) experimental or toxicological, and (2) observational or epidemiological. Experimental or toxicological studies involve either direct measurements of the effects of pollutants on health outcomes of human or animal subjects (see Figure 2-8 and Figure 2-9) or in vitro experiments in which the effects of pollutants on specific human or animal cells are examined.

In general, observational or epidemiological studies examine statistical relationships between the actual exposure of a population to a pollutant and some measure of adverse health effects in the population (for example, emergency room visits for asthma attacks or morbidity) over that same time

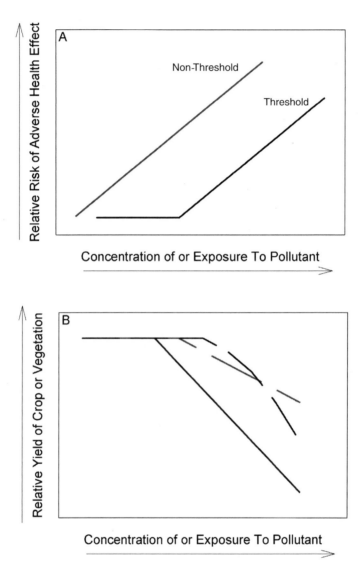

FIGURE 2-7   Schematic illustrating dose-response relationships between pollutant exposure and (A) human health effects and (B) crop or vegetation effects. In A, two types of dose-response relationships are illustrated: the upper one with no threshold for an adverse effect and the lower one with a threshold. In B, the different lines are to indicate that the response to pollutants typically varies substantially among plant species or among varieties within a given species. Dose-response relationships for health effects are usually plotted with risk increasing with increasing dose, but dose-response relationships for welfare effects are often plotted in terms of a diminishing return as a function of exposure.

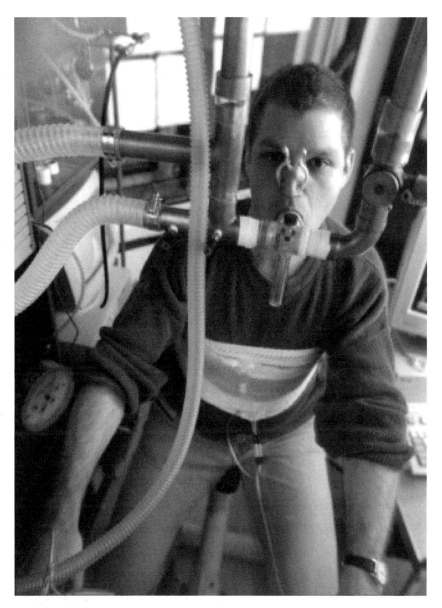

FIGURE 2-8   Exercising volunteer being exposed to ultrafine particles and monitored for health response. SOURCE: HEI 2000. Reprinted with permission; copyright 2000, Health Effects Institute.

a                                   b

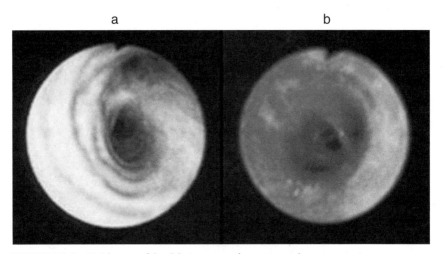

FIGURE 2-9   Evidence of health impact of ozone on human respiratory system based on an experimental study involving human subjects. (a) A healthy lung airway, and (b) the constricted opening of a lung airway inflamed from exposure to ozone. SOURCE: EPA 1999d.

period.   Population exposure is in turn estimated from observations of pollutant concentrations and, in some cases, information on activity patterns within the population.   To estimate the public health impact at any given level of exposure, relationships are derived from these studies.

In addition to observations of the concentration of the pollutant of interest, epidemiological studies require data on many other parameters that can affect health—such as meteorological condition and the concentrations of other pollutants—so that the influence of these potentially confounding influences on health outcomes can be accounted for.   As noted in Chapter 1, identifying and quantifying the health impact of a specific pollutant is a challenging task and typically requires the use of large sample sizes and sophisticated statistical methods.

Each approach has its advantages and disadvantages.   A major advantage of the toxicological approach is that, because the experiments can be carried out under controlled conditions, a much stronger cause-and-effect link exists between the administered exposure and the observed health effect.   For this reason, experimental studies are often considered more convincing than those based on epidemiological data.   Moreover, laboratory experiments provide an opportunity through ancillary observations to elucidate the physiological mechanism that results in the adverse health effect.   However, the experimental approach has drawbacks.   In addition to ethical questions that might be raised, there are a number of potential technical difficulties. For example, experimental conditions might not replicate the actual conditions in which population exposures occur (for example, meteorological conditions, mixtures of pollutants, or human activities). Further, animal experimental studies might not be relevant to humans, and human experimental studies typically have been unable, for ethical reasons, to include the frailest subgroups of the population that are suspected to be the most sensitive to air pollution effects. In contrast, epidemiological studies apply to the specific conditions and activities that exist at the time of exposure. Moreover, epidemiological studies are able to focus on specific segments of the population in terms of either physical condition (for example, elderly or people with asthma) or social condition (for example, the economically disadvantaged persons or those living near hot spots). Epidemiological studies have more difficulty characterizing the exposures of individual members of the population and distinguishing the effects of air pollution from a variety of other environmental, societal, and economic factors that can also affect health (for example, smoking behavior or socioeconomic status). Because epidemiological studies rely on ambient monitoring data, it must be assumed that the monitored pollutant concentrations actually reflect population exposure to the pollutant or pollutants. That assumption is not necessarily accurate and may explain the apparent inconsistencies in the findings of observational studies where one pollutant appears to have a stronger association with adverse health effects in one setting, but a different pollutant, or set of pollutants, is more strongly associated with effects in another setting (Samet et al. 2000).

In an attempt to address the limitations of the experimental and observational approaches, a hybrid approach involving the use of personal-exposure monitors has received increased attention in recent years. A personal-exposure monitor is essentially a miniaturized and automated air quality monitor that a person can carry during the course of his or her daily activities (see Figure 2-10). In this approach, selected members of a population are given personal exposure monitors to develop a quantitative record of their actual exposure to air pollutants and, at the same time, are carefully monitored for signs of any related adverse health effects. Because the approach includes clinical measurements of affected individuals, it (like the more standard experimental approach) has the potential to elucidate the specific physiological mechanisms that result in adverse health effects from air pollution. On the other hand, it does not have the disadvantage of the standard experimental approach of not being able to replicate the actual conditions under which the exposure occurs. In that regard, the use of personal-exposure monitors is even superior to the observational or epidemiological method, which must infer personal exposure from ambient measurements and personal-activity data. However, a major disadvantage is that only a small segment of the population can be studied at any one time.

FIGURE 2-10   Volunteer wearing a personal exposure monitor to measure actual exposures to PM and gases during daily activities. SOURCE: HEI 2000. Reprinted with permission; copyright 2000, Health Effects Institute.

The monitors also can seldom be used by themselves to determine the ambient contribution to personal exposure.

### Studies of Air Pollution Effects on Ecosystems

Standards and goals (such as the secondary NAAQS) designed to protect ecosystems are primarily based on four types of studies: (1) laboratory chamber studies, in which specific types of soil and vegetation are exposed to pollutants in greenhouses or carefully controlled (but artificial) environmental chambers (see Figure 2-11); (2) field studies, in which effects of air pollutants on ecosystems and the biotic components of these are monitored; (3) field studies, where a portion (for example, forest plot and stream reach)

FIGURE 2-11  Four-chamber greenhouse-based exposure system constructed to study effects of elevated $CO_2$ on plants. SOURCE: Photograph courtesy of Alberta Research Council Inc., Vegreville, Alberta, Canada.

of or entire (for example, lake and watershed) in situ ecosystems have been exposed to varying levels of acidity to demonstrate the effects of acidity in the otherwise natural environment (Hall et al. 1980; Schlinder et al. 1985; Hedin et al. 1990; Driscoll et al. 1996; Norton et al. 1994); and (4) hybrid studies, in which chamber experiments are conducted in the field using filters to scrub pollutants, such as $O_3$ so that exposure concentrations are less than those at ambient conditions, or pollutant concentrations are increased above ambient concentrations (see Figure 2-12). The first two approaches suffer from many of the same shortcomings discussed above for toxicological and epidemiological studies of effects on humans. Controlled experiments have the advantage of helping to establish cause-and-effect relationships. However, it may be difficult to relate these experiments to field conditions. Field measurements have the advantage of being representative of field conditions, but biotic response under field conditions is the integrated response of all stresses an organism is exposed to, air pollution being only one. Under field conditions, it is difficult to establish a cause-and-effect relationship. A significant shortcoming of these approaches is that they are limited to studying the effects experienced by a small number of species of organisms over a relatively short period of time. Although often adequate for agricultural crops, which are usually grown as isolated cultivars over one growing season, these approaches are not adequate for

FIGURE 2-12  Studies in open-top field chambers have shown the response of plants to ambient levels of $O_3$. Plants grown in chambers receiving air filtered with activated charcoal to reduce $O_3$ concentrations, do not develop symptoms that occur on plants grown in nonfiltered air at ambient $O_3$ concentrations. SOURCE: USDA-ARS 1998.

unmanaged ecosystems filled with a multitude of species, some of which have life cycles that span decades. These approaches also fail to address the indirect effects of air pollution on, for example, soils and aquatic ecosystems. As a result, the effects of pollutants on specific agricultural crops are generally better defined than the long-term effects of pollutants on unmanaged ecosystems, such as forests, grasslands, wetlands, lakes, and estuaries. One potential technique that might be useful for examining long-term effects of air pollutants on intact ecosystems has thus far been applied to studies of the long-term ecological effects of increased atmospheric carbon dioxide ($CO_2$) and is referred to as free air $CO_2$ enrichment (FACE) studies. In a FACE facility, scientists can increase the concentration of a trace gas, such as $CO_2$, in a controlled way in the air surrounding an intact ecosystem and measure plant and soil responses to the altered conditions over years to decades in accordance with the long life span of trees (see Figure 2-13 and Delucia et al. 1999). The FACE systems are not without fault but offer an alternative approach to enable long-term experimental exposure to air pollutants and evaluation of effects on whole ecosystems and their components under otherwise natural environmental conditions.

FIGURE 2-13   Free air $CO_2$ Experiment (FACE) is used to elucidate forest ecosystem responses to elevated $CO_2$. A similar type of experimental approach could be used to better understand the long-term ecological effects of elevated pollutant concentrations. SOURCE: DOE 2001.

The hybrid approach of integrating these types of studies has the advantage of investigating effects of air pollution under general conditions that closely approximate those in the real world, but also allowing investigators to carefully regulate the pollutant exposures experienced by the ecosystem being studied. This hybrid approach might include long-term measurements of air pollutants and their changes, associated ecosystem response to these changes, and coupled integrated experiments and model applications. An example of a site where this hybrid approach has been implemented is at the Hubbard Brook Experimental Forest in New Hampshire. At Hubbard Brook, long-term measurements of atmospheric deposition have been made (Likens and Bormann 1995). Associated with these measurements are long-term studies of soil, vegetation, and stream response to changes in atmospheric deposition (Likens et al. 1996; Driscoll et al. 2001a). These long-term measurements have been supported by integrated field experiments involving forest plots (Christ et al. 1999), streams (Hall et al. 1980; Hedin et al. 1990), and whole watershed manipulations and model development and application (Gbondo-Tugbawa et al. 2001; Gbondo-Tugbawa and Driscoll 2002, 2003).

## Scientific Basis for Setting Standards for the Nation's Air Quality Management

### Need for Additional Strategic Planning of Research That Underpins Health-Based Standards

At this time and most likely for the foreseeable future, no one single definitive method exists for establishing the dose-response relationships needed for setting health-based air quality standards. For that reason, standards and goals have been and should continue to be established using a combination of data from toxicological and epidemiological studies. A key question is whether there is an optimal mix of these studies that should be used to determine standards. In the past, the mix used by EPA has varied widely among the different criteria pollutants. Moreover, although EPA has normally identified a set of research priorities to fill key gaps at the end of each NAAQS review, those priorities have not explicitly defined the mix of toxicological and epidemiological data needed to establish pollutant standards and goals; nor have they, until recently, been part of a strategic research plan to develop such a consistent mix. For example, a large body of both animal and human experimental evidence on $O_3$ is available, and as a result, the $O_3$ NAAQS has been set with a great reliance on experimental data. For PM, on the other hand, direct human effects studies are scarce, and the bulk of the scientific evidence used to develop the PM NAAQS in 1997 came from epidemiological studies using mostly $PM_{10}$ data from atmospheric monitoring. This presents a series of challenges—most notably because $PM_{10}$ is a complex mixture and because toxicological information on the relative toxicities of its different components is sparse. As a result, in 1997 Congress appropriated funds for a major new PM research initiative, and directed EPA to work with the National Research Council to develop the first multiyear strategic research priorities for NAAQS research. In 1998, that committee identified 10 key priorities (NRC 1998b). Since that time, EPA has been investing substantial funds in implementing them, and the NRC committee is currently completing its evaluation of the program. Although EPA has begun developing similar strategies for other pollutants, no similarly comprehensive effort has been initiated to date.

Beyond the need to improve the science base for setting standards, there is also a need to improve how research results are summarized and synthesized in criteria documents. Criteria documents are designed to be comprehensive and have therefore become a useful but extensive catalogue of all recent research. A more systematic approach is needed to ensure that the criteria document includes all potentially relevant findings in the scientific review, identifies in an objective fashion the most valuable studies for assessing effects and setting standards, and does not overly emphasize findings of some studies without adequate consideration of their limitations.

### Accounting for Lack of Thresholds for Health Effects of Some Criteria Pollutants

Several recent epidemiological studies have introduced a new complication into the health-based standard-setting process. These studies suggest that there is no threshold concentration for $O_3$ (EPA/SAB, 1995), Pb (Canfield et al. 2003; Selevan et al. 2003), and PM (EPA 2002f) below which no observable health effects occur in the population (for example, see Figure 2-14). Although the validity of these findings still needs to be confirmed by additional research (EPA 2002f), the possibility that concentration thresholds may not exist for some pollutants raises serious questions about the technical feasibility of setting primary NAAQS that are consistent with the language in the CAA. In it, the EPA administrator is required to set primary NAAQS to protect public health with "an adequate margin of safety." Implicit in this

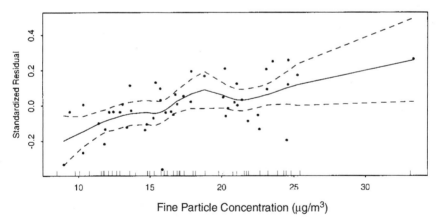

FIGURE 2-14   Concentration-response estimation from the reanalysis of the Pope/American Cancer Society Study on cardiopulmonary disease mortality (excluding Boise, Idaho). Each dot represents the risk of mortality for one of the cities studied as compared with the risk for a city at the mean PM concentration. (The risk is shown on the y axis as standardized residual where larger positive values indicate greater risk.) Note that the relative risk continues to decline even when annual average levels of fine particles ($PM_{2.5}$) are reduced to below the current NAAQS. SOURCE: Krewski et al. 2000a. Reprinted with permission; copyright 2000, Health Effects Institute.

instruction is the assumption that a NAAQS can be formulated by specifying a particular concentration below which the public health is protected from an adverse health effect of a pollutant. If a threshold does not exist, however, there might be no concentration below which the most susceptible members of the population are protected, raising the challenge for the administrator of how to arrive at an "adequate" margin of safety.

A variety of alternative approaches that would better reflect the observed dose-response relationships could be considered for pollutants without thresholds. One example would be the use of a "cumulative" form of the standard in which risk increases with concentrations accumulated over a period of time, and the standard would be set at some allowable level of risk. However, this approach would present problems. It would require incorporation of the concept of acceptable risk into the NAAQS-setting procedure, a controversial issue in its own right. It would also require accurate quantitative estimates of the nature and extent of adverse health affects in sensitive populations, and the models upon which such estimates might be based are currently clouded in considerable scientific uncertainty.

### Static List of Hazardous Air Pollutants

At the national level, toxic air pollutants are controlled when they are listed as hazardous air pollutants (HAPs), but the list of chemicals that fall under this regulatory regime has remained virtually unchanged since it was first developed in the CAA Amendments of 1990. That is the case despite the statute's directing the administrator to review the list periodically, to add a substance when it "is known to cause or may reasonably be anticipated to cause any adverse effects to human health or adverse environmental effects," and to delete a substance from the list when it "may not reasonably be anticipated to cause any adverse effects to human health or adverse environmental effects." At this time, no formal process seems to have been established at EPA for routinely reviewing the list of HAPs. In fact, EPA has encountered resistance—in the form of extensive regulatory debates, conflicting scientific evidence, lawsuits, and political pressure—when trying to add substances or remove them from the list. (One example is the recent debate over the potential removal of methanol from the list.)

The static nature of the HAPs list is problematic. In the United States, it is estimated that approximately 300 new chemicals are introduced into the environment each year by industry,[11] and yet not a single new air toxic has

---

[11]This information was obtained from the Notice of Commencement Database maintained by the inventory section in EPA's Office of Pollution Prevention and Toxics.

been added to the list in more than a decade. The Toxic Substances Control Act (P.L. 94-469) is intended to require chemical testing and, if necessary, impose controls on new chemicals that are manufactured or imported. However, implementation of this act has been difficult, and an analysis by EPA showed that for even those chemicals produced or imported at over 1 million pounds per year, only 7% had a full set of basic toxicity information (EPA 1998a).

There are two consequences of potential concern: (1) it is possible that the public health and welfare is not being adequately protected from all HAPs; indeed, without a formal procedure for reviewing the list of HAPs and assessing the risks associated with unregulated HAPs, it is difficult to assess the level of danger associated with this possibility; and (2) by leaving some harmful toxicants off the list of HAPs, the overall risks and costs to society of exposure to HAPs are probably underestimated. For example, although it is difficult to quantify the precise cancer risk from exposure to diesel exhaust (HEI 1999), some efforts to include diesel exhaust in estimates of cancer risk from exposure to HAPs have resulted in substantial increases in the overall estimates of risk (SCAQMD 2000).

Another problem with the current HAPs framework is that chemicals that do not meet the evidentiary threshold for regulatory action and that have only suggestive evidence of adverse health or environmental effects are not addressed by the federal regulatory system, and the resources needed to better quantify the risks of these chemicals are often not available. As an example, consider polybrominated diphenylethers (PBDEs). These compounds, widely used as flame retardants, have been found in dust emanating from old furniture and manufacturing facilities (Hermanson et al. 2003) and are now accumulating in the environment (Hale et al. 2001a,b; Ikinomou et al. 2002). PBDEs are structurally similar to polychlorinated biphenyls, which are regulated by EPA, and could elicit similar toxic effects (Hooper and McDonald 2000; Eriksson et al. 2001; McDonald 2002). Moreover, the concentrations of PBDEs in human tissue in the United States are on the rise (Mazdai et al. 2003; Schecter et al. 2003). However, because the toxicology of PBDEs is not well-documented, these compounds are unregulated air pollutants at the federal level in the United States.

In Chapter 7, suggestions are put forward for addressing many of the problems related to HAPs described above.

### Need for a Coordinated Strategic Program to Assess Ecosystem Effects

The nation has not had a continued strategic approach for conducting the research needed to characterize and understand the effects of air pollution on ecosystems, and that has made it extremely challenging to establish appropriate standards to protect ecosystems. Indeed, for the most part,

EPA has opted not to set unique secondary NAAQS for criteria pollutants. However, as noted earlier, there is growing evidence that tighter standards to protect sensitive ecosystems in the United States are needed, and an enhanced program of research on air pollution impacts on ecosystems is needed.

The development of a quantitative understanding of how air pollutants affect ecosystems is going to require the design and implementation of a substantially enhanced research strategy—one that monitors pollutants, as well as ecosystem structure and function, in a comprehensive and holistic way. A more detailed discussion of the monitoring networks that are in place and that would be needed to support such a research strategy is provided in Chapter 6.

### Need for Alternative Forms of Air Quality Standards to Protect Ecosystems

The CAA currently directs the administrator to protect ecosystems from criteria pollutants through the promulgation and enforcement of ambient-concentration-based standards (that is, the secondary NAAQS). However, concentration-based standards are inappropriate for some resources at risk from air pollutants, including soils, groundwaters, surface waters, and coastal ecosystems. For such resources, a deposition-based standard would be more appropriate. One approach for establishing such a deposition-based standard is through the use of so-called "critical loads." As described in Box 2-2, this approach has been adopted to protect ecosystems from acid rain by the European Union with some success.

### Limitations of Establishing Standards for One Pollutant at a Time

The CAA directs the EPA administrator to establish air quality standards for individual criteria pollutants and HAPs in isolation from other pollutants. That approach has contributed to the development of an AQM system in the United States that tends to focus on only one pollutant at a time, probably introduces inefficiencies into the pollution control program, and might even give rise to unintended consequences. Many air pollutants have common sources, and multipollutant strategies that take advantage of these common sources probably can achieve economies of scale that control strategies that target one pollutant at a time cannot. Moreover, pollutants can also be connected by similar precursors or chemical reactions once in the atmosphere. Thus, control strategies that target one pollutant may affect others, perhaps in unintended ways.

Because the standard-setting procedure for criteria pollutants and HAPs has also largely focused on individual pollutant effects, most health

---

**BOX 2-2   Critical Loads:**
**Europe's Approach to Setting Acid Rain Goals**

Critical loads came to the forefront in air quality policy development as part of the United Nations Economic Commission for Europe (UNECE) Convention on Long-Range Transboundary Air Pollution (LRTAP). LRTAP was signed in 1979 and entered into force in 1983; critical loads were adopted in 1988 as part of the protocol development process. Critical loads are scientifically determined estimates of the maximum long-term exposure to pollution that an ecosystem can withstand without significant harmful effects (Grennfelt and Nilsson 1988). Defining a critical load requires the integration of data on an ecosystem's soil type, land use, geology, rainfall, and other characteristics to classify its sensitivity to deposition of a pollutant or combination of pollutants. Initial efforts in Europe focused on sulfur deposition but have been expanded to include nitrogen and more recently heavy metals and persistent organic pollutants (POPs).

Three approaches have been used to calculate critical loads: (1) empirical, (2) mass-balance based, and (3) dynamic modeling. The empirical approach uses empirical relationships developed from experimental studies and field observations of soil and water chemistry across pollution gradients. The mass-balance approach involves calculations to determine an ecosystem's ability to neutralize inputs of acidity. These calculations assume that ecosystems are at steady-state conditions and that soil chemical pools do not evolve in response to changes in atmospheric deposition. Dynamic models take the mass-balance approach a step further by assessing the time-dependent response to changes in deposition (Jefferies 1997; Jenkins et al. 1998). Although the mass-balance approach is limited by the steady-state assumption, it requires far less information than dynamic models and has been widely embraced in Europe for critical-load calculations.

After the critical load is assessed for an area, it is possible to determine the exceedance level, which is the difference between the critical load and the actual deposition. The exceedance level indicates the emission reductions necessary to protect the ecosystem. Much like critical load values, exceedance levels can be calculated by different approaches, ultimately, shaping environmental policy. One method is to determine the land area of ecosystem types that exceeds the critical load. A more refined method, the average accumulated exceedance, produces a weighted average of the exceedance by the amount of area the ecosystems cover in the grid. In Europe, a 5-percentile critical load map was adopted in the second sulfur protocol in which a deposition level is considered to be less than the critical load if 95% of the ecosystems in the grid will not be harmed.

Once areas of excess deposition are identified, optimal emission-reduction strategies that take cost into consideration can be negotiated and implemented. The Convention on LRTAP determined that the cost to meet the 5-percentile goal was not economically feasible. Thus, policy makers identified "target loads" that accounted for economic and political considerations as an intermediate step to reducing emission levels to below critical load levels. The UNECE chose to set an interim load that would reduce the 1980 exceedance level by 60% by 2010. This objective is then revisited every 5 to 6 years to determine if additional measures need to be enacted.

*(continued on next page)*

**BOX 2-2**  continued

The critical-loads and target-loads approaches adopted in Europe have provided an objective framework for stakeholders to debate how the ecological effects of acid deposition and other deposition-based pollutants (for example, nitrogen, mercury) can be curtailed by air pollution control programs. It has resulted in a process that provides for ecosystem protection and has led to reductions in European emissions while accounting for variations in sensitivity between different ecosystems. Countries have typically implemented measures that provide for reasonable progress toward meeting the targets given cost considerations, but they typically have not adopted requirements stringent enough to meet the targets (Blau and Agren 2001). Due to its grounding in scientific assessments, the critical load process enjoys relatively strong political support from the majority of European countries. Nevertheless, although critical loads may provide a consistent scientific framework for evaluating emissions reductions, the calculated values can be highly uncertain, in part due to reliance on steady-state models. Further, the numerous methods for calculating both critical loads and exceedance levels allow for inconsistency in implementation. Another limitation in the approach is that the formulation is in terms of a single threshold for an entire ecosystem. In actuality, ecosystems vary in composition, and pollutant deposition probably has impacts on some species at even very low levels.

effects studies that form the current scientific basis for setting air quality standards have attempted to elucidate the effects of individual pollutants in isolation from others. That approach can pose significant challenges to investigators. In epidemiological studies, for example, identification of independent health effects of pollutants requires that a setting be chosen where only one of the pollutants is present (an unrealistic and somewhat artificial requirement) or, more commonly, that statistical modeling of the data be carried out to tease out individual pollutant effects. The latter approach can be hampered by the strong correlations that sometimes exist between air pollutants; the correlations can make it difficult or even impossible to separate the effects of one pollutant from another.

Health effects studies that focus on sources instead of individual pollutants offer one method for moving away from an AQM system focused on single pollutants to one focused on multipollutant controls. One approach that science is beginning to pursue is to move away from studies (and ultimately standards) based on concentrations of individual pollutants to studies of the effects of specific pollution sources. In a source-oriented method, scientists define exposure to a specific source and obtain information about the health effects of the exposure to the mix of pollutants from that source. Techniques are being developed to define pollution sources by using source "markers," which are indirect source measures (for example, distances from major roadways, determined by geographic information systems, to reflect motor vehicle pollutants), and factor analysis based on extensive chemical analysis of monitor filters (Laden et al. 2000). There also has been significant progress in developing exposure assessment techniques that determine the intake fraction—that is, that portion of a person's exposure that results from emissions from a particular source (Bennett et al. 2002). Finally, toxicological efforts are under way to systematically assess and compare the effects of well-characterized sources (Seagrave et al. 2002). These source-oriented approaches have potential advantages in focusing regulatory attention and research on potentially toxic emissions from specific sources and in directing public health initiatives to reduce emissions from specific sources rather than attempt to reduce general ambient concentrations of specific pollutants.

### Need to Address Health Risk Associated with Exposure in Hot Spots and Indoor Environments

There is a growing recognition within the scientific and regulatory communities of the potential importance of pollutant exposure in special microenvironments or hot spots. These environments may include highway toll plazas, truck stops, airport aprons, and areas adjacent to one or many stationary sources or busy roadways, as well as transit within vehicles and indoor environments. Pollution concentrations in hot spots may exhibit strong transient spikes, such as with the ebb and flow of traffic or as a result of off-normal (upset) conditions at stationary sources that might result in short-term emissions greater than those usually represented by typical operating conditions or by annual national averages (for example see NRC 2000c). Because of the small spatial scale and possible transient nature of hot spots, mitigation of air pollution within these microenvironments may not be adequately addressed in the nation's current AQM system, even though segments of the population may experience significant air pollution exposure within these environments (for example, California school bus study [Fitz et al. 2003]). In Chapter 7 of this report, recommendations are made to better address the problem of hot spots by encouraging states to include measures to mitigate hot-spot pollution in their implementation plans to meet NAAQS and other national goals and by evolving the AQM system toward a more risk-based multipollutant paradigm.

Within the general problem of hot spots, the health risk that may be associated with exposure to indoor air pollution is of particular concern, because most Americans spend more time indoors than outdoors (EPA 1987). For most Americans, exposure to the indoor environment dominates over the outdoor environment. The most vulnerable—children and older and infirm adults—generally spend over 90% of their time indoors.

Indoor air contains complex mixtures of contaminants. Some enter from the outside (for example, lead, mercury, and fine PM from contaminated soils, power plants, and other sources), and others arise from the presence and use of commercial, industrial, and household products, such as plastics, paint, solvents, pesticides, and stoves (see Figure 2-15). Biological agents— such as molds, dander, and dust mites—are another type of contaminant that can cause pulmonary reactions. The level of air pollution within any indoor environment depends on the specific types of products in use as well as the building's characteristics (for example, the ventilation and types of carpets) and the habits of the occupants.

In spite of the potential dangers, indoor air pollution remains largely unregulated in the United States, and to the extent that the public is protected, it is accomplished through a patchwork of information and some regulatory actions. In some instances, EPA has been able to exercise authority (for example, over asbestos). However, no federal statutes specifically give EPA authority to regulate indoor residential and commercial sources of air pollutants—perhaps because of a reluctance on the part of

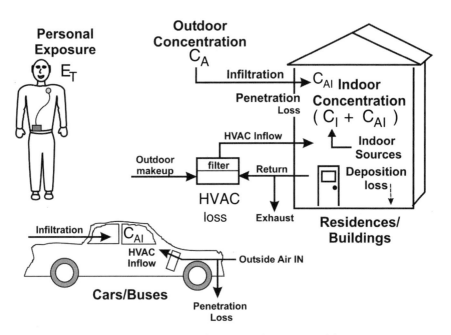

FIGURE 2-15 Schematic diagram illustrating the source of human exposure to indoor PM pollution. SOURCE: Rodes 2001. Reprinted with permission; copyright 2001, RTI International, Research Triangle Park, NC.

Congress to regulate any aspect of the inside of people's homes, public buildings, and offices. Regulatory power for some aspects of indoor air quality in occupational settings is given to the Occupational Safety and Health Administration (OSHA), and for some consumer products, to the Consumer Products Safety Commission (CPSC). However, in the case of OSHA, which is primarily responsible for protecting generally healthy worker populations, indoor air quality standards are generally not as protective as EPA's ambient (outside) air quality standards. In addition, some states and EPA, in regulating some products for VOC content (for example, surface coatings that are potential sources of VOC precursors to $O_3$ formation), have in effect also reduced exposure to those VOCs in indoor settings where those coatings are applied. For the most part, however, EPA has attempted to reduce indoor air pollution indirectly by identifying indoor air contaminants that pose significant risks (for example, environmental tobacco smoke and radon) and encouraging consumers, corporations, and other organizations to respond to that information or by regulating major sources of outdoor air pollutants that pose a health risk when transported indoors (for example, lead). In some cases, state and local governments have responded by improving their own standards, regulations, or ordinances. However, there is not at this time a coordinated federal program to ensure that the public is protected from unnecessary or avoidable health risks associated with indoor air pollution. A more coordinated approach would seem prudent.

## Risk Assessment and Priority Setting

There is a long-standing recognition of the need for robust quantitative estimates of the risks to human health and welfare associated with exposure to air pollutants. First, Congress expressly directed EPA to use an assessment of residual risk to design the second phase of controls on HAPs. Second, the final report of the National Commission on Risk Assessment and Risk Management, established in the CAA Amendments of 1990, found that risk assessment can be a powerful tool for setting priorities on resources for monitoring and regulating the myriad air pollutants to which humans, ecosystems, and materials are regularly exposed (NCRARM 1997). Moreover, under the present AQM system, setting priorities is done largely by statutory and/or agency fiat. For example, because of the detailed requirements specified for the regulation of criteria pollutants (as discussed in Chapter 3), this subset of pollutants generally receives a substantially larger share of the management effort and resources than do HAPs. Moreover, controls that most effectively reduce concentrations of pollutants in the ambient atmosphere tend to be favored over those that target pollution in hot spots. This emphasis on criteria pollutants in ambient air may or may not be justified on the basis of actual human health and ecosystem risk. Similarly, it is not clear if all of the currently regulated HAPs pose a greater risk to human health and welfare than many of the untested and unregulated air toxics known to be in the ambient air and at specific hot spots.

For the reasons noted above, it would be highly desirable for the nation's AQM system to have a robust risk-assessment capability that could reliably assess and set priorities on the relative risks posed by all pollutants in the atmosphere—in hot spots and microenvironments, as well as the ambient air. However, although the scientific community has learned a great deal about air pollution in recent decades, and there have been significant advances in the general field of risk assessment (NRC 1994), current knowledge is not yet extensive enough to rank pollutants comprehensively on the basis of risk. There is a lack of sufficient knowledge of the diversity of health and welfare effects associated with different pollutants, and, perhaps more important, with different mixtures of pollutants under environmental conditions. Another major deficiency is our inability to assess pollutant exposures accurately because of a lack of sufficient data on the distribution of pollutants. If these deficiencies are to be addressed, substantial investments over a substantial period of time will be needed for research on air pollution effects research and for more advanced systems to determine the spatial and temporal variability of air pollutants in specific hot spots and in indoor environments as well as the ambient air.

## SUMMARY

### Strengths of Goal-Setting Procedures

The establishment of the NAAQS has allowed for important and extensive input and feedback from the scientific and technical communities and has catalyzed additional research and understanding of the effects of air pollution.

- The standards-setting procedure for NAAQS has been responsive to new scientific information and has allowed for adjustments in the standards when scientific understanding so dictated.
- The establishment of NAAQS has provided targets for regulatory agencies and measures by which to assess improvement in air quality and the effectiveness of the AQM.

## Limitations of Goal-Setting Procedures[12]

• The funding of health effects research and the subsequent selection and review of this research in criteria documents and staff papers used to promulgate NAAQS often lack a coherent and standardized strategic plan.

• The CAA requirements for pollutant-specific air quality standards for criteria pollutants and HAPs have encouraged the evolution of an AQM system with control strategies that largely focus on one pollutant at a time. Changes that foster multipollutant approaches would be advantageous. One method that could be considered would be to assess the health and welfare effects of sources instead of pollutants.

• The list of HAPs (that is, those specifically regulated by EPA) has been essentially static for a decade and probably does not contain all the air toxics that pose a significant risk to human health and welfare.

• Current monitoring data and understanding are not sufficient to adequately assess the relative risks to human health and welfare posed by exposure to the myriad pollutants in the environment, as well as to the myriad microenvironments or hot spots in which these exposures may occur. Development of such a capability will be a major challenge and will require a substantial investment in resources for monitoring and effects research over a long period of time.

• Although progress has been made to improve exposure assessment and to link specific exposures and effects to specific sources, substantial additional work is needed.

• Indoor air pollution poses a significant health risk to humans and yet is not addressed comprehensively by any agency in the federal government.

• The current practice of letting the primary standard serve as the secondary standard for most criteria pollutants does not appear to be sufficiently protective of sensitive crops and unmanaged ecosystems. Moreover, concentration-based standards are inappropriate for some resources, such as soils, groundwater, surface water, and coastal ecosystems, that are at risk from the indirect effects that pollutants can foster (for example, eutrophication). A deposition-based standard would be more appropriate in some instances.

• It is a significant challenge to set ambient or emission standards to protect public health with an adequate margin of safety from harmful exposure to a pollutant if that pollutant does not exhibit a threshold concentration for an adverse health effect.

---

[12]Recommendations that address these limitations are provided in Chapter 7.

# 3

# Designing and Implementing Control Strategies Through the SIP Process

## OVERVIEW OF SIP PROCESS

The state implementation plan (SIP) and the tribal implementation plan (TIP)[1] are the central organizing elements in the management of criteria air pollutants (see Box 3-1 for a discussion of tribes and the Clean Air Act). The SIP defines the combination of local, state, and federal actions and emission controls needed for an area to bring about compliance with the National Ambient Air Quality Standards (NAAQS) and for an area that has attained the NAAQS to maintain compliance and prevent significant deterioration of air quality. However, the purpose of the SIP extends well beyond the attainment and maintenance of NAAQS. It provides the basic link between state regulations, EPA oversight of state actions, and federal enforcement. In addition to addressing criteria pollutants, SIPs are used by EPA to formally establish state and local agency obligations to meet emission standards and goals related to regional haze, acid rain, and hazardous air pollutants (HAPs).

When EPA approves a plan, the rules specified therein are federally enforceable. When EPA disapproves a plan or finds that the state is delinquent in implementing it, EPA can impose sanctions, such as the loss of highway funds. In extreme cases, EPA can prepare a federal implementation plan (FIP), which takes the role of a SIP.

---

[1]Hereafter, we will use the term state implementation plan (SIP) as shorthand to denote both a SIP and a tribal implementation plan (TIP).

SIPs are submitted by individual states in accordance with explicit requirements set forth in the Clean Air Act (CAA). The requirements have been interpreted via regulations and guidance promulgated by EPA. The CAA requires each state to produce a single comprehensive SIP; however, in practice, each state produces a separate implementation plan for each criteria pollutant, because of the timing and pollutant-specific requirements set

---

### BOX 3-1   Tribes and the Clean Air Act

There are more than 550 federally recognized tribes and more than 300 reservations in the United States. Indian tribes and individuals own approximately 55.4 million acres of land in the contiguous United States (Getches et al. 1998). The 1990 CAA Amendments authorized EPA to "treat tribes as states" for purposes of developing, administering, and enforcing air quality regulations within reservation boundaries, irrespective of land ownership (42 USC § 7601(d)(2)(B)). In doing so, Congress recognized the inherent sovereignty of tribes with respect to their land and members. Congress also delegated to the tribes regulatory authority over nonmembers operating on land within reservation boundaries.

Pursuant to the CAA, in February 1998, EPA promulgated its tribal authority rule, specifying requirements for tribal eligibility to administer air programs (40 CFR 49). To be eligible, a tribe must apply to the EPA regional administrator and demonstrate that it is "reasonably" capable of administering its program in a manner consistent with the terms and purposes of the CAA. Tribes may develop a full tribal implementation plan (TIP) and seek authority to carry out all the functions that states perform under the act, but they are not required to do so. EPA's regulations allow tribes to assume primacy over a subset of regulatory functions and to expand their authority gradually. EPA also has the flexibility to alter deadlines for plan submittal and other regulatory requirements.

In 1999, the Gila River Indian Community became the first tribe to become eligible for "treatment as a state" status. To date, 14 tribes have received eligibility to implement parts of the CAA. The Mohegan tribe submitted a TIP to EPA in FY 2002, while the Pequot, St. Regis Mohawk, and Gila River have TIPs in progress (EPA 2003d). The tribal council of the Gila River Indian Community adopted ordinances in March 2002 comprising the first section of its TIP.

On the basis of a telephone survey of 156 of the 237 federally recognized tribes in the Western Regional Air Partnership (WRAP) region, 60 tribes in this region have some form of air quality program. The scope of activities ranges from education and outreach to monitoring, emissions inventory development, and source permitting (ITEP 2001). Twenty-eight of the surveyed tribes had an emissions inventory and 51 tribes performed some air quality or meteorological monitoring. The survey suggested that if resources were available, the level of activity could double in the next few years. For example, 62 tribes indicated an interest in starting an air monitoring program in the next few years (ITEP 2001).

Tribes, like states, are eligible under the CAA for federal grants to support air quality monitoring and management efforts. EPA began seeking tribal participa-

*(continued on next page)*

**BOX 3-1**   continued

tion in its grants program in 1995, providing grants to about 20 tribes. In 2002, 121 tribes received air program grants (C. Darrel Harmon, EPA, personal commun., Feb. 19, 2003). The tribal share of EPA-administered state and tribal assistance grants (STAG) for air has been flat at about $11 million since 1999, while tribal interest in developing air programs has increased. Tribes seeking first-time grants are being turned away in Regions 9 and 10.[a] Additional support for tribal environmental programs is available through EPA's multimedia Indian environmental general assistance program (GAP) for which $63 million was included in the agency's 2004 budget request. However, GAP funding is used primarily for environmental infrastructure development, capacity building, education, and outreach and cannot be used to operate environmental management programs.

In addition to direct grants to tribes, STAG funding supports the Institute for Tribal Environmental Professionals (ITEP) at Northern Arizona University. This organization offers about 20 technical workshops each year for tribal environmental staff on monitoring and permitting, and other air quality management (AQM) topics and also serves as a forum for interaction and information exchange between tribes and with EPA. In partnership with ITEP, EPA also supports the Tribal Air Monitoring Support Center at the EPA Radiation and Indoor Environments Laboratory to provide air monitoring training and technical support to the tribes.

Until a tribe assumes control of its own air pollution programs, the responsibility and authority to regulate air pollution sources on tribal lands falls to EPA. States have no environmental regulatory jurisdiction on tribal lands. EPA administers a Title V operating permits program for major stationary sources on tribal lands along with a prevention of significant deterioration (PSD) preconstruction permit program for new major sources or source modifications in attainment areas. A new-source review (NSR) program under which individual permits can be issued for major sources in tribal lands within nonattainment areas and for minor sources in attainment areas does not exist. EPA is required to develop case-by-case federal implementation plans (FIPs) where these plans must go through the full and lengthy *Federal Register* rule-making process for each applicable facility. In addition, sources in tribal areas may be competitively disadvantaged by the lack of synthetic minor permitting programs that allow sources to be classified on the basis of actual rather than potential emissions.[b]

The National Tribal Environmental Council (NTEC) estimates that 80 Indian reservations are located within or partly within nonattainment areas for the 1-hr average ozone ($O_3$) standard and that the number is likely to be higher with the 8-hr standard. Some tribes have expressed concern about pending designations of nonattainment areas for $O_3$, because EPA guidance presumes that designations will be made based on metropolitan statistical area (MSA) or consolidated metropolitan statistical area (CMSA) boundaries (NTEC 2002). These default designations ignore the jurisdictional boundaries between tribes and states. EPA will consider recommendations from tribes that their lands be excluded from nonattainment areas, but the exemptions require detailed analysis that may tax tribal resources. In particular, air quality data for tribal lands are often lacking. Tribes are concerned that they will face the burdens of nonattainment designation, including offset requirements, even though they have historically borne little responsibility for air quality problems (or derived little economic benefit from the air pollution sources) in the adjacent metropolitan areas. Similar issues are likely to arise with respect to designations for particulate matter of 2.5 micrometers ($PM_{2.5}$).

Since EPA promulgated the tribal authority rule in 1998, tribes have demonstrated increasing interest in developing and administering their own air programs. Federal policy supports tribes' control over their own air resources and economic development. Tribal participation in regional organizations, such as the Western Regional Air Partnership, is vital for filling gaps in air quality and emissions data and for effective air quality management. However, lack of resources imposes severe constraints on air quality management activities for many tribes. As indicated in Chapter 7, more federal funding for tribal assistance grants is needed to support tribal self-determination and tribal participation in the national AQM system.

---

[a]EPA's FY 2004 budget request includes $11,050,000 for tribal grants and $228,550,000 for state and local assistance grants (EPA 2003e).

[b]Most states have operating permit programs for minor sources (those that fall under the relevant thresholds, for example, for maximum achievable control technology [MACT], NSR, and PSD). When such sources have the potential to emit more than the major source threshold, emission levels can be capped just below those thresholds. The sources (which are then called synthetic minors) do not have to go through the more onerous permitting processes required for major sources. At present, sources that might otherwise want to locate in Indian country cannot take advantage of these provisions. That can be a disadvantage for tribes in terms of economic development.

by EPA. Most SIPs are also continuously evolving to reflect new federal or state requirements, new information, or change in status of NAAQS attainment.[2] Thus, at any one time, a state may have a number of new and revised SIPs at various stages of development and federal review.

The requirements for any state's SIP depend on its air quality, which is determined by the attainment designations assigned to the individual areas[3] within that state. Section 107 of the CAA defines three designations for any area's compliance with NAAQS for criteria pollutants. They are as follows:

- *Nonattainment.* Any area that does not meet (or that contributes to ambient air quality in a nearby area that does not meet) a national primary or secondary ambient air quality standard for the pollutant.

---

[2]EPA is authorized to direct a state to revise its SIP if the agency finds that the SIP is substantially inadequate to attain the NAAQS. This procedure is referred to as a "SIP call." The nitrogen oxides ($NO_x$) SIP call, which is discussed later in this report, is intended to reduce the contribution of long-range transport to $O_3$ nonattainment by limiting $NO_x$ emissions over a large multistate airshed.

[3]The term area is used in this context to denote a locality, such as a metropolitan statistical area. An area typically consists of a county or parish or a combination of adjacent counties or parishes.

- *Attainment*. Any area that meets the national primary and secondary ambient air quality standard for the pollutant and does not contribute to the violation of a national primary or secondary ambient air quality standard in a nearby area.
- *Unclassifiable*. Any area that cannot be classified on the basis of available information as meeting or not meeting a national primary or secondary ambient air quality standard for the pollutant.

The CAA further classifies $O_3$ nonattainment areas as marginal, moderate, serious, severe, and extreme, and CO nonattainment areas as moderate or serious (see Table 3-1). The procedures used to determine the appropriate designation and classification for an area are described in Box 3-2.

The basic requirements for states in general and for those areas that are in nonattainment of one or more NAAQS are listed in Box 3-3. The requirements for a SIP grow in stringency and complexity as an area's designation shifts from attainment to nonattainment and, for ozone ($O_3$) and carbon monoxide (CO), to the more acute nonattainment classifications. Because the major efforts in air quality management (AQM) for criteria

TABLE 3-1　Classification of Nonattainment Areas for $O_3$ and CO Mandated in the CAA Amendments of 1990

| Area Classification | Design Value, ppm | Attainment Date for Primary Standard[a] |
|---|---|---|
| $O_3$ Nonattainment Areas | | |
| Marginal | 0.121–0.138[b] | Nov. 15, 1993 |
| Moderate | 0.138–0.160[b] | Nov. 15, 1996 |
| Serious | 0.160–0.180[b] | Nov. 15, 1999 |
| Severe-15[c] | 0.180–0.190[b] | Nov. 15, 2005 |
| Severe-17[c] | 0.190–0.280[b] | Nov. 15, 2007 |
| Extreme | 0.280 and above[b] | Nov. 15, 2010 |
| CO Nonattainment Areas | | |
| Moderate (low)[d] | 9.1–12.7[e] | Dec. 31, 1995 |
| Moderate (high)[d] | 12.8–16.4[e] | Dec. 31, 1995 |
| Serious | 16.5 and above[e] | Dec. 31, 2000 |

[a]The primary standard attainment date for $O_3$ is determined from the date of the enactment of the CAA Amendments of 1990.
[b]The classification scheme for $O_3$ was devised by Congress before EPA promulgated a new 8-hr standard, and, thus, the classification relates to the old 1-hr form of the standard.
[c]The requirements for severe-15 and severe-17 $O_3$ nonattainment areas are the same except for the attainment dates.
[d]Moderate CO nonattainment areas with design values of 12.7 ppm or less have reduced SIP requirements compared with those areas with design values above 12.7 ppm.
[e]These values for CO refer to a rolling 8-hr average.
Abbreviation: ppm, parts per million.

## BOX 3-2    Procedures Used to Designate an Area's Attainment Status

Designation of an area as an attainment, nonattainment, or unclassifiable area is made by the EPA administrator in consultation with the relevant state governors and is based on data gathered from the NAMS/SLAMS air quality monitoring networks (which are discussed in Chapter 6). The designation process engenders two fundamental assumptions: (1) that attainment and maintenance of a NAAQS can be established using data from a limited number of surface air quality monitoring sites, and (2) that geographical areas can be effectively regulated with a single, uniform designation and a common plan for achieving and/or maintaining an acceptable level of air quality.

An area's attainment or nonattainment status with regard to a criteria pollutant is determined by comparing the NAAQS with the area's "design value"[a] for the pollutant. The design value is derived from air quality monitoring data gathered by local or state authorities following guidelines specified by EPA. If the design value exceeds the NAAQS, the area is designated as being in nonattainment. For CO and $O_3$, an additional, more detailed classification scheme is used to indicate the severity of the nonattainment (see Table 3-1). Because the statistical form of the standard varies from one pollutant to another, the method used to determine the design value also varies by pollutant. For example, in the case of the old 1-hr $O_3$ standard, the design value is calculated as the fourth highest 1-hr averaged concentration observed at any monitoring site in the area over three consecutive years. The new 8-hr $O_3$ design value, on the other hand, is defined as the 3-year average of the annual fourth highest 8-hr averaged concentration observed each year. The new 24-hr averaged $PM_{2.5}$ design value is derived from the 3-year average of the 98th percentile measured concentrations in an area. The annual $PM_{2.5}$ design value is obtained from an average of all measurements made in the area. These various protocols involving multiyear and spatial averages and selecting concentrations below the absolute maximum observed have been chosen to limit the impact of statistical outliers and extreme and anomalous meteorological events. Nevertheless, as discussed elsewhere, the use of design values in the attainment demonstration of the SIP presents important challenges.

In addition to providing a mechanism for identifying and designating nonattainment areas, the CAA (in Section 175(a)) also provides a process by which a nonattainment area can be redesignated an attainment area. To be redesignated, an area must file a "maintenance SIP." In addition to showing that the area is in compliance with the NAAQS on the basis of relevant air quality monitoring data, the maintenance SIP must provide a plan for ensuring that the standard will not be violated in the future. Once such a plan is approved by EPA, the area is generally referred to as a maintenance area. EPA currently lists 62, 59, 30, and 16 maintenance areas for $O_3$, CO, $SO_2$, and $PM_{10}$ respectively. There are a number of other nonattainment areas that have demonstrated achievement of the standards, but either because they have not completed the process of filing a maintenance SIP or because EPA has not yet accepted it, they have not been redesignated.

---

[a]An area's "design value" is determined from the area's monitoring data. These values are determined differently for each of the criteria pollutants. For example, the design value for $O_3$ is the 3-yr average of the fourth highest 8-hr concentration observed each year. The derived design value is then compared with the NAAQS to determine if the area is in attainment.

## BOX 3-3   Clean Air Act Requirements for State Implementation Plans

### A. For all states[a]

SIPs must be submitted within 3 years of promulgation of new NAAQS and provide for "implementation, maintenance, and enforcement" of the standard. Among other things each SIP must

- Include enforceable emission limitations and controls as well as schedules and timetables to ensure compliance.
- Provide for the monitoring of ambient air quality.
- Include a program to enforce the emission limitations and control measures.
- Contain adequate provisions prohibiting emissions within the state to contribute significantly to nonattainment of NAAQS in any other state.
- Ensure that the state will have adequate personnel, funding, and authority to carry out the plan.
- Require stationary emission sources to monitor and provide periodic reports of their emissions.
- Meet requirements relating to consultation, public notification, and prevention of significant deterioration of air quality.
- Provide for air quality modeling and provide related data to demonstrate how emissions affect air quality.
- Require owners or operators of major stationary emission sources to pay fees to cover (1) reasonable costs of reviewing and acting upon permit applications, and (2) reasonable costs of implementing and enforcing the terms and conditions of the permit.
- Provide for participation by local political subdivisions affected by the plan.

### B. For nonattainment areas[b]

Attainment-demonstration SIPs must be submitted within 3 years of an area being designated a nonattainment area. In addition to the items listed in part A, the SIP must

- Provide a plan for the implementation of reasonably available control technologies (RACT) and attainment of primary NAAQS, for the offsetting of emissions of new or modified major stationary sources, and for the installation in major new stationary sources of technology capable of achieving the lowest achievable emission rate (LAER).
- Include a comprehensive emissions inventory for all relevant pollutants.
- Implement a new-source review (NSR) before construction, and for all new or modified stationary sources, implement a permit program that mandates use of control technologies that obtain the LAER and provides sufficient emission offsets from other sources in the area to ensure reasonable progress and attainment of NAAQS.
- Provide for the implementation of contingency measures in the event that the area fails to make reasonable progress or meet its attainment deadline.

For a nonattainment area to be redesignated an attainment area, a revised SIP must be submitted and approved. This plan[c] must

- Provide for the maintenance of NAAQS compliance for at least 10 years after the redesignation.
- Include additional measures, if any, to ensure such maintenance.

## C. For $O_3$ nonattainment areas[d]

In addition to the items listed in parts A and B, SIPs for marginal and above $O_3$ nonattainment areas must

- Include a vehicle emission-control inspection and maintenance (I/M) program.
- Include a volatile organic compound (VOC) and nitrogen oxides ($NO_x$) emissions inventory every 3 years for the area.
- Implement an NSR for VOC sources that includes an offset ratio of emission reductions to new emissions of at least 1.1:1.

In addition, SIPs for moderate and above $O_3$ nonattainment areas must

- Provide a plan for VOC emission reductions as specified in the CAA.
- Provide a plan for comprehensive introduction of RACT for specified VOC sources.
- Implement a vapor recovery program requiring gasoline service stations to install special refueling equipment to prevent the escape of VOCs.
- Implement an NSR for VOC sources that includes an offset ratio of emission reductions to new emissions of at least 1.15:1.

In addition, SIPs for serious and above $O_3$ nonattainment areas must

- Include an attainment demonstration using a photochemical grid model.
- Demonstrate that reasonable progress is being made through appropriate 3% per year reductions in VOC emissions (or its $O_3$-equivalent in $NO_x$ emissions) and submit triannual compliance demonstrations beginning in 1996 showing emission reductions are being met.
- Implement an NSR for VOC sources that includes an offset ratio of emission reductions to new emissions of at least 1.2:1.
- Implement a program of enhanced air quality monitoring.[e]
- Provide for an enhanced vehicle I/M program.
- Include a clean fuel (such as natural gas and propane) vehicle program for centrally fueled fleets.[f]
- Demonstrate conformity with regional transportation plans.[g]

In addition, SIPs for severe $O_3$ areas must

- Implement transportation control measures (TCM) to reduce single-occupancy-vehicle use through high-occupancy vehicle (HOV) lanes and car-pooling and van-pooling programs.
- Implement an NSR for VOC sources that includes an offset ratio of emission reductions to new emissions of at least 1.3:1 (or 1.2:1 if areawide best available control technology [BACT] is used).

(continued on next page)

**BOX 3.3** continued

• Implement a reformulated fuels program.

In addition, SIPs for extreme $O_3$ areas must

• Include a plan for use of clean fuels and advanced technology for electric utility, industrial, and commercial boilers.
• Implement an NSR for VOC sources that includes an offset ratio of emission reductions to new emissions of at least 1.5:1 (or 1.2:1 if areawide BACT is used).
• Implement a reformulated fuels program.

---

[a] As described in Section 110 of the CAA.
[b] As described in Section 172 of the CAA.
[c] As described in Section 175(a) of the CAA.
[d] As described in Section 182 of the CAA.
[e] This provision has resulted in the development and operation of the photochemical assessment monitoring (PAM) network described in Chapter 6.
[f] See Chapter 4 for discussion of this program.
[g] See Chapter 4.

pollutants are currently focused on nonattainment areas, our discussion here will be largely limited to the "attainment-demonstration SIP" submitted by an area after it has been designated a nonattainment area. As indicated in Box 3-3, the CAA requires that an attainment-demonstration SIP be submitted to EPA within 3 years of an area's being designated a nonattainment area. The agency or agencies responsible for preparing an attainment-demonstration SIP vary from state to state. Often, it is prepared by the relevant state or tribal authority; in other cases, a local or regional government is given primary responsibility.

## THE MAIN COMPONENTS OF AN
## ATTAINMENT-DEMONSTRATION SIP

The key elements of an attainment-demonstration SIP are the following:

• An emissions inventory.
• An analysis involving air quality model simulations as well as observational data and related evidence to determine the amount and types of emission reductions needed to bring about compliance by the appropriate date.
• A description of the emission-control strategies and enforcement measures to be adopted to achieve the required reductions.

Each of those components is described below.

## Emission Inventories

The first step in developing an emission-control strategy for a criteria pollutant is to develop an inventory of pollutant emissions that lists all sources of the pollutant or its precursor and the rate at which each source emits the pollutant to the atmosphere. EPA has specified a general procedure for emissions-inventory development that categorizes emissions into four source types: stationary, mobile, biogenic, and geogenic (EPA 2003f). Stationary sources are further divided into major stationary and area sources. Major stationary sources are defined as stationary sources having emissions that exceed a minimum or threshold level, which varies depending on the pollutant. Stationary sources that fall below the threshold level are merged as area sources. The reporting requirements for major stationary sources are more detailed than those for area sources (65 Fed. Reg. 33268 [2000]). Mobile sources include on-road and nonroad vehicles and sources such as lawn and garden, recreational, construction, and marine equipment; aircraft; and locomotives. Biogenic sources include plants, trees, grasses, and agricultural animals and crops, and geogenic sources include gas seeps, soil wind erosion, geysers, and volcanoes. Inventories are generally developed using a combination of the direct measurements and emission models described below (EPA 1998b).

### Continuous Emissions Monitoring Systems

The most direct way to determine the rate at which a pollutant is emitted into the atmosphere is through emissions monitoring. However, the large number and varying types of sources that generally exist for a given pollutant make this impractical and only the largest major stationary sources are generally equipped with continuous emissions monitoring (CEM) systems.

As a result of the requirements in the Acid Rain Title of the 1990 CAA Amendments, increasing numbers of electrical utility boilers are now using CEM systems. The data from them are posted quarterly on the internet, providing hourly emissions as well as values averaged quarterly and annually (EPA 2003g). Because of the use of CEM systems, the emission rates from these sources are generally viewed as being among the most accurate of all known rates in the United States.

### Emission Estimation Models

Models for estimating emissions have been developed and made available by EPA for selected sources, particularly area sources for which measurements are difficult to make. Emission models are used for estimating on-road emissions and air emissions from landfills, storage tanks, wastewater-collection and -treatment systems, wind erosion, fugitive dust from roads, material handling, agricultural tilling, and construction and demolition. These models generally estimate $E_i(s)$, the emission rate (in, say, tons per day) of pollutant $i$ from source $s$, as

$$E_i(s) = EF_i(s) \times A(s), \tag{3-1}$$

where $EF_i(s)$ is the emission factor (in units of tons of emissions per unit of activity) and $A(s)$ is the activity level for that specific emission. Depending on the source, various types of activity levels can be chosen—for example, the total amount of fuel used by the source, the amount of product produced or consumed, the population density, or the vehicle miles traveled. In some cases, an emission factor is derived for an uncontrolled source. In this case, the emission rate for a source with controls is corrected to allow for the fractional control of the emissions. The control efficiency $(CE)$ may be further modified by correction factors to take into account the rule effectiveness $(RE)$ and the rule penetration $(RP)$ (EPA 2001b). In this case, Equation 3-1 is modified as follows:

$$E_i(s) = EF_i(s) \times A(s) \, [1 - (CE)(RE)(RP)], \tag{3-2}$$

where $CE$ is the fraction of a source category's emissions that are controlled (for example, controlled by a control device or process change). $RE$ is an adjustment to the $CE$ to account for failures or uncertainties in the performance of the control. $RP$ is the fraction of the source category that is covered by the regulation or is expected to comply.

EPA compiles and periodically updates emission factors for a large number of sources (EPA 2002g). Emission factors are simply averages of all available data of acceptable quality and are generally assumed to be representative of long-term averages for all facilities in the source category (that is, a population average) (EPA 1995). For example, for mobile sources, emission factors are derived from measurements made on a selected set of vehicle types and ages deemed to be representative of the fleet of vehicles in use. In other cases, emission factors may be based on the properties of the fuel used. For example, sulfur emissions from vehicles are often estimated from fuel consumption and fuel sulfur concentration, which is determined by measurements made on a sampling of the fuel in use at the time.

An activity level can be estimated from a wide range of data and measurements. For area sources involving consumer products, it can simply be the population density or other relevant socioeconomic indicator (for example, the amount of a given item sold or consumed). Activity levels for mobile sources are often derived from estimates of vehicle miles traveled and fuel consumption statistics. Stationary-source activity levels can be based on economic indicators, fuel-usage statistics, and surveys and questionnaires.

### Critical Review of Emission Inventories

The emission inventories used in AQM and the methods used to develop them have been reviewed and critiqued extensively by previous investigators and committees (see NRC 1991, 1998b; Placet et al. 2000). Current emission inventories are generally held to have an uncertainty of about a factor of two or more (see NARSTO 2000), although, as discussed later, the uncertainty factor is poorly defined. The consequences of errors in emission inventories can be profound. For example, during the 1970s and 1980s, emission inventories for $O_3$ precursors, volatile organic compounds (VOCs), and nitrogen oxides ($NO_x$) had significant errors, the emissions of VOCs being greatly underestimated and the emissions of $NO_x$ from power plants being somewhat overestimated. Collectively, these errors probably contributed to the adoption of less-than-optimal control strategies for ground-level $O_3$ pollution in many regions of the United States (NRC 1991; NARSTO 2000).

A major contributor to the large uncertainties in current emission inventories arises from the use of emission models to derive the inventories (that is, estimating the emissions from a source as the product of an emissions factor and an activity level). The application of an average emissions factor based on the measurements of a small subset of a total population necessarily introduces uncertainty. Such application is especially problematic if the total emissions of a particular type are dominated by statistically anomalous members of the emitting population. For example, in the case of mobile emissions, a disproportionate amount of the emissions arises from a relatively small percentage of high-emitting motor vehicles (NRC 2001c and references therein; Holmes and Cicerone 2002 and references therein). Because it is not well-understood why some of these vehicles are high emitters and others are not, it is extremely difficult to identify a representative sample of motor vehicles that will provide a statistically valid representation of the total population. A related problem arises from the difficulty in accounting for large emission spikes from stationary sources related to so-called upsets, which are not captured in the average emission estimate for each source and which may result in increased overall emissions (TNRCC 2003). Additional uncertainty can arise when the design and operation of the sources that are measured to determine the emission factors are not representative of the design and operation of the entire population of sources. Beyond the uncertainties in estimating emission factors, the estimation of the activity levels for each source adds more uncertainty, some measures (such as miles traveled or numbers of vehicles in use) being relatively easy to estimate and others (such as industrial emissions tied to economic activity and fugitive dust tied to construction activity) being substantially more difficult to estimate. Further, other sources, such as wildfires, can contribute substantial quantities of air pollutants, but emission estimates are highly uncertain.

Another source of uncertainty in emission inventories arises from the "bottom-up" nature of the process; for example, an inventory is constructed by adding up the emissions from the known sources. With this process, there is no internal mechanism to ensure that important sources have not been overlooked. The importance of overlooked sources was highlighted recently by the measurement of emissions of particles and organics, including chlorinated dioxins and furans, from the burning of household waste in barrels. The research showed that this uncontrolled source of emissions could yield emissions comparable to those of a controlled combustor for a community of 30,000 people (EPA 1998c).

There are at least two steps that could be taken to address the difficulties discussed above and thereby improve the accuracy and utility of emission inventories in AQM. The first step involves the enhanced use of ambient air measurements in conjunction with diagnostic tests, such as source apportionment, to evaluate emission inventories independently (see Figure 3-1). It is often found that the process of reconciling emission estimates from source-oriented models with those from receptor-oriented models helps to identify problematic emissions and activity factors as well as important sources of emissions that had been heretofore neglected (Watson et al. 2002). The second step is the development and routine use of methods to estimate the uncertainty and determine the quality of the inventory data (see Frey et al. 1999). Currently, EPA grades inventory data from A to E based on the quality of the data, but no quantitative measure of uncertainty is maintained. As discussed later, a quantitative estimate of the uncertainty in an emissions inventory could then be used in a propagation of error analysis to advise policy-makers on the range of possible outcomes that could arise from any given emission-control strategy.

Although EPA is responsible for defining how to develop emission inventories, state and local agencies have most of the responsibility for gathering the data. To overcome some of the difficulties that this separation of responsibility generates, the Emission Inventory Improvement Program (EIIP)—a jointly sponsored effort of the State and Territorial Air Pollution Program Administrators and the Association of Local Air Pollution Control Officials (STAPPA-ALAPCO) and EPA—was established in the early 1990s. The goal of EIIP is to provide cost-effective, reliable inventories by improving the quality of emission information and develop-

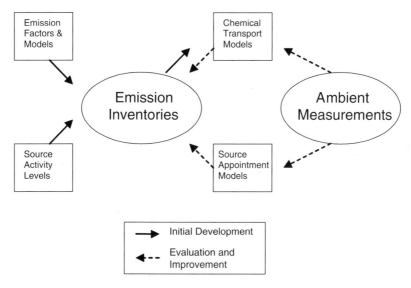

FIGURE 3-1  Emission-inventory development, evaluation, and improvement. Ambient measurements can be used to evaluate and improve emission inventories initially developed using emission factors and activity levels.

ing systems for collecting, calculating, and reporting emission data. EIIP is directed to set up protocols for data collection, transfer, and documentation to overcome some of the problems with data quality. EPA has indicated that documents prepared by EIIP should be used instead of existing federal guidance when appropriate.

## Mobile-Source Emission Inventories

The emission inventories for mobile sources are particularly noteworthy. Mobile emissions represent a major source of many pollutants in the United States.[4] Moreover, because of the large number of types, ages, and operating modes of mobile sources, their inventories are especially difficult to develop. As a result, uncertainties in mobile-source emissions can result in significant uncertainty in the overall emissions inventory for an area.

Mobile-source types range from lawn and garden equipment to motor vehicles, aircraft, and locomotives. EPA has developed two main computer tools for determining the emissions from mobile sources. The MOBILE

---

[4]A detailed discussion of the mobile-source emissions control program in the United States is presented in Chapter 4. Here, the discussion is limited to the emission inventories used to characterize their emissions.

computer program, in various versions, has been used to compute the emissions from on-road vehicles (cars, trucks, buses, and motorcycles) for over 30 years. The NONROAD program is used for the computation of emissions from other mobile sources.

As in all emission inventories, the accuracy of the estimated on-road and off-road emissions depends on two components: the emission factors and the activity levels. The MOBILE model provides emission factors for on-road vehicles. Local areas have to develop activity levels, such as vehicle miles traveled, to get a final emissions inventory in tons per day. In contrast, the NONROAD model contains both emission factors and activity levels. However, little is known about the accuracy or uncertainty of either component (Bammi and Frey 2001).

The National Research Council (NRC) has published a report on the modeling of mobile-source emissions, the major focus being the MOBILE model (NRC 2000b). A number of problems with MOBILE were identified, including emission factors based on laboratory testing and not actual driving conditions. A number of recommendations for improving that model were made. The NRC report also noted that nonroad (or off-road) sources were becoming an increasingly important part of the inventory (see discussion in Chapter 4) and that, because of a lack of data, the NONROAD model does not accurately estimate the emission inventories or the effect of controls on off-road sources. The report recommended that more attention be given to the NONROAD model.

At the time that the NRC report was issued, EPA was in the process of producing a new version of the MOBILE model, referred to as MOBILE6, which appears to address several of the concerns with previous versions of the model (NRC 2000b). However, a significant shortcoming remaining with MOBILE6 stems from its inadequate treatment of the effects of emission spikes that come from variability in engine loads and the importance that such spikes have in overall emission inventories (Barth et al. 1997; NRC 2000b; Hallmark et al. 2001). With the ultimate goal being to translate vehicle performance into their effects on air quality, a much tighter coupling between vehicle operating characteristics and emissions is being pursued through EPA's next-generation MOVES model (EPA 2002h). A tighter coupling of emissions to vehicle operations will aid not only in the simulation of regional emissions and air quality but also in the assessment of improvements that are more microscale in nature, such as the effects on emissions from changes in traffic-signal operations and traffic flow.

Another concern raised in a separate General Accounting Office report (GAO 1997) was how inspection and maintenance (I/M) programs are modeled in MOBILE. I/M modeling is an important element because local areas use this part of the model to derive mobile emission reductions arising from the implementation of I/M programs in their local areas. Another

NRC report on I/M programs concluded that such programs provide "much lower benefits" than those estimated by the MOBILE model (NRC 2001c).

## Air Quality Modeling

The second step in the development of an attainment-demonstration SIP is to determine the amount of pollutant-emission reductions that will be needed to bring about compliance with the NAAQS. Air quality models generally play the central role in that process. Typically a model is used to simulate one or more of the historical events that contributed to the area's "design value" (defined in Box 3-2) and, through these simulations, demonstrate that appropriate emission controls will prevent NAAQS violations in the future during similar events.[5] Because the model simulations are required to show compliance in the future, they must be run with future projections for a variety of hard-to-predict quantities that affect emissions (for example, socioeconomically driven activity factors, such as vehicle miles traveled and the effectiveness of emission-control technology). The projections and the attainment demonstration based on simulations of the extreme events that contributed to the design value introduce uncertainty into the process beyond that arising from intrinsic uncertainties in the emission inventories and models.

The air quality models used in these analyses are designed to allow policy-makers to link quantitatively pollutant emissions to concentrations of pollutants in the atmosphere. Models of several types have been in development and used for over 30 years and, through a close collaboration between the scientific, engineering, and regulatory communities, have continuously evolved in their capabilities and technical completeness (Russell and Dennis 2000). There are three major classes of air quality models: (1) statistical and empirical models that are based on observed relationships between pollutant concentrations and emission rates with little or no explicit consideration of the underlying physical and chemical processes that determine these relationships; (2) deterministic models that solve mathematical equations that describe the physics and chemistry of air pollutant emissions, formation, transport, and removal; and (3) a hybrid of the former two that, although essentially empirical or statistical in its approach, makes use of physically and chemically based algorithms. An example of the latter

---

[5]Some recent SIP revisions have not required the exercise of modeling ambient air quality. For example, modeling of ambient air quality was not required for the rate-of-progress SIP revisions that demonstrated that $O_3$ nonattainment areas classified as moderate or above would reduce VOC emissions by 15%. Although they were backed by regional-scale modeling carried out by EPA and the Ozone Assessment Transport Group (OTAG), the $NO_x$ SIP call and the I/M SIP revisions did not require air quality modeling by the states.

would be a grid-based model that uses inverse methods to derive pollutant emission rates from observed pollutant concentrations. This section briefly reviews and provides a critique of some of the key models commonly used in AQM.

### Empirical Rollback Model

The empirical rollback model is the simplest model available for AQM. This model assumes that an air pollutant's concentration in an airshed is directly proportional to the total emission rate of the pollutant in the airshed. In other words, if $P_o$ is the current concentration of a pollutant in an airshed, $E_o$ is the current rate of emission, and $E^l$ is the hypothetical emission rate after the introduction of emission controls, the new pollutant concentration, $P^l$, predicted by a rollback model would be

$$P^l = P_{background} + P_o (E^l / E_o),  \quad (3\text{-}3)$$

where $P_{background}$ is the so-called background concentration of the pollutant that would be found in the absence of any emissions within the airshed. From Equation 3-3, one can solve for $E_a$, the emission rate needed for attainment of $P_s$, the pollutant air quality standard or goal:

$$E_a = E_o (P_s - P_{background})/P_o. \quad (3\text{-}4)$$

Although easy to use, rollback models have distinct limitations. Because of the assumption of linearity between emissions and concentrations, these models are only suitable for primary pollutants or secondary pollutants with relatively simple chemical-production mechanisms. Because rollback models only calculate a single concentration for the pollutant in the airshed, they are unable to simulate or account for spatial and temporal variations in pollutant concentrations. As a result, these models are most useful for pollutants that tend to be uniformly mixed in an airshed. They were used extensively in AQM before the mid-1970s and have since been largely supplanted by more sophisticated models. They nevertheless continue to be used in some applications, most notably in the design of urban strategies to meet the NAAQS for carbon monoxide (CO) (NRC 2002b).

### Receptor Models

Receptor models are similar to rollback models in their dependence on observed concentrations and their neglect of chemical and meteorological processes. However, instead of being limited to assuming simple linear relationships between a pollutant concentration and its emission rate, receptor models can adopt more sophisticated statistical techniques to derive a more complex (and possibly more accurate) description of the relationship. They have been used most successfully and most widely in estimating the relative contributions of various emission sources to measured ambient air particulate matter (PM) composition at a monitoring site. The estimates are made by relating the measured elemental and organic tracer composition of ambient PM to the known elemental composition of the sources of PM in the region. Reviews of these methods can be found in Brook et al. (2003) and the references therein.

Although receptor models do not require emission inventories, some of them (such as chemical mass balance models) require source profiles, which define the detailed elemental or organic composition of all sources. In contrast, statistical factor analysis models do not require source composition data but use variations and covariations among species over time to identify sources. If the source profiles vary among sources in the same category, space, or time, a large measurement project is required to obtain profiles that adequately characterize all the source classifications that are to be used in the model. These models have a long history of applications and have in general provided valuable information for the design of primary PM-control strategies. Recent extensions of these models allow the identification of sources of primary organic PM as well as sources of inorganic particles. Their greatest weakness is the estimation of the contribution of secondary PM. To date, a combination of receptor models has been the most successful approach for PM nonattainment areas.

### Emissions-Based Air Quality Models

Emissions-based models for $O_3$ and CO are important tools used today in AQM to evaluate alternative emission-control strategies and estimate the amount of emission reductions required to meet a specific air quality goal or standard. The models use a mathematical representation of the relevant physical and chemical processes and then solve the governing equations (usually numerically) in time and space to determine the relationships between pollutant emissions and pollutant concentrations. Because these models require the input of pollutant emission rates, they are sometimes referred to as emissions-based models. One of the major advantages of these models for AQM is their predictive capability. Because they calculate pollutant concentrations as a function of pollutant emissions, they can be run in a prognostic mode to predict the air quality response to any hypothetical change in pollutant emissions.

Emissions-based air quality models can vary in complexity. The simplest are box models, which simulate the evolution of pollutant concentrations in an idealized well-mixed parcel of air. Dispersion models, such as

ISC3 and CalPUFF (EPA 1995a,b), simulate a plume parcel of air as it advects and mixes with ambient air. The most complex are state-of-the-science three-dimensional (3D) chemical transport models (CTMs), which attempt to recreate the observable distributions of chemical species in time and space. They are called 3D models (also, grid models) because they use a three-dimensional grid to represent space. In the chemical transport model, gases and particles are transported horizontally and vertically by winds and turbulence between adjacent grid cells of the model, and chemical reactions between gases and particles occur at rates governed by reactant concentrations in individual grid cells. These models generally require a massive amount of input data to operate (Russell and Dennis 2000). In addition to the emissions of all pollutants in the model domain, these models require the input of temporally (usually hourly) and spatially resolved information on the relevant meteorological conditions (wind speed and direction, temperature, relative humidity, mixing depth, and solar insolation) (Pielke and Uliasz 1998). They also require the initial concentration fields of all pollutants in the entire domain, plus temporally resolved concentrations at the domain boundaries. Collecting and processing all this input information usually requires many measurements and additional models (meteorological, emissions, and coarser resolution models for the boundary conditions) and is usually the most time-consuming and resource-intensive part of the modeling exercise. A relatively new but important component of some 3D models is the inclusion of algorithms to internally assess the statistical uncertainties in the model output.

### The First-Generation 3D CTMs—Urban-Scale Photochemical Grid Models for $O_3$

Ground-level $O_3$ is produced by the oxidation of VOCs in the presence of sunlight and $NO_x$. Control of ground-level $O_3$ pollution in an urban area can be achieved in principle by reducing the emissions of VOCs or $NO_x$ locally or in areas upwind of emission sources. However, the formation and transport of $O_3$ in the atmosphere is a complex chemical and physical process, and determining the most effective combination of VOC- and $NO_x$-emission controls to mitigate $O_3$ pollution has proved to be a challenging task (see Box 3-4). It is best accomplished, however, using so-called photochemical models capable of simulating the chemical reactions that lead to the production of $O_3$ in the presence of sunlight.

Since the 1970s, when the first NAAQS for $O_3$ was promulgated, photochemical models have evolved considerably in their scientific complexity and completeness as well as their numerical sophistication. Simple rollback and box models were used initially to develop $O_3$ mitigation strategies; by 1973, EPA had committed its research efforts to supporting the

## BOX 3-4   The VOC, $NO_x$, and $O_3$ Challenge

Since the first $O_3$ NAAQS were set in response to the 1970 CAA Amendments, significant progress in $O_3$ reduction has been made in some areas of the United States. These areas include Los Angeles, New York, and Chicago, where the average daily maximum 1-hr $O_3$ concentration decreased by about 15% over the decade 1986–1996. Despite almost three decades of massive and costly efforts to bring the $O_3$ pollution under control, however, the lack of $O_3$ abatement progress in a number of areas of the country has been discouraging and perplexing (NRC 1991). The complex $O_3$-formation process, which involves the interaction of $NO_x$, VOCs, and dynamic atmospheric processes, has probably contributed to the difficulties encountered in abating $O_3$ pollution in the United States.

Reduction of $O_3$ concentrations requires control of either $NO_x$ or VOC emissions or a combination of both. For the first two decades of $O_3$ pollution mitigation in the United States, VOCs were the primary targets for emission reductions. The initial decision to pursue VOC controls was based on results from federal smog chamber experiments and data collected at monitoring sites in several U.S. cities, as well as the recognition that VOC controls would reduce the concentrations of eye irritants. In 1971, EPA issued the so-called Appendix J curve (Figure 3-2) for use by state and local agencies in developing SIPs (Fed. Reg. 36 [1971]). The curve (essentially from a modified rollback model) was derived from observations in six U.S. cities. On the basis of the maximum $O_3$ concentrations observed at these cities and their estimated VOC emissions, the curve purported to indicate the percentage of VOC emission reduction required to reach attainment in an urban area as a function of the peak concentration of photochemical oxidants observed in that area.

In the late 1970s, a more sophisticated method involving the use of a photochemical box model was developed by EPA. The empirical kinetic modeling approach (EKMA) (Dimitriades 1977) used the improved chemical mechanisms that were under intense development in the late 1970s and early 1980s (Atkinson and Lloyd 1984) to simulate the production of $O_3$ within an idealized air parcel as it advected over an urban area with VOC and $NO_x$ emissions. Figure 3-3 illustrates the typical output from EKMA—the so-called Haagen-Smit or EKMA diagram, where lines of constant peak $O_3$ concentrations (called $O_3$ isopleths) are plotted as a function of VOC and $NO_x$ emissions. Inspection of Figure 3-3 reveals distinct regions of VOC and $NO_x$ sensitivity. For VOC-to-$NO_x$ ratios less than about 4:1, the atmosphere is VOC limited, the most expeditious path to $O_3$ control is VOC reduction, and a reduction in $NO_x$ could increase $O_3$. However, for VOC-to-$NO_x$ ratios greater than about 15:1, $NO_x$ controls provide the best way to reduce $O_3$.

EKMA plots, as shown in Figure 3-3, captured the major features and complexities of the $NO_x$/VOC/$O_3$ system. However, to effectively apply the information to an $O_3$ abatement strategy for a given nonattainment area, one must be able to accurately characterize where on the diagram an area is situated in terms of the VOC:$NO_x$ emissions ratio. Based on the available emissions inventories in the late 1970s and early 1980s, it appeared that most urban areas were near or above the ridge of the diagram, suggesting that VOC controls were the most effective path to attainment. Results using more sophisticated models, such as the UAM (urban airshed model), based on essentially the same emission inventories, generally

*(continued on next page)*

**BOX 3-4** continued

supported the conclusion reached by the applications of the EKMA plots—that VOC emission controls provided the best path to attainment of the $O_3$ NAAQS.

As a result of these analyses, the SIPs submitted in response to the 1970 and 1977 CAA Amendments focused on reducing VOCs to reduce ambient $O_3$, and from the early 1970s to the early 1990s, EPA and Congress promoted VOC control as the principal path to attaining the $O_3$ standard.[a] However, by the mid-1980s, it became clear that few areas would actually meet the December 31, 1987, attainment deadline for the $O_3$ standard required by the 1977 CAA Amendments; by 1990 some 100 areas were still classified as nonattainment for $O_3$. At the same time, there was growing appreciation for the importance of $NO_x$ emission reductions. In the mid-1980s California began to pursue $NO_x$ controls. Reports from both the OTA (1989) and the NRC (1991) identified $NO_x$ control as an essential new direction for reducing $O_3$ pollution. Although it has taken some time to implement this change, the late 1990s saw an increase in efforts to control $NO_x$, including much more stringent $NO_x$ controls on heavy-duty diesel vehicles and the $NO_x$ SIP call for the eastern United States.

A number of factors led to the slow pace of implementing substantial $NO_x$ emission controls to mitigate $O_3$ pollution. Both the J curve and EKMA model had deficiencies, but the major reason was the incomplete and often erroneous emission inventories used. It was not until the late 1980s, with the analysis of data gathered from various field experiments carried out in Southern California and the Southeast United States, that two deficiencies were identified in these inventories: a significant underestimate of VOC emissions from mobile sources (Pierson et al. 1990), and a failure to account for biogenic VOC emissions (Chameides et al. 1988). As these deficiencies were addressed, it became apparent that a greater emphasis on $NO_x$ controls was needed (NRC 1991). In retrospect, it is possible that more progress in mitigating $O_3$ pollution in the United States might have been achieved had there been a greater reliance on analysis of field data and the use of observation-based methods (Roth et al. 1993) to complement the more traditional emissions-based modeling approaches during the first two or three decades of $O_3$ mitigation efforts.

---

[a] Three main exceptions to the focus on VOC controls were (1) $NO_x$ reductions were required from motor vehicles; (2) the 1990 CAA Amendments allowed states some latitude in substituting $NO_x$ reductions for VOC reductions in some cases; and (3) the 1990 CAA Amendments required RACT and NSR controls for stationary-source $NO_x$ emissions.

---

development of an urban-scale 3D chemical transport (grid-based) model— the urban airshed model (UAM). At the same time, other urban-scale 3D models were developed and used by the scientific community (Reynolds et al. 1973). In the 1980s, the use of 3D models spread to other major metropolitan areas, and the 1990 CAA Amendments specifically called for

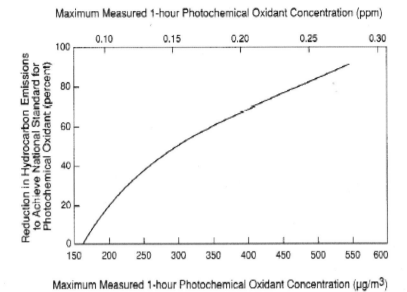

FIGURE 3-2   Appendix J curve.   Required hydrocarbon emissions control as a function of photochemical oxidant concentration.   SOURCE: EPA 1971.

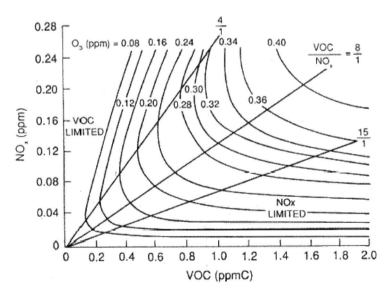

FIGURE 3-3   Empirical kinetic modeling approach (EKMA) diagram. SOURCE: NRC 1991, adapted from Dodge 1977.

the use of such models for all $O_3$ nonattainment areas classified as serious and above.

Today, the scientific and engineering community has developed more than 20 3D CTMs for $O_3$ in the United States and Europe, in addition to the UAM (see Russell and Dennis [2000] for a detailed review). These models continue to grow and develop as new scientific information becomes available and as regulatory needs dictate. For example, $O_3$ pollution was initially viewed as an urban problem, and the initial versions of the UAM were restricted to simulating individual urban areas. Today, it is recognized that $O_3$ pollution extends over multistate airsheds, and the current version of the UAM and many other contemporary 3D photochemical models used to simulate $O_3$ pollution are designed to treat this larger spatial domain.

*Modern Multipollutant, Multiscale CTMs*

Even as the UAM and other urban-scale CTMs were being developed and refined in the 1980s, the lessons learned from the application of these models were being applied to other types of CTMs. One of the most important of these efforts was the National Acid Precipitation Assessment Program's development of a regional (or multistate airshed) CTM able to simulate acidic deposition in the eastern United States and Canada (NAPAP 1991b). The resulting model, the regional acid deposition model (RADM), included state-of-the-science representations of the physical and chemical processes leading to the formation and deposition of acidic species over multistate geographic areas (Chang et al. 1987). The model (together with similar tools, such as the ADOM model developed for Canada) provided important insights into the source-receptor relationships in the acid deposition problem in the United States and Canada. Although these models ultimately played a relatively minor role in the development of the Acid Rain Program enacted in the CAA Amendments of 1990,[6] they represented a major step in the development of CTMs—namely, the expansion of the algorithms used in urban-scale CTMs for $O_3$ into a more comprehensive regional-scale model able to simulate processes related to the formation, transport, and deposition of PM as well as gaseous pollutants.

Just as the development of urban-scale CTMs for $O_3$ aided in the development of regional acid deposition models, such as RADM, the development of RADM provided the basis for significant advances in CTMs used for $O_3$ pollution. Most present-day $O_3$ models have regional rather than urban-scale coverage. Many have multiscale and nesting capabilities so that they can simulate a relatively large multistate area but also focus on

---

[6]A more-detailed discussion of the Acid Rain Program is presented in Chapter 5.

one or more localities of interest (for example, an urban center). Many of these models also have more advanced treatments of clouds and depositional processes, which can have a significant effect on $O_3$ and related gaseous pollutants.

The proposed fine particulate $(PM_{2.5})$[7] NAAQS and the need for models to carry out the related SIP attainment demonstrations will present an opportunity to build further upon the experience gained from modeling both $O_3$ and acid deposition. Fine particulate matter comes not only from direct emissions but also from atmospheric reactions that transform primary pollutants such as sulfur dioxide $(SO_2)$, $NO_x$, ammonia, and VOCs into particulate sulfate, nitrate, carbonaceous material, and ammonium. Some of the processes that transform gases into PM take place in clouds and are therefore closely tied to the processes that lead to acidic precipitation. As a result, models for $PM_{2.5}$ require a conjoining of processes relevant to the formation of ground-level $O_3$ and to the formation of acid deposition. Because of this complexity, most models capable of simulating PM can also be used to address other air pollutant problems $(O_3, acidity,$ and visibility reduction), and for that reason, they are also referred to as multipollutant or unified air quality models. A prime example of such a model is EPA's new Models-3, which is available to both the scientific and regulatory communities in various forms. Seigneur (2001) has published a review of the strengths and weaknesses of these tools.

*Observation-Based Model for $O_3$*

One significant problem with emissions-based air quality models for $O_3$ is their dependence on uncertain VOC and $NO_x$ emission inventories (see Box 3-4). In recognition of that limitation, additional diagnostic tools have been developed to complement the air quality models. These tools, sometimes referred to as "observation-based methods," use ambient air quality measurements rather than emission inventories to determine the relative effectiveness of VOC and $NO_x$ emission reductions on $O_3$ pollution mitigation. The advantage of these methods is that they are not affected by uncertainties in emission inventories. On the other hand, they have two distinct disadvantages: (1) they require high-quality data on the ambient air concentrations of relevant chemical constituents, and much of those data are not gathered in routine monitoring networks; and (2) they are diagnostic and not prognostic (predictive). The latter limitation means that observation-based models can usually be used to estimate the types but not the amounts of emission reductions needed to meet a specific air quality

---

[7]$PM_{2.5}$ refers to particulate matter with an aerodynamic equivalent diameter of 2.5 micrometers $(\mu m)$ or less.

goal or standard (NARSTO 2000). For that reason, observation-based models provide a useful adjunct to emissions-based models in an attainment demonstration but are not a substitution for them.

### Air Quality Models: Lessons Learned

Scientists have had over three decades of experience in developing and applying a number of models for air quality management, and in that time, a number of insights into what is working well and what is not working well have been gained.

*What Is Working Well*

• The dynamic partnership existing between the technical and regulatory communities promotes continuous improvement and advancement in air quality modeling and management. On the one hand, the evolving needs of the regulatory community promoted and catalyzed the scientific and engineering communities to develop increasingly sophisticated air quality models; on the other hand, insights gained from advanced models have prompted members of the regulatory community to rethink their approaches to air quality management.

*What Is Not Working Well*

• As a general rule, models should be subjected to comprehensive performance evaluations using detailed data sets from the atmosphere before using them in regulatory applications (Roth 1999; Seigneur et al. 2000). Unfortunately, because of the lack of adequate atmospheric data (such as those on pollutant concentrations), that rule is all too often overlooked by the regulatory community.[8] The consequences of this practice can be and, in some instances, have been the promulgation of less than optimal or even ineffective control strategies (see Box 3-4).

• Model results are sometimes used inappropriately by regulators. One example is the practice in attainment demonstrations of using a model to predict pollutant concentrations years or even decades into the future without recognizing that input parameters, which are based on estimates of future socioeconomic factors, for the simulation are highly uncertain.

• Substantial delays occurred in incorporating new scientific insights from models into policy design. One example is the reliance on VOC

---

[8]In a small number of cases, EPA has required mid-course review, but in most cases, no provision is made for such a review.

controls for $O_3$ pollution long after science suggested that this strategy was not effective.

• The ability to quantify uncertainty in the predictions of state-of-the-science chemical transport models remains inadequate, although improvements have been made.

• In a number of areas of the country, there have been inadequate resources (financial and personnel) for the correct application of these tools to local and regional problems.

Beyond the specific concerns listed above, there are two broader concerns—model uncertainty and over-reliance on models for $O_3$ SIPs—and one important opportunity—multipollutant models—that bear more detailed discussion.

### Model Uncertainty

An assessment of the uncertainties in model output has not been a routine component of the information used by regulators. Unfortunately, the uncertainty in model predictions depends on a variety of factors, including the type of prediction (such as variables and averaging time), the application, the uncertainties associated with the model's input and internal parameters, and the model itself. As a result, a universal, comprehensive statement cannot be made about the uncertainties of any model simulation (see Oreskes et al. 1994). Literature estimates for individual components of an air quality model—emissions, chemistry, transport, vertical exchange, deposition—typically indicate uncertainties of 15–30%, but when the supporting data sets are weak, the uncertainties can be significantly higher. A consideration of all factors reveals that model uncertainty can be both significant and poorly characterized (NARSTO 2000). It follows, therefore, that relying solely on the output of an air quality model to resolve emission-control issues or to demonstrate attainment of an air quality standard or objective is problematic.

Despite these limitations, air quality models remain the only tools available for quantitatively simulating or estimating future outcomes. Although challenging to use, air quality models are essential to the current AQM system. Their use can be optimized by collecting and analyzing appropriate data sets, carefully assessing model sensitivity and uncertainty, and avoiding inappropriate applications.

### Over-Reliance on Models for $O_3$ SIPs

As discussed in more detail later in this chapter, SIPs developed for $O_3$ nonattainment areas have relied heavily on so-called attainment demonstrations, in which an air quality model is used to determine the amount of emission reductions needed to reach attainment by a specified date. In general, these attainment demonstrations have tended to be overly optimistic, and $O_3$ concentrations in these areas have tended to exceed the values projected by the models (API 1989).

### An Opportunity: Emerging Multipollutant Models

The growing development and use of multipollutant models suggests that a multipollutant approach to AQM is now viable. The current approach in the United States tends to ignore the interrelationships among pollutants. For the most part, planning and regulatory efforts have occurred without serious consideration of the linkages among pollutants and the commonality of their sources (the one exception being particles and visibility). At the very least, the effort to solve each air quality problem in isolation from the rest has probably resulted in missed opportunities to address different pollutants simultaneously. Future multipollutant modeling efforts should be enhanced to support strategies for simultaneous reduction of multiple pollutants.

### Using the Weight-of-Evidence Approach
### in the Attainment Demonstration

In recognition of the uncertainties inherent in the use of the air quality models, EPA now encourages states to perform complementary analyses using available air quality, emission, and meteorological data along with outputs from alternative receptor- and observation-based models before arriving at a final target for the emission reductions that will be needed to meet attainment. These complimentary analyses can be used to assess the reasonableness of the results obtained from the air quality model simulations or as a weight-of-evidence determination that a given amount of emission reductions will be adequate to meet the NAAQS attainment standard.

The inclusion of a weight-of-evidence analysis in attainment-demonstration SIPs should be a positive development from a scientific and technical point of view. It implicitly acknowledges the limitations of air quality model simulations and allows planners to use information and insights from existing data and other analytical procedures to develop a more comprehensive and conceptual understanding of the relationship between pollutant emissions and the concentration of a criteria pollutant. In principle, such an understanding can make it possible for SIP developers to arrive at a more robust estimate of the emission reductions that will be needed to reach attainment. However, for this approach to work, the weight-of-evidence analysis must be applied in an unbiased manner and not simply to justify lower emission reductions than those indicated by air quality model simulations. The introduction of the weight-of-evidence approach in the 1996 guidance document for preparing new SIPs for the 1-hr $O_3$ standard appeared to have invited such a biased application, and bias in using the weight-of-evidence approach has been alleged in legal challenges to SIPs for a number of states. More recent guidance documents for SIPs for the 8-hr $O_3$ standard (EPA 1999e) and for $PM_{2.5}$ and regional haze (EPA 2001c) appear to have a more balanced approach. Time will tell how states will respond to this guidance.

Finally, the inclusion of the weight-of-evidence analysis does not eliminate a major problem in the current SIP process. This problem—an overemphasis on the emissions target obtained in the attainment demonstration—is discussed in more detail later in this chapter.

### Emission-Control Strategy Development in an Attainment-Demonstration SIP

When the total amount of emission reductions required to reach attainment of the NAAQS has been determined though modeling and weight-of-evidence analysis, air quality planners must devise a strategy of emission controls and enforcement to bring about these reductions by the required date. In principle, these reductions can be derived from measures designed to change sociological and behavioral factors that influence pollutant emissions as well as from technological changes that directly affect emissions. In practice, however, AQM in the United States has emphasized technological solutions (see Box 3-5).

The emission-control strategy in an attainment-demonstration SIP is typically developed in stages, as the emission reductions from federal measures, mandatory local measures, additional local measures, and possible regional multistate measures are assessed consecutively.

### Federal Measures

The CAA empowers EPA to impose nationwide emission-control measures on selected industries (for example, motor vehicle manufacturing facilities). When developing a SIP, state and local agencies can include the emission reductions anticipated from these measures. The specific amount of emission-reduction credit an area is allowed from each federal measure is determined from EPA-supplied guidelines and models. For most states, emission-reduction credits from federal control measures have represented a major fraction of the emission reductions in their respective SIPs (see Box 3-6); thus, the existence of federally mandated emission-control measures

---

### BOX 3-5   Technological Change versus Social or Behavioral Measures

One of the challenges facing the SIP process and the attainment of NAAQS has been the inability of air quality management in the United States to rely on social and behavioral measures in addition to technological innovations to improve air quality. The 1970 CAA Amendments were passed at a time of great environmental activism. In addition to technological requirements, the amendments contained provisions to obtain emission reductions from actions to change behavior, such as reducing vehicle miles traveled or changing land-use planning. Early efforts to implement these measures (for example, through increased tolls, parking freezes, and other measures to discourage travel) were met with such intense political resistance that EPA was expressly prohibited in the 1977 CAA amendments from imposing the more stringent measures on any area. The 1977 CAA Amendments did include, however, the requirement that transportation-control measures be enacted in certain nonattainment areas. In the 1990 CAA Amendments, Congress added the employee commute options requirement. Even these less stringent measures have met with resistance from those who do not believe that implementation of the CAA should result in changes in life style. In 1995, Congress repealed the employee commuter requirements and placed limitations on the use of enhanced I/M requirements. As a result, most emission reductions continue to come from technological changes. In some instances, the inability of air quality managers to affect social and behavioral patterns has significantly degraded the impact of the technological changes that were imposed to lower pollutant emissions (see, for example, the discussion on mobile emissions in Chapter 4).

---

have eased the burden of state and local authorities in developing attainment SIPs.

### Mandatory Local Measures

Once an area has accounted for emission reductions based on federal measures, the area must then plan and account for local emission-control measures that are specifically mandated by the CAA. The extent of these mandatory measures depends on the pollutant and the classification of the area. As detailed in Box 3-3, the measures range from the imposition in all nonattainment areas of reasonably available control technology (RACT) on large stationary sources, such as electricity-generating and industrial combustion facilities, to the promulgation of the reformulated gasoline program in severe and extreme $O_3$ nonattainment areas. As in the case of the federal measures, the credits for emission reductions based on each mandatory measure are determined by EPA-supplied guidelines and models. The emission reductions allowed by EPA for some of these measures (for

## BOX 3-6 Relative Roles of Federal, State, and Local Controls in Illustrative SIPs

When an area is found to be in nonattainment with a NAAQS, the relevant state (and/or local) agency must submit an attainment-demonstration SIP to EPA. Among other things, the SIP must contain a list of emission-control measures that, when implemented, will bring the area into compliance. The control measures may include those promulgated at the national level by the federal government as well as those promulgated at the state and local level. To illustrate how the mix of federal, state, and local emission programs have been adopted, the $O_3$ attainment-demonstration SIPs from four nonattainment areas are examined here:

1. Houston-Galveston, TX: SIP submitted in December 2000 (TNRCC 2000) and September 2001(TNRCC 2001).[a]
2. South Coast Air Basin, CA: included in the California SIP submitted in November 1994 (CARB 1994).
3. Southeast Wisconsin: SIP submitted in December 2000 (WDNR, 2000).
4. Philadelphia, PA, Trenton, NJ, and Wilmington, DE: SIP submitted in August 1998 (CPDEP 1998).

Although some of these areas have submitted SIP revisions more recently, the ones chosen were those with the most comprehensive discussion of emission-reduction strategies since passage of the 1990 CAA Amendments and are thus most appropriate for our discussion here. Tables 3-2 and 3-3 list the federal, state, and local emission-control measures and anticipated emission reductions for the two $O_3$ precursors, VOCs and $NO_x$, for each of the SIPs. The identification of a reduction as a federal, state, or local measure in the SIPs is determined by the states themselves and is therefore not consistent among the states. For example, stringent motor vehicle standards initially implemented in California and then later extended to other states in a federal regulation are counted as a state measure in California and a federal measure elsewhere. In addition, most states claim I/M programs as a state initiative rather than a federal control, even though the CAA Amendments of 1990 require areas in moderate $O_3$ nonattainment to have basic I/M programs and areas in serious and above $O_3$ nonattainment to have enhanced I/M programs.

The data in the tables suggest that federal programs often play a major role in the emission reductions assumed in SIPs. Federal measures are responsible for the majority of VOC emission reductions in all but one of the examples. California is the exception in that it claims that nearly all of its emission reductions result from state and local initiatives. In the case of $NO_x$ emission controls, the federal measures result in essentially no reductions in Southern California and an almost 30% decrease in Wisconsin. In the Philadelphia-Trenton-Wilmington area, multistate measures, specifically the $NO_x$ SIP call, account for about 30% of the anticipated $NO_x$ emission reduction.

---

[a]The Houston-Galveston SIP did not demonstrate sufficient reductions for either $NO_x$ or VOCs. In theory, EPA should have bumped up this area to an extreme nonattainment area.

TABLE 3-2  Federal, State, and Local VOC Emission-Reduction Measures in Four Illustrative SIPs[a]

| | Houston-Galveston, TX | South Coast, CA | Southeast Wisconsin | Philadelphia, PA; Trenton, NJ; Wilmington, DE |
|---|---|---|---|---|
| Attainment date | 2007 | 2010 | 2007 | 1999 |
| 1990 base year inventory | 945 | 1,517 | 407 | 616 (grown to 625 in 1999) |
| Inventory for attainment | 694 | 323 | 222 | 431 |
| Reductions needed | 251[b] | 1,194 | 185 | 164 |
| Reductions from: | | | | |
| Existing state and local programs | — | 463 (39%) | — | — |
| Existing federal programs | 133 (81%) | 0 (0%) | 110 (57%) | 40 (23%) |
| New local measures | 2 (1%) | 453 (38%) | 0 | 18 (10%) |
| New state measures | 30 (18%) | 231 (19%) | 12–17 (7%)[c] | 59–77 (39%)[c] |
| Mobile | 12 | 114 | 6 | — |
| Enhanced I/M | 18 | 26 | 4 | 59 |
| Stage-2 vapor recovery | — | — | 2 | 0–18[d] |
| Pesticides | — | 2 | — | — |
| Consumer products | — | 89 | — | — |
| Industrial controls | — | — | 0–5[e] | — |
| New federal measures | 0 (0%) | 47 (4%) | 67–72 (36%)[c] | 40–58 (28%)[c] |
| Off-road vehicles | — | 43 | — | — |
| RACT | — | — | 67–72[e] | 10 |
| On-board vapor recovery | — | — | — | 0–18[d] |
| Other | — | 4 | — | 30 |
| Total | 165[b] | 1,194 | 194[c] | 175[c] |

[a]VOC emissions and emission reductions in tons per day (tpd, Texas and California) and tons per summer day (tpsd, Wisconsin and Pennsylvania).

[b]The VOC emission-reduction measures described in the Houston-Galveston SIP (that is, 165 tpd) are less than the state claims are needed for attainment (that is, 251 tpd).

[c]Mid-point of the range was used in calculations.

[d]Documentation is unclear about contribution of Stage 2 and on-board vapor recovery systems.

[e]Documentation is unclear about contribution of state measures and federal RACT.

TABLE 3-3  Federal, Multistate, State, and Local $NO_x$ Emission-Reduction Measures in Four Illustrative SIPs[a]

| | Houston-Galveston, TX | South Coast, CA | Southeast Wisconsin | Philadelphia, PA; Trenton, NJ; Wilmington, DE |
|---|---|---|---|---|
| Attainment date | 2007 | 2010 | 2007 | 1999 |
| 1990 base year inventory | 1,236 | 1,361 | 397 | 439 |
| Uncontrolled inventory for attainment year | 1,375 | | | 455 |
| Inventory for attainment | 414 | 553 | 297 | 373 |
| Reductions needed | 961[b] | 808 | 100 | 82 |
| Reductions from: | | | | |
| Existing state and local programs | 0 (0%) | 429 (53%) | 0 (0%) | 0 (0%) |
| Existing federal programs | 209 (23%) | 0 (0%) | 27 (26%) | 17 (21%) |
| New local measures | 24 (3%) | 43 (5%) | 0 (0%) | 0 (0%) |
| New state measures | 674 (74%) | 227 (28%) | 75 (74%) | 32 (39%) |
| Point source | 586 | — | 53 | — |
| Enhanced I/M | 36 | 31 | 17 | 32 |
| Mobile (other) | 35 | 196 | 5 | — |
| Other | 17 | — | — | — |
| New multistate measures | 0 (0%) | 0 (0%) | 0 (0%) | 27 (33%) |
| $NO_x$ SIP call | — | — | — | 27 |
| New federal measures | 4 (<1%) | 109 (14%) | 0 | 6 (7%) |
| Off-road vehicles | — | 87 | — | — |
| Other | 4 | 22 | — | 6 |
| Total | 911[b] | 808 | 102 | 82 |

[a]$NO_x$ emissions and emission reductions in tons per day (tpd, Texas and California) and tons per summer day (tpsd, Wisconsin and Pennsylvania). [b]The $NO_x$ emission reduction measures described in the Houston-Galveston SIP (that is, 911 tpd) are less than the state claims are needed for attainment (that is, 961 tpd).

example, I/M and oxygenated fuels) have been criticized as being unrealistically large (NSTC 1997; NRC 2001c, 1996).

## Additional Local Measures

As envisioned in the CAA, if the emission reductions projected by federal measures, regional measures (see discussion in next section), and local mandatory measures are not sufficient to bring about compliance, the air quality planners must identify additional local emission controls to make up the shortfall. The only specific requirements on these additional emission reductions are that they be sufficient to meet the areawide target reduction indicated by the attainment demonstration and that the measures are verifiable and enforceable. They can involve deeper emission reductions from sources already regulated via federal or mandatory local measures and emission controls on previously unregulated sources, such as lawnmowers, small industrial and consumer products, open burning, and construction equipment. Although EPA cannot consider costs in setting NAAQS, states and local authorities can and do consider costs and cost-effectiveness when identifying the mix of local emission controls to be included in the SIP. Inclusion of economic incentive programs, such as cap-and-trade programs discussed in Chapter 6, and consideration of environmental justice, as discussed in Chapter 2, are also emerging as important drivers in the process.

## Multistate Regional Measures

At the time of the initial passage of the CAA, it was thought that nonattainment within a given area was largely caused by emissions within that area and could be mitigated by local emission controls. It was reasonable, therefore, to place the final responsibility for devising a plan to mitigate this pollution on local and state authorities. By the end of the 1980s, however, it had become apparent that some air quality problems had a larger, multistate component and that a substantial contribution to an area's pollution could arise from upwind emission sources. In response, Congress included provisions in the 1990 CAA Amendments for implementing multistate air pollution mitigation strategies through the creation of regional planning organizations (RPOs). Although these RPOs cannot be vested with regulatory authority, in principle, their strategies can be implemented and legally mandated by EPA's issuance of an appropriate multistate rule and by the voluntary inclusion of the measures in the SIPs prepared by the states. As a result, a number of statutory as well as ad hoc RPOs have emerged. In the eastern half of the United States, where regional $O_3$ pollution has been an important concern, organizations such as the Ozone Transport Commission (OTC) and the ad hoc Ozone Transport Assessment Group (OTAG) were established. In the West, where degrading visibility in Class 1 areas has been a growing problem, the Grand Canyon Visibility Transport Commission and Western Regional Air Partnership (WRAP) was established. A brief overview of those three efforts and their relationship to two major regional rules promulgated by EPA (the $NO_x$ SIP call and the regional haze rule) is provided below.

## The $O_3$ Transport Region and $O_3$ Transport Assessment Group

In recognizing that $O_3$ and its precursors can be transported over large distances and that $O_3$ nonattainment areas, especially in the eastern United States, can be affected significantly by upwind sources, the 1990 CAA Amendments created the Ozone Transport Region (OTR) to help states in the northeastern United States develop a coordinated strategy. The amendments also authorized EPA to create other $O_3$ transport regions.

The states from Virginia to Maine and the District of Columbia make up the OTR. Their representatives and the EPA offices constitute the OTC, which Congress created to formulate solutions to regional $O_3$ pollution. OTC recommends a variety of VOC- and $NO_x$-reduction measures, which the state governments then make enforceable by incorporating them into their respective SIPs. The major initiatives adopted thus far by OTC to address regional $O_3$ pollution include the California low-emissions vehicle (LEV) program and $NO_x$ budget trading program for electric utilities and large industrial boilers. Phase I of the program included the installation of RACT, which was required in the 1990 CAA Amendments of all OTR states. Phases II and III committed OTR states to further regional $NO_x$ emission reductions in 1999 and 2003 via an integrated interstate emissions-trading program.[9]

OTAG was established in 1995 by EPA and the 31 states east of the Mississippi River. The major impetus for establishing OTAG was the inability of $O_3$ nonattainment areas in the eastern United States to submit complete SIPs by the 1994 deadline mandated in the 1990 CAA Amendments. The incompleteness of the SIPs arose from the inability of state and local regulators to develop viable plans to reach attainment on the basis of local emission contributions alone. Therefore, OTAG was formed to assess the role of more distant sources in causing $O_3$ nonattainment in localized, largely urban areas and to develop strategies for mitigating their contributions.

In contrast to the OTC formation, the affected states played a major role in initiating OTAG. In addition to EPA and the relevant state agencies,

---

[9]This program is discussed in more detail in Chapter 5.

affected industrial and environmental groups, the academic research community, and Canadian representatives participated in the deliberations. The work and deliberations carried out by OTAG produced a general consensus recommendation to EPA for a new regulatory initiative to reduce $NO_x$ emissions throughout the region. In response, EPA proposed a $NO_x$ SIP call that would cap $NO_x$ emissions in 22 states and the District of Columbia. The caps were imposed on states on the basis of their estimated impact on regional $O_3$ concentrations and not on the basis of their own attainment status with regard to the $O_3$ NAAQS. Following a series of court challenges by midwestern states and industry groups to some of the details of EPA's proposed cap, the $NO_x$ SIP call rule was adopted with some changes and will become effective in 2004. The lengthy regulatory and legal challenge process substantially delayed the implementation of the regional $NO_x$ SIP call program and the associated air quality benefits. The OTAG process will have taken 9 years from its initiation to deliver its initial $NO_x$ reductions via the $NO_x$ SIP call. · Under the OTC $NO_x$ budget program, which relied on more explicit authority in the 1990 CAA Amendments, the initial $NO_x$ reductions were achieved in only 5 years.

OTC and OTAG have provided effective mechanisms for facilitating coordination among states faced with air quality problems that have a significant multistate or regional component. However, the actual emission-reduction strategies developed by those groups have yet to be fully implemented; therefore, it is too early to determine whether these strategies will bring about the desired $O_3$ reductions.

### EPA's Regional Haze Rule

Section 169(a) of the 1990 CAA Amendments required EPA to establish regulations to ensure reasonable progress in improving visibility in 156 national parks and wilderness areas (Class I areas) in the United States. In response, EPA issued a regional haze rule in 1999. The rule sets specific visibility improvement targets for the nation (described in Chapter 2) and then requires all states to develop plans to achieve "reasonable progress" toward those goals.

A notable aspect of the regional haze rule is that even states that do not have visibility-degraded Class I areas are required to submit SIPs to reduce the sources of pollution from the states that contribute to visibility degradation elsewhere. Thus, like the $NO_x$ SIP call, EPA's regional haze rule represents a program that attempts to address a regional air pollution problem by making all contributors to the problem contribute to the solution even if they do not suffer significantly from the consequences of this pollution.

Although the rule requires all states to participate, it does not impose specific intra- or interstate emission controls or limits. Instead, all but nine western states are required to develop long-term strategies for achieving the visibility improvement goals set by EPA and to submit these strategies in the form of a regional haze SIP to EPA for approval and review. The nine western states exempted from this procedure are Arizona, California, Colorado, Idaho, Nevada, New Mexico, Oregon, Utah, and Wyoming. As described below, these states, whose pollution has been deemed to contribute most significantly to visibility degradation in the Grand Canyon National Park, are treated in a separate section of the regional haze rule that allows them to implement the strategies recommended by the Grand Canyon Visibility Transport Commission (GCVTC).

The regional haze rule gives states the option of developing their own implementation plans but encourages them to work collaboratively with other states by forming RPOs. Today there are five regional planning organizations addressing regional haze: WRAP is working to implement the recommendations of the GCVTC (see below), the Central States Regional Air Partnership (CENRAP), the Midwest Regional Planning Organization (Midwest RPO), the Mid-Atlantic/Northeast Visibility Union (MANE-VU), and the Visibility Improvement State and Tribal Association of the Southeast (VISTAS).

Problems have arisen despite the authority of EPA to promulgate multistate regulations to address visibility degradation. For example, when developing their long-term implementation strategies, states are required by the regional haze rule to identify all major sources to which the statute's best available retrofit technology (BART) requirement can be applied. However, the Court of Appeals for the District of Columbia Circuit has set aside this aspect of the rule on the grounds that it impermissibly constrains the states' authority to make individualized BART determinations (American Corn Growers Ass'n v. EPA, 291 F.3d 1 [D.C. Cir. 2002]). To help avoid such problems in the future, a recommendation is made in Chapter 7 to enhance EPA's responsibility for multistate air pollution problems.

### Grand Canyon Visibility Transport Commission and the Western Regional Air Partnership

Along with the OTC, the 1990 CAA Amendments also instructed EPA to form GCVTC to deal with fine-particle haze that impaired visual air quality in the Grand Canyon National Park (42 U.S. Code § 7492(f)). The purpose of GCVTC, which EPA officially established in 1991, was to develop consensus recommendations on measures to protect visual resources in the national parks and wilderness areas of the Colorado plateau. Voting members of GCVTC included representatives from eight western states, four tribes, the Columbia River Inter-Tribal Fish Commission, and five federal agencies. Over a 5-year period, GCVTC and its committees performed scientific, technical, and economic assessments of existing and projected visibility in the region and obtained public input on air quality management alternatives. GCVTC reported its final recommendations to EPA in 1996. The regional haze rule that EPA issued in 1999 included separate provisions and deadlines for the western states and tribes to pursue GCVTC's recommendations (40 CFR § 51.309).

WRAP was formed in 1997 as the successor to GCVTC. Eleven western tribes and nine states are listed as WRAP members (WRAP 2003). In September 2000, WRAP submitted an annex to the 1996 GCVTC report, proposing measures to implement GCVTC recommendations and meet the requirements of the 1999 regional haze rule (WRAP 2000). The annex addresses the period through 2018 and features a shrinking emissions cap-and-trade program for $SO_2$ emissions in the GCVTC region. WRAP expects the emissions cap to be met through voluntary measures; thus, the trading program serves as a "backstop" in case the voluntary measures are not sufficient. EPA proposed to approve the annex on May 6, 2002 (67 Fed. Reg. 30418 [2002]).

### Institutional Accountability in the SIP Process

The original 1970 CAA Amendments required SIPs to demonstrate that the primary NAAQS would be achieved in every area within a state by 1977 at the latest. When most urban areas of the country failed to meet that deadline for one or more pollutants, Congress amended the statute in 1977 to require states to demonstrate that attainment of NAAQS would be achieved in every area by 1987 at the latest. In addition, the 1977 Amendments provided for sanctions that EPA could impose upon states that failed to submit adequate plans and for states that did not attain the standards by the deadlines. In especially difficult cases, EPA could take action by writing a federal implementation plan (FIP). Possible sanctions included the withdrawal of federal highway funds for all but the most critical safety-related highway projects in the area and a ban on construction of new major stationary emission sources in the area. In the mid-1980s, EPA instituted sanctions on a few areas that failed to meet the 1983 deadline for PM and $SO_2$, but the agency refrained from administering sanctions in areas in states that had submitted plans for attaining the $O_3$ standards by the 1987 deadline plans that later proved to be inadequate. In addition, EPA rarely wrote FIPs for states that had submitted SIPs that did not appear adequate to achieve NAAQS attainment.

In response to the failure to meet the 1980s deadlines, the 1990 CAA Amendments specified attainment deadlines that depended on how far an area was from reaching attainment in 1990 (see Table 3-1) and a new "bump-up" provision for dealing with areas that failed to meet their attainment deadlines. In this provision, the EPA administrator is directed to re-classify any $O_3$ nonattainment area that is ranked below severe that fails to attain the NAAQS by its attainment date to "the higher of (i) the next higher classification for the area, or (ii) the classification applicable to the area's design value." Since the SIP requirements for $O_3$ nonattainment areas increase dramatically as the classification increases, the automatic bump-up provision provides a sanction of sorts for areas that fail to reach attainment by the required dates. However, EPA has only rarely complied with its statutory obligation to make timely findings that areas have not achieved attainment by the statutory deadlines and thereby bump up their non-attainment status. In some cases, EPA has applied a "downwind extension policy" under which areas that did not achieve attainment by the relevant deadlines because of emissions or transport of emissions from other areas would not automatically be bumped up. Nevertheless, one district court ordered EPA to make the required nonattainment finding in a timely fashion, and three courts of appeals found the downwind extension policy to be invalid.[10]

In addition to the bump-up provision, the 1990 CAA Amendments provide other more tangible penalties and sanctions for states that fail to submit a SIP or submit an inadequate SIP. These include the authority for EPA to write a FIP as well as to impose two types of sanctions: (1) cutting off federal highway funds for the area, and (2) limiting the construction or modification of major new sources in the area by requiring two-for-one offsets from other sources of the same pollutant in the same or (in some cases) in adjacent areas (42 USC § 7509(a), (b)). A new aspect of the sanctions provision in the 1990 CAA Amendments is the requirement for automatic imposition of sanctions on deficient areas. According to the statute, once an area has been found by EPA to be deficient because of a failure to submit an adequate SIP, that area has 18 months to correct the deficiency. At the end of this 18-month "sanctions clock," EPA must impose at least one of the two sanctions on the area and keep the sanctions in effect until the deficiency has been corrected.

At the time of the writing of this report, EPA has had to impose one or more of the sanctions mandated in the CAA Amendments of 1990 in 28 instances. An additional 39 areas under the 18-month sanctions clock are facing the imposition of sanctions in the near future unless they can correct the relevant deficiency or deficiencies. For the most part, when sanctions have gone into effect, states have acted rapidly to have them removed, and as a result, they have tended to remain in effect for relatively brief periods

---

[10]Sierra Club v. EPA, 2002 U.S. App. LEXIS 25289 (5th Cir. 2002); Sierra Club v. EPA, 311 F.3d 853 (7th Cir. 2002); Sierra Club v. EPA, 294 F.3d 155 (D.C. Cir. 2002); Sierra Club v. Browner, 130 F. Supp. 2d 78, 95 (D.D.C. 2001).

(in many cases, sanctions were lifted within weeks or even days of being imposed). In virtually all cases where highway sanctions were imposed, they did not result in an actual loss of highway funds, because no highway projects were being proposed in the areas during the brief periods that the sanctions were in effect. (The two exceptions were East Helena, MT, and Iron County, MO.) The rapidity with which states and areas acted to remove sanctions that had been imposed by EPA might indicate that the requirement for mandatory sanctions in the 1990 Amendments did in fact provide a positive impetus for compliance with the CAA and the SIP requirements. However, it is beyond the scope of this report to determine whether that was the case. The imposition of sanctions appeared to have little or no direct economic impact on the affected areas and states.

## THE EFFECTIVENESS OF THE SIP PROCESS

Two basic metrics can be used to assess the effectiveness of the SIP process. The less stringent metric is based on an assessment of whether implementation of SIPs has resulted in a general decrease in pollutant emissions and concentrations and, more specifically, in a decrease in criteria pollutant concentrations in nonattainment areas. As discussed in more detail in Chapter 6, data from the nation's air quality monitoring networks suggest that the SIP process has resulted in considerable progress in improving air quality on the basis of this metric.

The more stringent metric is based on whether implementation of federal, regional, and local programs through the SIP process has resulted in the attainment of the NAAQS for criteria pollutants. On that basis, the effectiveness of the SIP process is less apparent. For example, in Table 3-4, where the number of nonattainment areas in the United States in 1992 and 2002 are listed for each of the criteria pollutants, significant progress is clearly shown. The number of nonattainment areas for CO, lead (Pb), $SO_2$, and nitrogen dioxide ($NO_2$) have decreased substantially. On the other hand, with the exception of $NO_2$, nonattainment areas for each of these pollutants still exist. In the case of $O_3$, there has also been a substantial decrease in the number of nonattainment areas. However, most of the areas that have been redesignated as being in attainment had a moderate or marginal classification. None of the serious $O_3$ nonattainment areas, which were scheduled to reach attainment in 1999 in the 1990 CAA Amendments, have been redesignated. Similarly, all the areas classified as severe and extreme remain on the nonattainment list; however, they are not scheduled to reach attainment until 2005 and 2010, respectively. Thus, the effectiveness of the current SIP process in addressing serious and above $O_3$ nonattainment has yet to be established. Of even greater concern is the lack of progress in alleviating nonattainment of the $PM_{10}$ NAAQS, especially given

TABLE 3-4    Classifications and Number of Nonattainment Areas in 1992 Remaining in Nonattainment as of February 6, 2003

| Pollutant | Classification (ppm) | 1992 | 2003 |
|---|---|---|---|
| CO | Serious ($\geq$ 16.5) | 7 | 6 |
| | Moderate  (12.7 to 16.4) | 4 | 1 |
| | Moderate (9.1 to 12.7) | 32 | 4 |
| | Not classified[a] | 33 | 2 |
| | TOTAL | 76 | 13 |
| Pb | — | 13 | 3 |
| $NO_2$ | — | 1 | 0 |
| $O_3$ | Extreme ($\geq$ 0.280) | 1 | 1 |
| | Severe-17 (0.190 to 0.280)[b] | 5 | 5 |
| | Severe-15 (0.180 to 0.190)[c] | 5 | 5 |
| | Serious (0.160 to 0.180) | 14 | 15 |
| | Moderate (0.138 to 0.160) | 32 | 8 |
| | Marginal (0.121 to 0.138) | 44 | 21 |
| | Section185A[d] | 11 | 3 |
| | Other[e] | 2 | 1 |
| | TOTAL | 137 | 73 |
| $PM_{10}$[f] | Serious | 8 | 8 |
| | Moderate | 78 | 58 |
| | TOTAL | 86 | 66 |
| $SO_2$ | Primary[g] | 48 | 21 |
| | Secondary[h] | 6 | 6 |

[a]"Not classified" is an area designated a CO nonattainment area as of the date of enactment of the CAA Amendments of 1990, which did not have sufficient data to determine if it is meeting or is not meeting the CO standard.

[b]"Severe-17" nonattainment areas have 17 years to attain standards.

[c]"Severe-15" nonattainment areas have 15 years to attain standards.

[d]"Section 185A" of the CAA (previously called transitional) is an area that has designated an $O_3$ nonattainment area as of the date of enactment of the CAA Amendments of 1990 and that has not violated the national primary ambient air quality standard for $O_3$ for the 36-month period beginning January 1, 1987, and ending December 31, 1989.

[e]This category includes areas that violate the $O_3$ standard and have a design value of less than 0.121 ppm. That occurs when the exceedance is higher than the $O_3$ standard exceedance rate of 1.0 per year, even though the estimated design value is less than the standard.

[f]This is not an official list of nonattainment areas. See the *Code of Federal Regulations* (40 CFR Part 81) and pertinent *Federal Register* notices for legal lists and boundaries (http://www.epa.gov/oar/oaqps/greenbk/pfrnrpt1.html). For an area designated as being in nonattainment of the NAAQS for $PM_{10}$, Section 188 of the CAA outlines the process for classification of that area and establishes that area's attainment date. At the time of designation, all $PM_{10}$ nonattainment areas are initially classified as moderate by operation of law. A moderate area can subsequently be reclassified as serious either before the applicable moderate area attainment date, if EPA determines the area cannot "practicably" attain the $PM_{10}$ NAAQS by this attainment date, or after the passage of the applicable moderate area attainment date, if EPA determines the area has failed to attain the standards.

[g]"Primary" standard based on health-related effects.

[h]"Secondary" standard based on welfare-related effects.

Abbreviation: ppm, parts per million.

SOURCE: EPA 2003h.

the likelihood that the promulgation of the new $PM_{2.5}$ NAAQS will give rise to a large number of PM nonattainment areas.

## Critical Discussion of the SIP Process

The SIP process is an important and essential component of the nation's AQM system. It allows state and local agencies to take into account emission controls adopted at the federal and multistate levels and then choose a suitable suite of additional local emission-control measures to reach attainment. On balance, this process should provide an appropriate division of responsibility. It can also provide the basis for a constructive partnership between the federal and state governments that steadily improves air quality on the local, multistate, and national levels. Air quality monitoring data confirm that such improvements have occurred in the past two decades (see Chapter 6). Nevertheless, important adjustments to the SIP process are needed if the difficult challenges ahead are to be effectively addressed. Some of the major concerns are discussed below. Recommendations for addressing these concerns are presented in Chapter 7.

### An Overly Bureaucratic Process

The SIP process now mandates extensive amounts of local, state, and federal agency time and resources in a legalistic, and often frustrating, proposal and review process, which focuses primarily on compliance with intermediate process steps (see Box 3-7). This process probably discourages innovation and experimentation at the state and local levels; overtaxes the limited financial and human resources available to the nation's AQM system at the state, local, and federal levels; and draws attention and resources away from the more germane issue of ensuring progress toward the goal of meeting the NAAQS.

### Overemphasis on Attainment Demonstrations

An attainment-demonstration SIP is statutorily required to demonstrate through air quality model simulations and other weight-of-evidence analyses that the relevant nonattainment area will reach attainment by a certain date as a result of the specific pollution-control measures proposed in the SIP. Such an exercise provides policy-makers with critical information: estimates of the magnitude of emission reductions that will be needed to reach attainment and assessment of the efficacy of various options that could be adopted. That information is essential to the development of an implementation plan and should be retained in the AQM plan.

**BOX 3-7 The State Implementation Plan:**
**Examples of Bureaucratic Overload**

• Every SIP revision requires a rule-making or legislative action at the state level. That process can take months or even years. Moreover, once approved at the state level, the revision must then be subjected to a full formal review by EPA and EPA's own federal rule-making process. A major overhaul of the process in 1989 encouraged the use of "direct final notices" and "conditional approvals" to reduce the time needed to approve a state submittal, but the process remains largely duplicative, resource intensive, and time consuming.

• The CAA uses the nonattainment status of areas to impose numerous prescriptive measures on those areas (see Box 3-3). From the sector-specific RACT requirements to, for example, vehicle I/M programs, the measures must be implemented according to strict EPA regulations or guidance, thus reducing the flexibility with which areas can design and implement emission-reduction measures. The rigidity of federally mandated requirements for SIPs containing areas that are classified as serious and above may represent a congressional recognition of the failure of more flexible SIP requirements to achieve attainment during the 20 years before the enactment of the 1990 CAA Amendments. Although federally mandated sector-specific measures reduce local flexibility, they should have the benefit of accelerating the EPA approval process, but in general, they have not done so. In some cases, even minor differences between EPA's language and a state's language have resulted in a full-scale SIP review. (A notable exception is the procedure used by the California South Coast Air Quality Management District [SCAQMD] to adopt New Source Performance Standards [NSPS]; the SCAQMD regulations refer to the language of the *Code of Federal Regulations* rather than the language of local rules.)

• The CAA and its associated amendments specify a number of deadlines that proved to be unrealistic. A prime example is the specification of attainment deadlines that proved to be infeasible for $O_3$ in the CAA Amendments of 1970 and 1977. (It remains to be seen if the more liberal attainment deadlines specified in the CAA Amendments of 1990, which extend to 2010, are feasible.) Setting unrealistic deadlines can lead to frustration for local and federal agencies that do not see any reasonable way to achieve the requirements of the act. It can also introduce an aura of fiction to the entire SIP process as agencies endeavor to meet the letter of the law by promulgating attainment demonstrations that have little likelihood of accurately forecasting future air quality trends.

• The actual emission reductions that can be achieved from specific measures are difficult to predict before they have been fully implemented and tested. Experience shows that many of the emission reductions claimed in SIPs prove to be too high. That has been especially true of the credits allowed by EPA for programs to reduce mobile-source emissions, such as the I/M and oxygenated fuels program (NSTC 1997, NRC 2002a). It is possible that the requirement to demonstrate attainment in a SIP inadvertently encourages the regulatory community to be overly optimistic when considering the benefits of specific measures. It is also possible that, in some cases, EPA has allowed local and state agencies to take large emission credits for specific programs to encourage program use and propagation. Finally, it is possible that EPA has allowed some to take overly generous emission credits to put off rancorous policy disputes.

However, the use of the attainment demonstration as a one-time robust prediction of how air quality in a given area will evolve over a multiple-year to a decadal time scale does not take into account the significant modeling, socioeconomic, and control-technology uncertainties implicit in such a process, and thus improperly applies the scientific and technical tools used in the demonstration (see, for example, NARSTO 2000). Moreover, the attainment demonstration can provide a false sense of assurance, which can discourage a review of the underlying assumptions of the plan until attainment has not been achieved after the prescribed time, such as 5 years or more. Finally, although mid-course reviews of SIPs do occur and although it should be possible to amend a SIP as new information and updated modeling simulations become available, in practice this task is very difficult, because the CAA requires that any such changes be subjected to the full and complex review and approval process used for the original SIP. In Chapter 7, the committee recommends implementation of a more iterative process, which retains the attainment demonstration but places greater emphasis on tracking and measuring progress and performance.

### Single-Pollutant Focus of SIPs

Air pollutants occur in complex mixtures, and yet SIPs are constrained to address only individual criteria pollutants. As a result, the entire, relatively cumbersome SIP process must be undertaken for a pollutant such as $O_3$ and then again for PM in a separate process and on a different timetable, despite the fact that the exposures are simultaneous, the sources are often the same, and the two pollutants share many common chemical precursors. One result of this separation is that facilities and other emitters of air pollution may be faced with multiple requirements over time to deal with similar pollutants. This process is made more problematic by the inability to consider key HAPs at the same time unless they are VOCs and therefore also precursors of the other pollutants. Some of the pollutants that states have sought to control in the past had limited sources and could be addressed individually (for example, Pb and CO). However, the major air pollution challenges today, which involve multiple emissions from common mobile and stationary sources, can be more effectively addressed using a multipollutant approach. Such an approach can simultaneously seek reductions of pollutants posing the most significant risks. It can also focus on achieving the most cost-effective mix of emission reductions of key pollutants from any one source rather than asking that source to separately address reductions of different pollutants at different times in response to different SIPs.

### Barriers to Addressing Multistate Airshed Pollution

Over the past two decades, the AQM system has had to grapple increasingly with air pollution phenomena that extend over multistate airsheds—phenomena whose effective control requires coordination across state boundaries and participation by states that may contribute to serious air quality problems but not experience them. For example, for acid rain, Congress chose to prescribe a national rule; for regional haze, Congress attempted to provide EPA with the authority to develop multistate strategies; and for $O_3$, OTC (mandated by the CAA) and OTAG (developed by EPA and the states) were formed to develop suitable multistate strategies.

The current form of the CAA does not provide EPA and states a clear mandate and procedure for regularly analyzing and identifying when and how to implement multistate efforts. With the exception of the OTC, the multistate regulatory approach for $O_3$ and regional haze has been linked to the traditional SIP, requiring that a set of multistate controls be identified first and then incorporated voluntarily into the SIPs of all the states in the given region. This indirect and controversial process has been time consuming and fraught with legal uncertainties. The SIPs of the future will require a more effective mechanism for identifying and linking multistate control strategies with local measures. Of particular importance will be the need for mechanisms to induce states upwind of emission sources to take actions that have little direct benefits for them but that are needed for successful attainment of the NAAQS in states downwind of the sources.

## SUMMARY

### Strengths of the SIP Process

- The SIP process provides a reasonable mechanism for state and local agencies to take into account emission controls adopted at the federal and multistate levels and then to choose a suitable suite of additional local emission-control measures to attain the NAAQS.
- The existence of federally mandated emission-control measures has eased the burden of state and local authorities in developing attainment SIPS.
- The requirement for emission inventories in SIPs has facilitated the development of a uniform methodology for quantifying pollutant emissions in the United States.
- The requirement for modeling analysis in SIPs has promoted the development of increasingly sophisticated air quality models that link pollutant emissions to pollutant concentrations in the atmosphere.

- The SIP process has resulted in a general decrease in criteria pollutant concentrations in the United States and, in some areas, has resulted in NAAQS attainment.
- The sequencing of attainment dates for $O_3$, based on nonattainment classification, provides a more reasonable and flexible timetable for state and local agencies to come to address this persistent air pollution problem.

## Limitations of the SIP Process[11]

- Implementation of federal, regional, and local control measures through the SIP process has not resulted in attainment for $O_3$ and PM in many areas in the United States.
- The SIP process has become overly bureaucratic and draws attention and resources away from the more germane issues of tracking progress and assessing performance.
- The attainment-demonstration SIP places too much emphasis on uncertain emissions-based modeling simulations of future air pollution episodes.
- SIPs must be developed individually for each criteria pollutant, making it difficult for states and local agencies to consider potentially more cost-effective and more protective multipollutant strategies.
- The SIP process lacks sufficient mechanisms and governmental infrastructure for addressing multistate airshed aspects of air pollution.

---

[11]Recommendations are provided in Chapter 7.

# 4

# Implementing Emission Controls on Mobile Sources

## INTRODUCTION

Since the earliest days of the Clean Air Act (CAA), mobile sources have been recognized as one of the most important sources of air pollution and, as a result, have been a prime target for emission control. Despite almost three decades of increasingly tight and wide-ranging regulations, emissions from mobile sources are still a major source of air pollution in the United States (see Figure 1-2 in Chapter 1). The persistence of the need to address mobile emissions is not, however, an indication that the largely technological controls applied to mobile sources have been ineffective; indeed, emission rates from individual vehicles have decreased substantially since enactment of the CAA. Instead, human behavior and other social factors, such as increases in vehicle miles traveled (VMT) and the growing popularity of sport utility vehicles (SUVs), which had been regulated at a less stringent level, appear to have offset many of the gains achieved from the imposition of technological controls.[1] Difficulties in identifying and repairing high-emitting vehicles was probably also a contributing factor.

Currently, a wide variety of mobile sources are subject to control under the CAA. These are broadly divided into on-road and nonroad sources (see Table 4-1). On-road sources include light-duty vehicles (LDVs, also referred to as passenger cars), light-duty trucks (LDTs), heavy-duty vehicles

---

[1]As discussed later in the chapter, new SUVs will soon be required to meet the same gram-per-mile emission standards as cars.

TABLE 4-1  Types of Vehicles and Engines Regulated by AQM in the United States

| Source Category | Status (Date Promulgated; Effective Date) |
| --- | --- |
| **On-road sources** | |
| Light-duty vehicles (LDVs or passenger cars) | Tier 2 regulations require about 99% reduction over previous control levels for cars (2000; 2004–2007) |
| Light-duty trucks (LDTs) (pickup trucks, minivans, passenger vans, and sport-utility vehicles with a gross weight of less than 8500 lb) | Must comply with Tier 2<br>Lightest truck (2000; 2004–2007)<br>Heavier light trucks (2000; 2008–2009) |
| Medium-duty passenger vehicles (MDPVs) (vehicles with gross weight 8,500–10,000 used for personal transportation, such as larger sport utility vehicles [SUVs] and passenger vans) | Tier 2 standards to be phased in beginning with the 2008 model year (2000; 2008–2009) |
| Heavy-duty vehicles (HDVs) (vehicles with gross weight greater than 8500 lb used commercially, such as large pickups, buses, delivery trucks, recreational vehicles [RVs], and semi-trucks. | Highway Heavy-Duty Rule requires new vehicles to meet substantially lower PM and $NO_x$ standards (>90% reduction) (2001; 2007–2010)<br>Also strict fuel sulfur levels (15 ppm) (2001; 2006) |
| Motorcycles | New standards to reduce emissions by 80% (based on California) (proposed 2002; to take effect 2006–2010) |
| **Nonroad Sources** | |
| Nonroad, spark ignition (gasoline) engines | For handheld engines, new rules require 70% $NO_x$ plus hydrocarbon reduction (2000; 2002–2007)<br>New standards promulgated for larger spark ignition engines (2002; 2004–2007) |
| Nonroad recreational vehicles and engines | New rules for variety promulgated to require between 55% and 80% reduction in emissions (depending on pollutant) (2002; 2006–2009, depending on type of engine) |
| Nonroad compression ignition (diesel) engines | Nine size ranges, including agricultural and construction equipment; existing rules promulgated in 1998 (1998; 1999–2008)<br>New standards comparable to Highway Heavy Duty Rule proposed 2003 |

TABLE 4-1    continued

| Source Category | Status (Date Promulgated; Effective Date) |
|---|---|
| Marine engines | New recreational engines required to meet new standards with 55–80% reduction in emissions (depending on pollutant) (2002; 2006–2009) Largest maritime commerce engines regulated to international standards for $NO_x$ (not PM) (2003; 2004) Smaller maritime commerce engines regulated for $NO_x$, PM (1999) |
| Aircraft | Initial standards for smoke only; new standards adopting International Civil Aviation Organization rules requiring 20% reduction in $NO_x$ emissions (1997; full effect in 1999) |
| Locomotives | Regulations applying to new locomotive engines and to remanufactured engines from 1973 and later locomotives; ultimately (2040) expected to reduce $NO_x$ by 60%; PM by 46% (1997; Tier 0 for engines 1973–2001; Tier 1 for engines 2002–2004; Tier 2 for engines 2005 and later) |

(HDVs), and motorcycles that are used for transportation on the road. On-road vehicles may be fueled with gasoline, diesel fuel, or alternative fuels, such as alcohol or natural gas. Nonroad sources refer to gasoline- and diesel-powered equipment and vehicles operated off-road, ranging in size from small engines used in lawn and garden equipment to locomotive engines and aircraft.

In principle, the mobile-source emissions can be controlled with three types of strategies:

- New-source certification programs that specify emission standards applicable to new vehicles and motors.
- In-use technological measures and controls, including specifications on fuel properties; vehicle inspection and maintenance programs; and retrofits to existing vehicles.
- Nontechnological (for example, behavior modification) measures to control usage or activity (for example, via management of transportation).

This chapter summarizes how each of those strategies is used in the United States and concludes with a critical discussion.

## CONTROLLING EMISSIONS THROUGH CERTIFICATION STANDARDS ON NEW VEHICLES AND MOTORS

In the United States, the first mobile-source emission reductions were aimed at lowering the emissions on new vehicles and engines. As older model cars were retired, the in-use fleet would become increasingly populated with regulated vehicles, and emissions would steadily decrease. The first regulations were imposed on passenger cars[2] and then were applied to other on-road vehicles, such as trucks and buses, and most recently, to nonroad sources, such as tractors and construction equipment.[3,4] The specific emission standards listed in Table 4-1 for mobile-source categories have sometimes been set directly by Congress in an amendment to the CAA, but more typically, the EPA administrator has set such standards. Except for California, states do not have independent authority to set new emission standards (see Box 4-1).

At the same time, as standards have been set for vehicle emissions, vehicle manufacturers have also been required to comply with requirements for Corporate Average Fuel Economy (CAFE) standards, which were initiated by Congress in the Energy Policy and Conservation Act of 1975. CAFE established mandatory fuel efficiencies in the form of required miles per gallon (mpg) goals for fleets of passenger cars and light-duty trucks.[5] These standards were enacted in the wake of a petroleum supply interruption and were less concerned with air quality improvements. Nevertheless, for a given vehicle technology, reduced fuel consumption per mile would, in principle, result in lower engine-out emissions, and, therefore, result in less burden on the afterengine emission control system to meet a given emission per mile standard. In practice, implementation of the relatively modest CAFE standards appears to have helped in meeting some hydrocarbon emission standards. However, future application of some especially fuel-efficient engines—which have higher engine-out emissions of some pollutants (for example, $NO_x$)—will likely require development of substantially more effective emission control technologies in order to simultaneously improve fuel economy and emissions (NRC 2002c).

---

[2]Although the first federal controls on mobile sources began with regulations on passenger cars for the 1968 model year, California began mandating emission controls on passenger cars in the early 1960s.

[3]An exception is aircraft engines, which have been subject to emission regulations since 1974. The earliest regulations were for smoke from turbojet engines.

[4]Because nonroad sources have been essentially uncontrolled until recently, their contribution to total mobile-source emissions has grown considerably since passage of the CAA (see Table 4-2).

[5]The current CAFE program requires vehicle manufacturers to meet a standard in miles per gallon (mpg) for the fleet they produce each year. The standard is 27.5 mpg for automobiles and 20.7 mpg for light-duty trucks.

---

### BOX 4-1   California's Unique Role in Controlling Mobile Emissions

The CAA expressly precludes any state except California from setting its own motor vehicle emission standards. Because of California's experience with the most severe air pollution in the nation, it has historically assumed a leadership role in promoting the application of new control technologies for automobiles. By the 1950s, researchers in California had been able to establish a cause and effect linkage between gaseous emissions from motor vehicles and the formation of photochemical smog with its concomitant high concentrations of $O_3$ and PM. In response to that finding, California enacted legislation in 1961 that established statewide new-vehicle emission standards beginning with the 1966 model year. The federal government followed suit in 1965 with the Motor Vehicle Pollution Control Act, which set similar standards for the entire nation beginning with model year 1968. That pattern has been repeated on numerous occasions, as stricter emission standards have been enacted first in California and then propagated to the entire nation by direct congressional mandate or rule-making by the EPA administrator (see Figure 4-1A,B,C). Although states are not allowed to set their own emission standards, the CAA permits them to choose California's stricter standards (typically as part of their state implementation plans). In addition to emission standards, states can opt for other aspects of California's more aggressive program to control mobile-source emissions; these include programs on fuel composition, regulation of individual motorists' use of their automobiles, and controls on transportation infrastructure planning and investment. As a result, California regulations and programs on mobile emissions have played an important role in state implementation plans throughout the nation (see, for example, the discussion on the Ozone Transport Commission in Chapter 3).

---

### Emission Standards for Light-Duty Vehicles and Light-Duty Trucks

The CAA Amendments of 1970 required auto manufacturers to reduce LDV and LDT emissions by 90%. That reduction was to be achieved for carbon monoxide and hydrocarbon emissions by 1975 and for nitrogen oxides ($NO_x$) emissions by 1976 (Jacoby et al. 1973). However, these standards were not fully implemented until the 1980s. Claiming that the initial compliance dates for a 90% reduction in LDV and LDT emissions were infeasible, the auto industry achieved only a partial implementation of the mandated emission-reduction goals by 1975–1976 (Howitt and Altshuler 1999). The CAA Amendments of 1977, extended the emission standards deadlines for carbon monoxide and hydrogen until 1983 and 1980, respectively. The amendments also extended the deadlines for $NO_x$ to 1982 and changed the standard from a 90% to a 75% reduction. In addition, the amendments allowed the EPA administrator to relax the $NO_x$ requirement selectively until 1984 for automotive technologies that promised better fuel economy (Crandall et al. 1986).

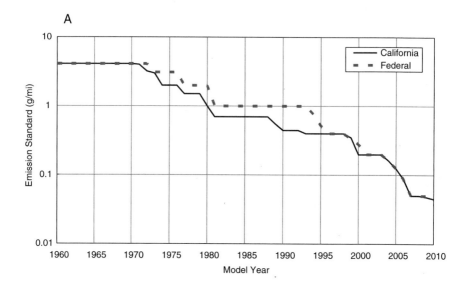

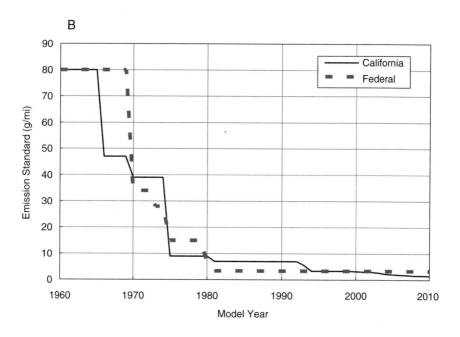

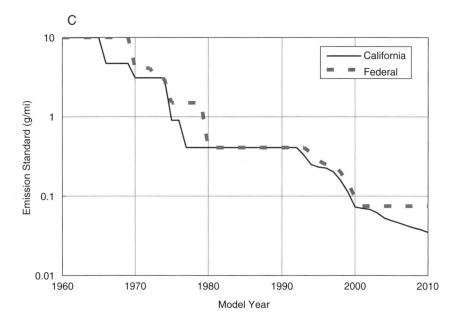

FIGURE 4-1   The evolution of California and federal tailpipe standards on passenger car exhaust emissions since the 1960s.   (A) $NO_x$ emissions, (B) CO emissions, and (C) VOC emissions. NOTE: A, B, and C are not completely consistent. There have been changes in test methods that are accounted for in an approximate manner.   Current emission standards for VOC are defined in terms of nonmethane organic gases (NMOG).   In addition, the most recent emission standards are based on vehicle categories and an average over these categories.   For the California program, manufacturers must meet a fleet average standard for NMOG; for the federal program, manufacturers must meet a fleet average standard for $NO_x$.   The "standards" for CO and $NO_x$ in California for 1999 and later and the federal standards for NMOG and CO in model years 2004 and later are based on an assumed distribution of standards used for compliance.   In addition, vehicles currently have to meet additional useful life standards and standards for supplemental test procedures to test operation under off-cycle driving conditions and under air-conditioning operations.   Thus, the actual progression of standards has been more stringent than shown in the figure.   These figures are for exhaust emissions only. For the uncontrolled vehicles in 1960, there were also VOC emissions from crankcase blowby, which have been completely eliminated, and evaporative VOC emissions, which have also had a high degree of control.   Note that the vertical axis on the VOC and $NO_x$ charts is a log scale.   The federal Tier II standards apply to a full useful life of 120,000 miles.   Those standards have been adjusted to equivalent 50,000-mile-useful-life standards using data from CARB, which has standards for both 50,000 and 120,000 miles.

Despite the difficulties and delays in implementation, passage of the emission standards from LDVs and LDTs represented a watershed in AQM in the United States: Congress's adoption of a "technology-promoting" strategy for lowering mobile-source emissions and, over time, industry's response by developing and installing new and innovative pollution-control technologies for passenger cars (for example, catalytic converters) (see Box 4-2). As this new technology was developed, refined, and installed on new LDVs and LDTs in the 1980s, the emission characteristics of new vehicles sold in the United States steadily and dramatically improved. As higher-polluting vehicles were replaced by newer ones, emission controls became widespread in the U.S. automotive fleet (Howitt and Altshuler 1999). The catalytic-control technology developed in the United States to meet the emission standards mandated in the CAA is now broadly used throughout the world. In China, for example, all cars operated in the Beijing metropolitan area are required to have catalytic controls either as factory-installed equipment or as a retrofit (Liu et al. 1999; Webb 2001).

Despite those technological advances, many areas could not meet the National Ambient Air Quality Standards (NAAQS) for ozone ($O_3$). That was due to challenges in reducing emissions from all sources, but among mobile sources, there was continued growth in VMT and new scientific and technological information that some emissions, especially evaporative emis-

---

### BOX 4-2   Technology Innovation and Emission Controls

The development and widespread application of pollution-control technologies have permitted reductions in criteria pollutant emissions even while vehicle miles traveled has continually increased. Early pollution-control technologies included positive crankcase ventilation valves to direct crankcase blowby emissions into the engine; charcoal canisters to sequester volatile hydrocarbons for later burning in the engine, exhaust gas recirculation valves to reduce $NO_x$ formation during fuel combustion; and catalytic converters designed to oxidize partially combusted hydrocarbons and CO to $CO_2$. Today, vehicles are being driven by cleaner fuels (for example, the removal of fuel sulfur to extend the life of catalytic converters and the further development and improvement of emission-control technologies including high-performance and three-way catalytic converters capable of reducing $NO_x$ to nitrogen gas, hydrocarbon adsorbents, coupled engine-exhaust controls that optimize air-to-fuel mixtures, and leak-free exhaust systems). These technologies are also being expanded for use on heavy-duty vehicles and nonroad equipment. The recent introduction of new automotive technologies, such as electric-gasoline hybrid vehicles with lower fuel consumption, will further decrease emission levels. The continued development of new technologies and application of current technologies to unregulated or less stringently regulated sources is expected to continue to drive decreases in criteria pollutant emissions.

sions, were not being fully controlled. In addition, HDVs and nonroad vehicles were important contributors. The CAA Amendments of 1990 mandated emission reductions (referred to as Tier I controls) for LDVs by 1994. In addition to calling for further reductions in $NO_x$ and volatile organic compounds (VOCs), the 1990 requirements tightened significantly the controls on evaporative emissions, especially during refueling.[6] Also, the 1990 CAA Amendments authorized the EPA administrator to establish more stringent Tier II controls in 2004 if they were judged to be needed, technically feasible, and cost-effective (Howitt and Altshuler 1999). The Tier II regulations have since been promulgated. Besides the tightening of $NO_x$ and VOC emission standards, Tier II includes a limit on the sulfur content of fuels (to extend the effective lifetime of catalytic converters) and regulations on medium-duty passenger vehicles (MDPVs), such as the largest SUVs,[7] as well as a provision that allows manufacturers to use fleet averaging to meet the $NO_x$ standards. The entire set of regulations will be phased in between 2004 and 2009.

The CAA authorizes California to set stricter vehicle emission standards because of the magnitude of its air pollution problems. California required manufacturers to achieve average emissions that were lower than those mandated by the federal Tier I regulations, beginning with the 1994 model year. The state also defined a family of low-emissions vehicles: the transitional low-emissions vehicle (TLEV), the low-emissions vehicle (LEV), and the ultra-low-emissions vehicle (ULEV). In addition, California required manufacturers to offer consumers so-called zero-emission vehicles, or ZEVs,[8] by 1998 and achieve a 10% statewide market share for ZEVs by 2003 (Sperling 1991). California has delayed and modified its requirement for ZEVs and has recently made additional modifications in response to a June 2002 federal district court injunction that prohibits implementation of

---

[6]An enhanced aspect of the emission standards mandated in the CAA Amendments of 1990 was the inclusion of tighter standards on evaporative emissions as well as the more conventional tailpipe emission standards. The addition of tighter evaporative emission standards was in direct response to research in the late 1980s that pointed to these emissions as an important and growing problem (NRC 1991).

[7]The increasing popularity of SUVs has caused an unintended emissions increase because these vehicles had been governed by emission regulations for trucks. The truck regulations are intended to provide for emissions control while allowing the truck to operate under heavy loads. The new Tier 2 regulations define a new class, medium-duty passenger vehicles (MDPV), which must meet the same gram-per-mile emission standards as cars, starting with the 2008 model year.

[8]ZEVs are so-called because they are electric-powered cars and as such have no tailpipe emissions. Nevertheless, the name is misleading. If the electricity used to power a ZEV is generated from burning fossil fuel, then its operation will certainly result in pollutant emissions at the power plant.

the 2001 Amendments to the ZEV program for the 2002 and 2003 model years.[9]

The CAA authorized states other than California to adopt the California standards. After the 1990 CAA Amendments set more stringent requirements for nonattainment areas, Maine, Massachusetts, New York, and Vermont chose to include California mobile emission standards in their state implementation plans (SIPs). In 1994, the Ozone Transport Commission (OTC) (described in Chapter 3) petitioned EPA to impose these automotive technology standards on the entire region, including the states that had not done so individually (Howitt and Altshuler 1999). EPA complied, but their decision was overturned when the District of Columbia circuit court determined that EPA lacked the authority to do so.

A series of negotiations among concerned states, automobile manufacturers, environmental groups, and EPA that took place between 1995 and 1998 led to the national low-emission vehicle (NLEV) program. EPA set regulations for this voluntary program; these regulations came into effect only when individual states and automobile manufacturers opted into the program. ZEVs, and the ability of individual states to require them by electing to apply California vehicle standards, were a major issue in these negotiations. Ultimately, Massachusetts, New York, and Vermont retained the requirement for California vehicles, including ZEVs, while the NLEV program was applied in all other states.

With the implementation of the Tier II program, the difference in standards between the California program and the federal program will be substantially reduced, one exception being the ZEV mandate in California, Massachusetts, New York, and Vermont.

### Emission Standards for Heavy-Duty Vehicles

LDVs have traditionally been the target of new-vehicle emission standards. However, as emission rates from LDVs declined and the use of on-road freight increased, HDVs became responsible for an increasing fraction of the overall mobile-source emissions of $NO_x$ and particulate matter of up to 10 micrometers in diameter ($PM_{10}$) (EPA 2001a, 2003a). The differentiation between LDVs and HDVs historically has been 8,500 pounds gross vehicle weight (the weight of the vehicle plus the weight of the rated load-hauling capacity). In response to this trend, EPA began regulating HDVs in the 1980s and adopted new-vehicle emission standards at several junctures.

---

[9]The basis of the injunction was not directly related to the emission standards but rather to language in the ZEV requirement that the court found was a form of "fuel economy" mandate that federal law reserves for the federal government alone (Central Valley Chrysler-Plymouth, et al. v California Air Resources Board, et al. E.D. Cal. 2002).

EPA issued new, even more stringent regulations on emissions from HDVs in early 2001 (65 Fed. Reg. 59896 [2000]; 66 Fed. Reg. 1535 [2001]). These new regulations are similar to the Tier 2 standards for LDVs discussed in the previous section in that they require both a tightening of emission certification standards and a decrease in the fuel sulfur content. The regulations, to be phased in between model years 2004 and 2010, will reduce PM and $NO_x$ emissions by at least 90% from current standards. To meet these more stringent standards for diesel engines, the sulfur content of diesel fuel will be reduced by 97% from its current level of 500 parts per million (ppm) to 15 ppm beginning in 2006 in most cases.

The HDV emission standards are technology-promoting in that they will require the use of new exhaust-after-treatment technologies for diesel-powered HDVs, as well as substantial requirements for control technology life (up to 435,000 miles in some vehicles). In contrast to LDVs, which have included after-treatment technologies (catalytic converters) since the mid-1970s, previous emission standards on HDVs have only required modifications of engine operations. However, meeting the new $NO_x$ and PM standards will require diesel-powered HDVs to further refine engine operations, as well as include control technology for $NO_x$ and PM.

### Emission Standards for Nonroad Engines

Compared with the long history of regulation of LDV emissions, nonroad mobile-source emissions, like HDV emissions, have been relatively unregulated and, as a result, represent a growing fraction of overall mobile-source emissions (see Table 4-2). In fact, with the set of currently enacted controls, nonroad emissions already exceed on-road emissions of PM and are projected to exceed on-road mobile emissions of VOC and carbon monoxide (CO) in the next two decades. Regulation of these emissions presents some specific challenges to air quality management (AQM). For

TABLE 4-2 Contribution of Nonroad Emissions to Mobile-Source Total and to Manmade Total

| Pollutant | Nonroad Emissions (1,000s of tons per year), yr | | Nonroad as a Percent of Mobile-Source Total, yr | | Nonroad as a Percent of Manmade Total, yr | |
|---|---|---|---|---|---|---|
| | 2000 | 2020 | 2000 | 2020 | 2000 | 2020 |
| HC | 3,488 | 3,139 | 47.7 | 58.0 | 19.1 | 20.3 |
| CO | 25,843 | 37,331 | 34.2 | 43.3 | 26.4 | 34.0 |
| $NO_x$ | 5,447 | 4,164 | 40.6 | 67.0 | 22.2 | 25.7 |
| PM | 466 | 510 | 66.0 | 77.9 | 15.0 | 16.8 |

SOURCE: EPA 2002r.

example, nonroad emissions are emitted from a wide variety of engines, including land-based diesel engines (tractors, backhoes, and generators), land-based spark ignition engines (chain saws, lawn mowers, airport ground-service equipment, and motorcycles), marine engines, jet and propeller aircraft engines, and diesel locomotives. In most cases, individual emission-control strategies need to be devised for each engine type and for widely varying use and performance cycles.[10] Although control of many new nonroad engines, such as portable power equipment and construction vehicles, can follow implementation and jurisdiction patterns similar to those used for on-road vehicles, the control of substantial emissions from aircraft and engines used in international maritime commerce is made much more difficult by the diverse international jurisdictions under which these vehicles fall.

The 1990 CAA Amendments directed EPA to prepare a study of the scope and sources of nonroad emissions and to regulate them if they were found to make a substantial contribution to $O_3$ or CO nonattainment. The EPA report did not make a formal determination of a significant effect, but it contained an inventory of emissions from nonroad sources and concluded, "because nonroad sources are among the few remaining uncontrolled sources of pollution, their emissions appear large in comparison to the emissions from sources that are already subject to substantial emission-control requirements" (EPA 1991). EPA regulations for some of these engines and vehicles started in 1995, and subsequent regulations continue to bring new classes of engines and vehicles under regulation while more stringent regulations are being prepared to replace existing control requirements. In 2002, EPA finalized regulations for a number of nonroad engines and vehicles, including large spark-ignition engines and recreational vehicles (67 Fed. Reg. 68242 [2002]). In April 2003, the agency proposed a national program to reduce emissions from nonroad diesel engines by combining engine and fuel controls. It is expected that engine manufacturers will meet proposed emission standards by producing new engines with advanced emission-control technologies. Because these control devices can be damaged by sulfur, EPA is also proposing to reduce the allowable level of sulfur in nonroad diesel fuel. The proposed exhaust emission standards would apply to diesel engines used in most kinds of construction, agricultural, and industrial equipment and are expected to reduce emissions by

---

[10]A performance or test cycle is where a vehicle or engine is run over a prespecified range of conditions to create repeatable emission-measurement conditions. Test cycles, such as the federal test procedure, are linked to the certification process to verify and ensure compliance with new-equipment emission standards. As described in this chapter, it is difficult to ensure that all possible driving conditions are represented within the new-vehicle emissions certification process.

greater than 90% as compared with today's engines. The proposed standards would take effect for new engines starting as early as 2008 and be fully phased in by 2014. The regulations plan to lower sulfur concentrations in diesel fuel from the current uncontrolled level of 3,400 ppm to 500 ppm beginning in 2007 and then to 15 ppm in 2010 (EPA 2003i; 68 Fed. Reg. 28328 [2003]).

### Control of Mobile-Source Air Toxic Emissions

According to the 1996 National Toxics Inventory (NTI), major stationary sources account for approximately 25% of hazardous air pollutant (HAP) emissions, mobile sources contribute 50%, and area sources (for example, commercial dry cleaning facilities) and miscellaneous sources contribute the remaining 25% (EPA 2001a).

Section 202 (l) of the CAA required EPA to study emissions of air toxics from mobile sources and fuels and set standards for benzene and formaldehyde at least. EPA issued a first draft of the study in 1993. The agency's final rule on control of emissions of air toxics from mobile sources, incorporating the final version of the study, was issued in March 2001 (66 Fed. Reg. 17230 [2001]). The rule was issued in accordance with a judicial consent order that mandated EPA action once it had missed the CAA deadline.

The rule identifies 21 air toxics associated with mobile sources (including nonroad mobile sources as well as LDVs and HDVs). The agency projects that existing tailpipe, evaporative, and fuel formulation regulations on vehicles and fuels will reduce emissions of benzene, acetaldehyde, and formaldehyde by approximately 70% in 2020 as compared with 1990 levels and that emissions of diesel PM from on-road vehicles will be reduced by 90% (EPA 2000d). The rule requires gasoline refineries to maintain their average 1998–2000 levels of toxic-emission control in response to existing toxic-emission performance standards for reformulated gasoline (RFG) and gasoline. Because of the substantial progress projected from the other emission-control standards and from those soon to be promulgated (as discussed previously in the chapter), no other new standards are set for either vehicles or fuels. A lawsuit contending that the rule is inadequate was filed against EPA on May 24, 2001, by the attorneys general of New York and Connecticut and a consortium of environmental groups.

### Implementation of Emission Standards for New Mobile Sources

Motor-vehicle emission standards are implemented through a certification procedure in which manufacturers provide EPA or the California Air Resources Board (CARB) with prototype data showing that the vehicles meet the emission standards for the legally mandated vehicle lifetime. Following a satisfactory agency review of these data, the manufacturer is issued a certificate of compliance, which allows the production of vehicles to proceed. Produced vehicles are then tested to ensure that the new vehicles meet the applicable emission standards. In addition, in-use vehicles are tested to ensure that the vehicles maintain their performance over time. A failure to maintain in-use performance can result in a recall. Manufacturers are also required to provide a warranty on the emission controls in the vehicle.

The initial exhaust-emission test used to certify cars, referred to as the federal test procedure (FTP), was based on measurements that sought to replicate driving during a typical vehicle trip in downtown Los Angeles during the 1960s. In this procedure (known as a test cycle), the vehicle exhaust is continuously collected and quantified as the vehicle is started and run through a series of operating modes (for example, cold start, accelerations, and decelerations). However, the FTP had a number of shortcomings when it was used to forecast in-use mobile emissions: (1) driving behavior and conditions vary so widely that it is difficult to represent all; (2) because of limitations in the dynamometers used at the time the test procedure was developed, the maximum acceleration during the test had to be limited unrealistically to 3.3 miles per hour per second (mph/sec); and (3) most of the vehicle speeds in the FTP are below 30 mph, a speed that is not reflective of the actual speeds used by most motorists in the United States. Thus, "off-cycle" emissions occurred during vehicle-operating modes that were not tested in the FTP. Although vehicles were certified to appropriate standards, their real-world emissions were higher than would be expected by the standards.

Congress sought to remedy the off-cycle emissions problem in the 1990 CAA Amendments by requiring EPA to control off-cycle emissions. To do so, EPA developed the supplemental federal test procedure (SFTP), which tests cars over an extended range of speeds and accelerations. Certification to the SFTP will be fully phased in by the 2004 model year. However, questions persist whether the SFTP fully captures the large increases in emissions that occur due to rapid accelerations and other changes in the vehicle load.

In addition, EPA took steps to address the issue of evaporative emissions. The test cycle for evaporative emissions was originally developed around 1970, and initial controls on evaporative emissions used a test procedure to measure emissions that would occur from a parked car (called diurnal emissions) and from a car that had just ceased operation (called hot-soak emissions). Like FTP for exhaust emission, these tests proved to be poor predictors of actual mobile evaporative emissions. Between 1987 and 1993, EPA developed a new test procedure for evaporative emissions that measured evaporative emissions under a more demanding set of operating conditions.[11] Under the test procedure (59 Fed. Reg. 16262 [1994]), evaporative emissions are collected by placing a vehicle in an air-tight enclosure and measuring the released hydrocarbons. The test is called the sealed housing evaporative determination (SHED) test. Under this test, the activated carbon canister that traps evaporative emissions vented from the tank is initially loaded with fuel vapor. That is followed by a period of driving that purges the fuel vapor from the canister and into the engine. Next, emissions are measured from a parked car during a simulation of repeated hot days. Finally, evaporative emissions are measured during vehicle operation to assess running losses.

In addition to controlling evaporative emissions vented from the vehicle fuel tank via onboard refueling vapor recovery (ORVR) controls, EPA regulations also require gasoline-filling stations to recover gasoline vapors emitted from filler pipe during refueling operations using so-called stage-two systems.[12] California, per that state's stationary-source regulations, has used these systems to control refueling emissions since the late 1970s. The 1990 CAA Amendments required stage-two systems in all except marginal $O_3$ nonattainment areas. There is a possibility that stage-two and ORVR controls can interact to reduce the effectiveness of both. Because the ORVR system removes fuel vapors displaced from a tank during refueling, the stage-two system would force air, rather than fuel vapors, into the underground gasoline storage tanks, potentially leading to increased gasoline evaporation and fugitive vent emissions. CARB initially considered seeking a waiver from ORVR regulations for California vehicles, but later adopted ORVR systems to promote a consistent vehicle design for all 50 states and to reduce the testing burden for vehicle manufacturers. CARB has since promulgated regulations for enhanced stage-two vapor recovery, which includes measures to ensure compatibility with ORVR systems. The on-board recovery systems were phased in on passenger cars between 1998 and 2000; they will be phased in on LDTs between 2001 and 2006. After these systems are in widespread use (probably sometime after 2010), EPA intends to drop the requirement for stage-two systems (EPA 1998d).

Emissions control for HDVs applies emission standards to the engines, which are tested on an engine dynamometer before being installed in the HDV. The test cycle is based on engine torque and rotational speed. The emission standards are written in terms of the emission rate per unit power

---

[11]Language was incorporated into the Clean Air Act Amendments that directed EPA to achieve reductions in evaporative emissions during certain operating conditions.

[12]Stage-one vapor recovery systems control evaporative emissions when fuel is delivered to a service station by a tank truck.

output, typically in grams per brake horsepower-hour (g/bhp-hr) or grams per kilowatt-hour (g/kW-hr).

## CONTROLLING IN-USE MOTOR-VEHICLE EMISSIONS

### Light-Duty Vehicle and Truck Emissions Inspection and Maintenance Programs

Characterizing emissions from in-use mobile sources has long been a controversial issue. Emission measurements in the field have shown that vehicles in use sometimes do not perform as well as those tested by the manufacturers during certification. High-emitting vehicles (commonly referred to as high emitters) appear to be a major factor. A small fraction of the fleet of LDVs and LDTs in the United States are responsible for a disproportionately large fraction of the total mobile-source emissions (NRC 1991, 1999b, 2001c; Holmes and Cicerone 2002). In some cases, the high emitters have very high evaporative emissions, most likely because of leaks in the fuel system and, in other cases, high tailpipe emissions, due to faulty emission-control systems, or poor engine maintenance.

The initial response to these problems was the development of inspection and maintenance (I/M) programs. In implementing the 1977 CAA Amendments, EPA determined that many states needed annual or biennial I/M programs to ensure that the tailpipe emission-control devices it had mandated were in place and operating properly. In some cases, motorists appeared to intentionally disable these devices or damage them by using leaded gasoline (Howitt and Altshuler 1999). Even though I/M programs did not affect everyday travel behavior and could often be combined with vehicle-safety inspections already required by some state and local authorities, a number of states adopted programs only when the federal government invoked sanctions authorized in the CAA and suspended much of their highway funding (Howitt and Altshuler 1999).

The 1990 CAA Amendments strengthened EPA's position with regard to I/M by specifically mandating enhanced I/M programs in a number of nonattainment areas. In response, EPA developed an enhanced I/M program with dynamometer testing by using a special cycle known as IM240.[13] Although the enhanced I/M requirements in principle allowed states a choice for their program, the requirements in the 1990 Amendments and the stringent criteria that EPA developed for program effectiveness gave states little choice but to accept EPA's version. For example, EPA's model (MOBILE)

---

[13]The IM240 is a transient loaded-mode emissions test that lasts for 240 seconds, is run on a dynamometer, and is intended to simulate a range of normal driving conditions (NRC 2001c).

for estimating emission-reduction benefits from I/M for SIP development greatly discounted benefits from programs that did not meet the EPA criteria. As a result, implementation of enhanced I/M has been highly contentious, with California and, to a lesser degree, other states fighting to implement alternative plans (see, for example, IMRC 1995a,b). In response to the controversy, Congress provided states with greater latitude in a provision of the 1995 National Highway System Designation Act (Bennett 1996). Most states subject to the requirement are currently proceeding with some form of enhanced I/M, but few have been willing to adopt the full set of practices that EPA prescribed in its original enhanced I/M regulations (Howitt and Altshuler 1999).

In addition to the political and regulatory controversies associated with I/M, serious technical questions persist. Recent scientific and technical reviews of I/M have concluded that these programs have been less effective than originally forecasted in identifying vehicles with faulty and/or noncompliant emission-control devices (NRC 2001c; Holmes and Cicerone 2002). Thus, there is a continuing need for an effective regulatory program that can identify and then facilitate the repair or removal of high-emitting vehicles from the fleet (NRC 2001c). It is possible that tailpipe testing programs will be eventually supplanted by other more effective mechanisms for ensuring in-use compliance with vehicle emission standards. First, the in-use performance of 1990 and later model-year vehicles appears to have been significantly better than that of vehicles of earlier model years, perhaps because of the maturing of emission-control technology and the requirements for extended warranties. Second, remote-sensing technologies (see discussion below) are becoming increasingly sophisticated and could provide accurate measurements of in-use vehicle emissions under actual driving conditions. Finally, on-board diagnostic (OBD) systems,[14] specifically the OBDII system which has been required on vehicles since 1996, periodically check many emission-control functions (with oxygen sensors, for example). If a problem is detected that could cause emissions to exceed 1.5 times the emission standards, the OBDII system illuminates a malfunction indicator light, known as the "check engine" light on vehicle dashboards. However, broad implementation of OBD in state I/M programs is still in early stages. Because the real-world effectiveness of OBDII in I/M programs has not yet been demonstrated, the transition from direct tailpipe measurement to reliance on these alternative procedures in I/M programs should proceed cautiously, retaining levels of direct tailpipe measurement to confirm that such systems are functioning effectively. Further, it remains

---

[14]An OBD system contains electronic sensors and actuators that monitor and modify specific motor-vehicle components, as well as the diagnostic software in the on-board computer.

to be seen whether those procedures will be acceptable to the American public.

## Remote Sensing of In-Use Vehicle Emissions

A few states have adopted I/M programs that assign a supplementary role to on-road emissions testing by using remote-sensing technology (TNRCC 2003a; Fresno Bee 2002; Laris 2002). In some states (for example, Texas), roadside enforcement officers use remote sensors to identify vehicles that have malfunctioning emission-control systems (similar to the way officers use radar to identify vehicles exceeding the speed limit) and then record license plate numbers. The owners of the vehicles can then be contacted and required to take appropriate corrective actions. Other states, such as Colorado and Missouri, use a "clean screening" program, in which roadside remote sensing is used to exempt vehicles from central testing requirements.

In principle, the use of remote sensing has a number of advantages over a test-center-based inspection system. Remote sensing provides a method for identifying certain types of high tailpipe emitters that periodic inspection at a test facility using a test cycle might not capture; it also can identify high-emitting vehicles that are not showing up for testing (NRC 2001c). At the same time, challenges remain for remote sensing. First, further controlled testing of the technology for quality assurance and quality control, as well as development of technologies for other mobile-source pollutants, such as PM, will be important to its expanded use. In addition, because remote sensing does not monitor a vehicle over its full range of operating conditions and is not yet able to monitor evaporative emissions, it is probably best used at this time as an adjunct to annual or biennial inspections and on-board diagnostics. Other technical issues include site selection of the sensing equipment and the effect of weather conditions on the equipment.

## In-Use Emissions from Heavy-Duty Vehicles and Engines

Ensuring that in-use emissions of HDVs (both on-road and nonroad) meet intended standards is considerably more challenging than it is for LDVs and LDTs. First, the durability of HDVs (Davis and Diegel 2002), and the large initial capital investment to replace them, means that older vehicles and engines remain in service much longer, delaying the benefits expected from the introduction of newer, cleaner vehicles and engines (see Box 4-3). Second, it is difficult to conduct in-use emissions testing for HDVs; unlike cars, they cannot be simply placed on a dynamometer to check their emissions through a full driving cycle. Although a number of states have implemented "smoke" regulations that remove gross diesel emit-

---

**BOX 4-3 Emissions from Heavy-Duty Vehicles**

Emissions from HDVs are substantially higher in older models than in newer models. Table 4-3A and 4-3B shows the average PM and $NO_x$ emissions for HDVs by model year from pre-1976 to 1976 through 2000 and the percent total $PM_{2.5}$ and $NO_x$ emitted in each category for the percent of vehicle miles traveled (VMT). Emissions were calculated for the national fleet traveling on U.S. freeways in July 2000. The three classes of trucks make up most of the heavy-duty fleet. Substantial decreases in PM emissions have been obtained with newer vehicles. Diesel trucks built before 1980 emit 10 times the PM emissions in grams per mile than do trucks built after 1996. The higher ratios of percent total PM emitted in category to percent VMT indicate that older trucks generally are substantial PM emitters despite being used less than newer trucks. Although trucks built before 1981 represent only 4% of VMT, they are responsible for about 11% of the PM emissions from trucks. $NO_x$ emissions have not shown as large a decrease because hydrocarbon and CO standards that went into effect during the 25-year period increased combustion temperatures, resulting in increased $NO_x$ emissions. In addition, the use of $NO_x$ "defeat devices"[a] has resulted in an increase in $NO_x$ emissions in newer trucks. However, diesel trucks built before 1977 emit twice as much $NO_x$ as those built after 1996, and those built between 1977 and 1980 emit 50% more than those built after 1996. $NO_x$ emissions from trucks built before 1981 represent about 5% of the total from trucks.

---

[a]$NO_x$ defeat devices were used to shut off emission-control systems during steady-state-operation models, such as cruising on the freeway but were mostly inactive during transient operation, such as acceleration.

---

ters from the highway, they are only sporadically enforced in some states, and do not necessarily identify the emitters of high concentrations of fine particles and $NO_x$, which cannot be correlated with smoke. As a result, a large number of older high-emitting vehicles remain on the road, especially for short-haul service in urban areas where the greatest population exposures can occur. A 1997 census of truck use in the United States (DOC 1999) showed that, although most trucks are used for local trips (less than 50 miles/day), the newest trucks tend to be used for the longest trips (Figure 4-2). Some states (for example, California, replacing public school buses) and recently EPA have initiated programs to accelerate the replacement and retrofit of older vehicles, but to date these programs have not been systematic and have relied to some extent on the availability of state funds, which in the current economy have become increasingly scarce.

Perhaps even more challenging is the evidence that at least some of the newer, purportedly cleaner vehicles were operating with emissions in excess of the standards for which they were certified. In 1998, EPA undertook enforcement actions against six diesel engine companies for producing ve-

TABLE 4-3A    Average PM$_{2.5}$ Emissions by Vehicle Model Years for Medium- and Heavy-Duty Trucks

| Model Years | PM$_{2.5}$ g/mile | VMT (millions of miles)[a,b] | Total Emissions for Vintage (10$^6$ g) | % Total Emitted in Category/ % VMT[a,b] |
|---|---|---|---|---|
| | Gasoline Trucks (8,001–10,000 lb GVWR) | | | |
| 1976 and earlier | 0.16 | 1,760 | 282 | 2.22 |
| 1977–1980 | 0.16 | 2,200 | 353 | 2.22 |
| 1981–1985 | 0.16 | 4,230 | 677 | 2.22 |
| 1986–1990 | 0.08 | 9,630 | 808 | 1.16 |
| 1991–1995 | 0.06 | 19,300 | 1,120 | 0.81 |
| 1996–2000 | 0.06 | 40,500 | 2,350 | 0.81 |
| | Diesel Trucks (33,000–60,000 lb GVWR) | | | |
| 1976 and earlier | 1.91 | 346 | 661 | 3.74 |
| 1977–1980 | 1.92 | 439 | 843 | 3.76 |
| 1981–1985 | 1.83 | 1,260 | 2,306 | 3.58 |
| 1986–1990 | 1.57 | 3,130 | 4,914 | 3.07 |
| 1991–1995 | 0.41 | 7,800 | 3,198 | 0.80 |
| 1996–2000 | 0.19 | 16,500 | 3,135 | 0.37 |
| | Diesel Trucks (>60,000 lb GVWR) | | | |
| 1976 and earlier | 1.90 | 1,700 | 3,230 | 3.39 |
| 1977–1980 | 1.91 | 2,040 | 3,897 | 3.40 |
| 1981–1985 | 1.83 | 5,420 | 9,919 | 3.26 |
| 1986–1990 | 1.60 | 12,400 | 19,840 | 2.85 |
| 1991–1995 | 0.41 | 28,200 | 11,562 | 0.73 |
| 1996–2000 | 0.19 | 55,300 | 10,507 | 0.34 |

[a]
$$\frac{\% \text{ total PM}_{2.5} \text{ emitted in category}}{\% \text{ VMT}} = \frac{PM_{2.5\text{vintage}} / \sum_{i=\text{pre-1976}}^{2000} PM_i}{VMT_{\text{vintage}} / \sum_{i=\text{pre-1976}}^{2000} VMT_i}$$

where $i$ = vintage year.
[b] These columns are calculated within the specific truck category (8,001–10,000 lb, gasoline; 33,000–60,000 lb, diesel; more than 60,000 lb, diesel).
Abbreviations: GVWR, gross-vehicle-weight rating; VMT, vehicle miles traveled; g, gram.
SOURCE: Data from S. Srivastava, EPA, personal communication, June 14, 2002.

hicles that operate to maximize fuel economy in a way that resulted in unacceptably higher NO$_x$ emissions (63 Fed. Reg. 59330 [1998]). EPA also initiated more procedures for in-use testing, and the 2007 rules, discussed earlier in this chapter, have substantially increased durability requirements for control systems—to as much as 435,000 miles (66 Fed. Reg. 5002 [2001]).

TABLE 4-3B   Average $NO_x$ Emissions by Vehicle Model Years for Medium- and Heavy-Duty Trucks

| Model Years | $NO_x$ g/mile | VMT (millions of miles)[a,b] | Total Emissions for Vintage ($10^6$ g) | % Total Emitted in Category/ % VMT[a,b] |
|---|---|---|---|---|
| Gasoline Trucks (8,001–10,000 lb GVWR) | | | | |
| 1976 and earlier | 6.33 | 1,760 | 11,200 | 1.17 |
| 1977–1980 | 6.60 | 2,200 | 14,600 | 1.22 |
| 1981–1985 | 6.64 | 4,230 | 28,100 | 1.22 |
| 1986–1990 | 7.63 | 9,630 | 73,500 | 1.40 |
| 1991–1995 | 5.54 | 19,300 | 107,000 | 1.02 |
| 1996–2000 | 6.65 | 40,500 | 188,000 | 0.85 |
| Diesel Trucks (33,000–60,000 lb GVWR) | | | | |
| 1976 and earlier | 48.5 | 346 | 16,800 | 1.67 |
| 1977–1980 | 39.1 | 439 | 17,200 | 1.35 |
| 1981–1985 | 29.9 | 1,260 | 37,700 | 1.03 |
| 1986–1990 | 27.8 | 3,130 | 87,000 | 0.96 |
| 1991–1995 | 30.8 | 7,800 | 240,000 | 1.06 |
| 1996–2000 | 27.6 | 16,500 | 457,000 | 0.95 |
| Diesel Trucks (>60,000 lb GVWR) | | | | |
| 1976 and earlier | 62.4 | 1,700 | 106,000 | 1.86 |
| 1977–1980 | 47.4 | 2,040 | 97,000 | 1.41 |
| 1981–1985 | 30.9 | 5,420 | 168,000 | 0.92 |
| 1986–1990 | 30.7 | 12,400 | 380,000 | 0.91 |
| 1991–1995 | 38.2 | 28,200 | 1,080,000 | 1.14 |
| 1996–2000 | 30.7 | 55,300 | 1,700,000 | 0.91 |

[a]

$$\frac{\text{\% total } NO_x \text{ emitted in category}}{\text{\% VMT}} = \frac{NO_{x\,\text{vintage}} \Big/ \sum_{i=\text{pre-1976}}^{2000} NO_{x_i}}{VMT_{\text{vintage}} \Big/ \sum_{i=\text{pre-1976}}^{2000} VMT_i}$$

where $i$ = vintage year.
[b]These columns are calculated within the specific truck category (8,001–10,000 lb, gasoline; 33,000–60,000 lb, diesel; more than 60,000 lb, diesel).
Abbreviations: GVWR, gross-vehicle-weight rating; VMT, vehicle miles traveled; g, gram.
SOURCE: Data from S. Srivastava, EPA, personal communication, June 14, 2002.

## Regulating the Content of Gasoline and Diesel Fuels

For most of the first 20 years of implementing the CAA, mobile-source emissions were controlled through technological changes to engines and exhaust systems. With the exception of lead (see Box 4-4), fuel was not

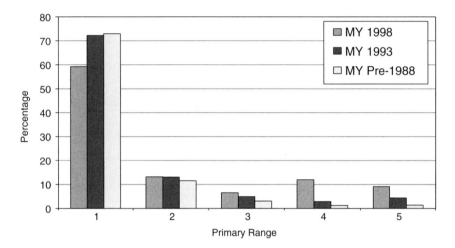

FIGURE 4-2 Percentages of U.S. trucks within selected model years (MY) used for various primary daily driving ranges: (1) up to 50 miles, (2) 51 to 100 miles, (3) 101 to 200 miles, (4) 201 to 500 miles, and (5) more than 500 miles. The survey includes light-, medium-, and heavy-duty trucks. The survey excludes vehicles owned by federal, state, and local governments; ambulances; buses; motor homes; and farm tractors. SOURCE: Data from DOC 1999.

regulated for emissions control. Beginning in the late 1980s, however, a more balanced strategy began to take shape that combined regulations on vehicle performance with regulations on the fuels used by those vehicles (see Table 4-4). This strategy was driven by a desire to control emissions from the entire vehicle fleet—both old and new vehicles—at one time and by emerging control technologies that were especially sensitive to some fuel elements (especially sulfur). The result has been a number of regulations that control fuel content and formulation by EPA under the CAA, as well as by states.

In addition to these regulations, there have been various proposals and requirements in the CAA, state rules, and national energy legislation for the promotion of alternative fuels to reduce the emissions from motor vehicles. Fuels proposed for this purpose include methanol, ethanol, natural gas, liquefied petroleum gas (LPG), and hydrogen. Although these fuels have some limited use, fuel and distribution infrastructure costs have prevented their widespread adoption. Electric vehicles using batteries or fuel cells have also been required or promoted. Current battery technology provides vehicles with limited range, and although they have been introduced into the market, they have not been received well. Vehicle manufacturers are hoping to introduce a limited number of fuel-cell vehicles over the next

---

**BOX 4-4    Getting the Lead Out of Gasoline:**
**Intended and Unintended Consequences**

Most of the federal regulations on gasoline aimed at improving air quality were implemented following the CAA Amendments of 1990. One notable exception is the requirement in the 1970s to remove tetraethyl lead from fuels, an action that led to one of the best documented benefits to human health from AQM in the United States through an innovative market-based approach that allowed petroleum refineries to trade and bank lead-reduction credits to lower implementation costs (see Chapter 5). As a result of the removal of lead from gasoline, the concentration of lead in the atmosphere dropped precipitously, and that in turn resulted in a very significant and measurable decrease in blood concentrations in Americans. It is estimated that the drop in blood concentrations in children (see Figure 4-3) has resulted in 30,000–60,000 fewer individuals with IQs below 70 each year (EPA 1997).

However, the decrease in blood concentrations was not the primary rationale for the decision to phase out lead in gasoline. This decision was precipitated by the requirement for catalytic controls on LDVs and LDTs to meet the emission standards mandated in the 1970 CAA Amendments. Because lead contaminates metal catalysts, rendering them permanently ineffective, lead had to be phased out of gasoline. As a result of the lead phase-out and introduction of catalytic controls, substantial reduction in pollutant emissions from LDVs and LDTs was achieved.

There was also an unintended negative consequence of the lead phase-out. Tetraethyl lead was originally introduced into gasoline as an octane enhancer to improve vehicle performance. To maintain octane levels after the phase-out of lead, refiners blended higher amounts of light hydrocarbons and aromatics, such as benzene, into the fuel. The blending had the unintended consequence of increasing evaporative VOCs and air toxic emissions (EPA 1973; Johnson 1988; NRC 1991).[a] At the same time, one of the responses to replacing lead, especially in premium gasoline, was the beginning of the use of small quantities of MTBE. The requirements in the 1990 CAA Amendments to control evaporative emissions from LDVs and implement the reformulated gasoline program were in large part an effort to reverse the increase in evaporative VOCs. The 1990 Amendments also required a substantial increase in oxygen content, resulting in a later unintended consequence—groundwater contamination from the high use of MTBE.

---

[a]For an informative history of lead in fuels, see http://www.epa.gov/history/topics/lead/03.htm.

---

5–10 years; however, fuel-cell vehicles currently are an order of magnitude more costly than gasoline-powered vehicles, and if adopted, an infrastructure would be needed to provide the hydrogen fuel.

One adaptation to the technological limitations to date has been the growing introduction of hybrid vehicles fueled by gasoline (or possibly diesel fuel) and driven by a combined gasoline (or diesel) engine and electric powertrain. These vehicles are only moderately more expensive than

TABLE 4-4　Timeline of Significant Federal and State Regulations for Motor Vehicle Fuels

| | |
|---|---|
| Early 1970s | Production of unleaded gasoline begun in anticipation of introduction of catalysts in 1975 (see Box 4-4) |
| 1985–1986 | Full phase-out of lead in gasoline |
| 1988 | Colorado established oxygenated fuels requirements (1.5% oxygen by weight, starting January 1, 1988) |
| 1988 | California Clean Air Act |
| 1989 | Summer volatility (Reid vapor pressure) regulations, Phase 1 |
| 1990 | Clean Air Act Amendments |
| 1992 | California reformulated gasoline, Phase I |
| 1992 | Summer volatility regulations, Phase 2 |
| 1992 | Winter Oxygenated Gasoline required 2.7% minimum oxygen content in about 40 CO nonattainment areas |
| 1995 | Federal reformulated gasoline, Phase I |
| 1996 | Completion of phase-out of lead in gasoline |
| 1996 | California reformulated gasoline, Phase II |
| 2000 | Federal reformulated gasoline, Phase II |
| 2002 | California Phase III gasoline effective December 31, 2002 |
| 2003 | California regulations for reformulated diesel fuel effective |
| 2004 | Phase-in of federal low-sulfur gasoline requirements begins |
| 2006 | Phase-in of federal low-sulfur diesel requirements begins |

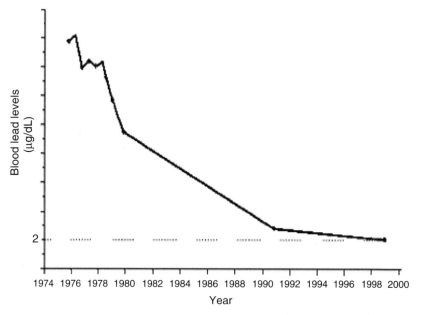

FIGURE 4-3　Blood lead concentrations in the U.S. population from 1976 to 1999. SOURCE: Adapted from CDC 2002.

conventional vehicles, use current technology, and are no more limited in range than current vehicles.

Recently, a much greater use of diesel engines in passenger cars has also been promoted in part as a way to get better fuel economy. Diesel engines have been greatly improved in recent years in noise and smoke production and in cold-starting ability. However, they continue to produce higher $NO_x$ and PM emissions than gasoline engines. Great efforts are now under way to devise emission controls to treat these emissions and meet Tier II emission standards.

Beyond the lead phase-out, several states preceded the federal government in regulating motor-vehicle fuels. In the early 1970s, California imposed limits on the Reid vapor pressure (RVP,[15] a measure of volatility) of gasoline sold in some parts of the state. RVP affects evaporative emissions of gasoline from vehicles as well as from storage tanks and distribution facilities. California was also the first state to impose limits on the sulfur content of gasoline.

In the years preceding adoption of the 1990 CAA Amendments, increased action by states was focused on the need to regulate fuels in addition to motor vehicles. In 1988, Colorado pioneered oxygenate requirements for gasoline—specifying a minimum oxygen content of 1.5% by weight during the winter—to reduce CO emissions. Several other western states soon followed suit.

In 1989, first the northern states and then the federal government took another step in fuel property regulation, imposing RVP limits for gasoline. At the time, RVP limits, as compared with other various mobile- and stationary-source controls, were judged to be by far the most cost-effective means available for reducing VOCs and, in turn, $O_3$ which had become an important issue because of the pervasive violations of the $O_3$ NAAQS that occurred during the summer of 1988 (OTA 1989). Average RVP levels had steadily risen during the 1980s, in part because of the use of light hydrocarbons to replace tetraethyl lead as octane enhancers (see Box 4-4).

The 1988 California Clean Air Act charged the Air Resources Board with attending to both vehicles and their fuels in pursuit of the maximum possible emission reductions of VOCs and $NO_x$. The Atlantic Richfield Company (ARCO) introduced the first reformulated gasoline (RFG), known as EC-1, in Southern California. EC-1 contained reduced amounts of olefins, aromatics, lead, and sulfur as compared to regular unleaded gasoline; had a lower RVP; and had methyl *tert*-butyl ether (MTBE) added to raise the oxygen content.

---

[15]RVP is the vapor pressure of volatile petroleum products, as measured under a specified protocol at 100 °F. RVP values are reported in units of pounds per square inch (psi); typical values for unregulated gasoline range from 9 to 11 psi.

In 1989, a major collaborative program between the automobile manufacturers and the oil industry, known as the Auto/Oil Air Quality Improvement Research Program (AQIRP), tested a variety of fuel and engine combinations designed to reduce emissions (Auto/Oil AQIRP 1993, 1997).

With these precedents, fuel regulations became a significant element of the motor-vehicle emission-control provisions (Title II) of the 1990 CAA Amendments. They mandated the use of RFG during the summer in nine major metropolitan areas with a severe or extreme $O_3$ nonattainment status and allowed additional areas to opt into the RFG program. Winter oxygenated gasoline requirements (oxygen content 2.7% by weight) were mandated for about 40 CO nonattainment areas. The 1990 CAA Amendments also established a nonattainment area fleet program and a California pilot program to encourage the development and use of clean-fueled vehicles. As described in Chapter 2, the RFG program was also intended to reduce emissions of benzene, one of the principal mobile-source air toxics.

The federal RFG program described in Table 4-5 includes both performance requirements, specified in terms of emission reductions, and content

TABLE 4-5    Part 1:  California and Federal Reformulated Gasoline Programs[a]

| California RFG Program, Phase 1 (1992–1996) | Federal RFG Program, Phase I (1995–1999) |
|---|---|
| • Effective January 1, 1992. | • Mandated in 42 U.S.C. 7545 as a result of language in Section 211(k) of the CAA Amendments of 1990. |
| • Set gasoline RVP limit at 7.8 psi. | |
| • Required detergent additives and no lead in gasoline. | • Effective beginning January 1, 1995, in the 9 metro ozone-nonattainment areas with population of 250,000 or greater classified as "extreme" or "severe" as of November 15, 1990: |
| • *No explicit oxygen requirement for summer gas.* | |

California RFG Program, Phase 2 (1996– )
• Effective with beginning of 1996 ozone season.
• Set flat limits for the following properties:

| • RVP: | 7.0 psi (gauge) |
| • Sulfur: | 40 ppm (vol) |
| • Oxygen: | 0–2.7% (wt) |
| • Olefins: | 6.0% (vol) |
| • Aromatics: | 25% (vol) |
| • Benzene: | 1.0% (vol) |

• Temperature at which 50% of fuel is distilled/vaporized ($T_{50}$): 210°F.
• Temperature at which 90% of fuel is distilled/vaporized ($T_{90}$): 300°F.

Federal RFG Program, Phase I (continued):
• Los Angeles (South Coast Air Basin)
• San Diego
• Baltimore/Washington
• Hartford-New Haven-Waterbury, CT
• New York/New Jersey/SW Connecticut
• Philadelphia/Wilmington/Trenton
• Chicago/NW Indiana
• Milwaukee/Racine, WI
• Houston/Galveston/Brazoria, TX
(Sacramento, CA was later added)
• Specified content criteria for gasolines to be sold in these areas primarily during the *summer* ozone season: oxygen minimum of 2.0% by wt; benzene maximum of 1.0% by vol; aromatics maximum of 25.0% by

## TABLE 4-5    continued

• Meets federal Phase II RFG specification and performance requirements (see Table 4-5 Part 2) *except* that oxygenate content requirement may be waived if a refiner demonstrates, through emissions test results for 20 vehicles in four technology classes, that a fuel's exhaust-emissions performance targets can be achieved without it.

• Properties *may* be measured according to an average limits provision, as long as the flat limits are met on average over a specified period of time.

• RFG performance relative to that of a specified base fuel *for exhaust emissions only* is calculated with the Predictive Model, which California developed using approximately the same data base that EPA used in developing the Complex Model.

vol; must contain detergent additive; must exclude heavy metals.

• Per-gallon performance requirements: 15.0% reduction in toxics; at least 15.6% northern states; 35.1% southern states; reduction in VOC relative to specified baseline gasoline, as computed by Complex Model (Simple Model valid until January 1, 1998).

• Average performance requirements (across all RFGs from a refiner): at least 16.5% reduction in toxics; at least 17.1% northern states, 36.6% southern states; reduction in VOCs as computed by Complex Model (Simple Model valid until January 1, 1998).

• RVP limits based on 40 CFR 80.28 standards, which cover all gasolines sold. Other areas may opt into program irrespective of ozone attainment status and may opt out if alternative means of attaining (and maintaining) ambient ozone standards are demonstrated.

[a]Unless otherwise stated, standards for the first phase of both programs carry forward to the second phase.
SOURCE: NRC 1999b.

## TABLE 4-5  Part 2:  Future Reformulated Gasoline Program

Federal RFG Program, Phase II
(2000– )

• Effective January 1, 2000.

• Revises per-gallon performance criteria for gasolines to be sold in covered and opt-in areas during the ozone season:  at least 20% reduction in toxics; at least 25.9% (northern states) and 27.5% (southern states) reduction in VOCs; and at least 5.5% reduction in $NO_x$ (which was not previously controlled) for VOC-controlled areas; relative to specified baseline gasoline, as computed by the Complex Model.

• Similarly, if a refiner opts to meet performance criteria on a pooled average (rather than per-gallon) basis as described in 40 CFR 80.67, targets are at least 21.5% reduction in toxics; at least 27.4% (northern states) and 29.0% (southern states) reduction in VOCs (but 23.4% and 25.0%, respectively, for any individual gallon sampled); and at least 6.8% reduction in $NO_x$ for VOC-controlled areas; all relative to specified average baseline gasoline, as computed by the Complex Model.

• For areas not designated VOC-controlled, the pooled average $NO_x$ reduction standard for RFGs is 1.5%.

• Per-gallon oxygen minimum requirement relaxes to 1.5% by wt as long as an average oxygen content across all RFGs produced by a refiner for a given area is 2.1% or higher.

• Per-gallon benzene maximum requirement relaxes to 1.3% by wt, as long as an average benzene content across all RFGs produced by a refiner for a given area is 0.95% or lower.

SOURCE: NRC 1999b.

requirements by weight, including a requirement that the fuel contain 2% oxygen (NRC 1999b). The RFG program provides two compliance options to refineries. One is a per-gallon performance requirement ensuring that all fuel sold in program areas meets the standards. The other is a nominally more stringent performance requirement to be met on average across all of the RFG produced by a refinery. To determine whether a specific fuel complies with the performance standards, EPA developed two regression models from laboratory tests of vehicles operated on fuels with a wide range of compositions and properties. These models, known as the simple and complex models, estimate exhaust and evaporative emissions based on fuel properties such as RVP, oxygen content, sulfur content, and the fuel's distillation curve. Although these models provide a straightforward regulatory framework for refiners and air quality managers to determine whether a given fuel blend meets the specifications required for the RFG program, the output of these models might not accurately reflect the actual performance of the fuels when used with the fleet of cars in operation (NRC 1999b).

Phase II of the federal RFG program took effect in January 2000. It tightens the performance standards, as shown in Table 4-5, allowing refineries to choose between per-gallon or on-average compliance options. Currently, the RFG program is required in 10 metropolitan areas: Los Angeles, San Diego, Baltimore-Washington, Hartford, New York, Philadelphia, Chicago, Milwaukee, and Houston, as specified in the 1990 CAA Amendments, and Sacramento, which was added when it was reclassified as a severe nonattainment area in 1995. Thirteen metropolitan areas or states are voluntarily participating. In addition, Phoenix has its own, more stringent RFG program (EPA 2002j).

In addition to the federal RFG program requirements, Table 4-5 shows the California RFG program's distinct requirements. That program was implemented on an accelerated schedule: Phase 1 of the California program started in 1992 and Phase 2 in 1996. Phase 3, which includes a ban on the use of MTBE, was originally scheduled to start on December 31, 2002, but has been postponed until December 31, 2003 (CARB 2003a), because of delays in removing the CAA oxygen requirement for RFG.

Oxygen-content requirements for RFG (2% by weight) and oxygenated fuels (2.7% by weight) mandated by the 1990 CAA Amendments have been controversial. Although the motivation for including them in the 1990 Amendments was ostensibly to aid in the reduction of emissions, they were also motivated strongly by a desire on the part of the sponsors from farm states to increase the use of ethanol in the fuel supply for purposes of farm policy and for decreasing reliance on foreign sources of oil (EPA 1999f). Ethanol and MTBE are the two additives most widely used to meet the oxygen content requirement, MTBE being used most heavily outside the

Midwest. The addition of ethanol to gasoline can increase its RVP, in turn, increasing evaporative emissions from vehicles as well as from the fuel storage and distribution system. The net impact on air quality of adding ethanol to fuel is small but under some circumstances might be negative. MTBE does not increase RVP, but leaking gasoline storage systems have contaminated groundwater and raised health concerns (NSTC 1997). An NRC (1999b) report concluded that although adding oxygenates to gasoline reduces CO in vehicle exhaust, as well as the evaporative and tailpipe emissions of some air toxics, it has little effect on exhaust emissions of VOCs and may increase exhaust emissions of $NO_x$. Because of concerns about the oxygen additives, CARB requested a waiver from EPA from the oxygen content requirements of the RFG program, arguing that the requirement impeded its efforts to reduce $NO_x$ emissions. A blue ribbon panel convened by EPA to review the matter in 1999 recommended a substantial reduction in the use of MTBE and the removal of the RFG oxygen content requirements (EPA 1999f). Although EPA denied California's request, it has supported changes to the program. Congress is considering amendments to energy legislation that would ban MTBE and replace the oxygen requirement with a renewable fuels requirement that would likely substantially increase ethanol use.

For CO emissions, the benefits of oxygenated fuel use are most pronounced in older vehicles. New vehicles with properly functioning closed-loop control systems show relatively little benefit.[16] Thus, the effectiveness of the oxyfuels program has diminished as the vehicle fleet has turned over. The CAA Amendments of 1990 do not have a sunset provision for the oxygenated fuels requirement, but in effect this requirement has been diminishing as a large portion of locales with high ambient CO concentrations have attained or are close to attainment of the CO NAAQS.

In addition to the emission standards promulgated in the new Tier 2 emission standards for passenger vehicles, including SUVs, minivans, and other LDTs, EPA is imposing new restrictions on sulfur in gasoline, primarily to prevent sulfur-poisoning of catalytic converters (65 Fed. Reg. 6698 [2000]). In 2004, corporate average sulfur concentrations of 120 ppm with a cap of 300 ppm will be required. By 2006, the corporate average limit will be reduced to 30 ppm and the cap to 80 ppm. The regulations will be phased in more slowly for gasoline produced for sale in the western United States and for small refiners. They also allow refiners to generate credits for

---

[16]Closed-loop systems control the engine air-to-fuel ratio from an oxygen sensor in the exhaust stream and are able to oscillate it within a narrow window about the stoichiometric ratio. This process ensures that a three-way catalyst maintains a high conversion efficiency for VOCs, CO, and $NO_x$.

reductions achieved ahead of schedule; refiners may then apply these credits toward meeting their regulatory requirements or transfer them to other refiners or importers.

Like the Tier 2 LDV standards, EPA's new emission standards for heavy-duty diesel engines, published in January 2001, are also accompanied by tightened limits on fuel sulfur content to prevent catalyst damage (66 Fed. Reg. 5002 [2001]). Starting in 2006, the new sulfur limit for most diesel fuel will be 15 ppm. Through 2009, up to 20% of diesel may still be produced with a sulfur content of up to 500 ppm, but it may only be sold for use in pre-2007 engines. As with the gasoline rule, small refineries have a slower phase-in schedule. Refineries producing diesel fuel and gasoline for sale in the western states may stagger their compliance schedules for the two fuels. The regulation also includes an averaging, banking, and trading component. Here, refineries can generate credits on the amount of 15-ppm sulfur diesel fuel that exceeds their required 80% production. These credits may be averaged with another facility owned by that refinery, banked, or sold to another refinery.

## BEHAVIORAL AND SOCIETAL STRATEGIES TO REDUCE MOBILE-SOURCE EMISSIONS

Although individual vehicle emissions have been reduced substantially over the past 30 years, those improvements have been offset at least partially by continued increases in vehicle miles traveled (VMT). Thus, controls on vehicle activity and transportation represent a potentially important avenue for reducing emissions from mobile sources.

### Regulation of Motorists' Vehicle Use

Under the CAA Amendments of 1970, EPA required the states to develop transportation control plans (TCPs) for their air pollution control areas (normally, metropolitan regions). The TCPs were to be incorporated into the overall SIPs for attaining and maintaining compliance with the NAAQS. Because the 1970 CAA Amendments required attainment of the standards by 1975, it appeared that stringent transportation controls would be required in many areas; for example, parking supply restrictions, high taxes or surcharges on parking, and downtown access restrictions. These policies proved highly unpopular, however, and most states refused to submit TCPs (Howitt and Altshuler 1999).

In 1973, EPA promulgated federal TCPs for 19 major metropolitan areas that included the types of policies that the states had chosen not to impose on their own authority (Altshuler et al. 1979; Suhrbier and Deakin 1988). Congress then restricted EPA's authority to require price disincentives (such as road-use tolls and parking surcharges) or to restrict parking at all, and EPA effectively abandoned efforts to enforce the federal TCPs (Altshuler et al. 1979; Howitt and Altshuler 1999).

The 1977 CAA Amendments did not mandate restrictions on personal travel, although they permitted states to adopt restrictions if they wished. They also authorized the federal government to withhold most federal highway funds from any state failing to submit an acceptable SIP to EPA (Howitt and Altshuler 1999). Although the states did submit their SIPs, very few proposed or implemented controls on personal travel (Horowitz 1978; Deakin 1978; National Commission on Air Quality 1981; Yuhnke 1991).

The CAA 1990 Amendments again left the decision of whether to adopt transportation control measures to state and local officials, with one major exception: mandated employer trip reduction programs (the employee commute options [ECO] requirement) in the 10 most severely polluted nonattainment areas. Some of the 10 affected areas initially proceeded with EPA's ECO program, requiring employers to reduce the amount of automobile commuting to their work sites. Beginning in 1994, EPA found it difficult to defend the ECO program against growing resistance from business groups because few emission-reduction benefits were expected from the program. Congress made the program voluntary in December 1995 (Public Law 104-70).

Outside the United States, there have been a number of efforts to try to address motorist behavior. These include strict restrictions in Singapore (Chia and Phang 2001), alternate day vehicle restrictions in a number of cities, and recent efforts to impose a user fee for drivers into central London. Although these programs may provide some useful insights, some have been implemented in very different governmental circumstances (for example, the authoritarian practices of the Singapore government). Also, others have taken place in the context of broader government policies that have imposed a much higher cost of fuel than is politically realistic in the United States. Still others, such as the London experiment, are not substantially different from programs already in place in the United States. For example, the cost of crossing the Hudson River into New York City is already nearly as high as the new charge imposed in London.

Regulatory and other efforts under the CAA to reduce motor vehicle use have proved to be politically infeasible. On the other hand, some efforts to promote voluntary reductions in vehicle use, such as ride-sharing programs, enhancement of existing transit service, compressed work weeks, and telecommuting, have won political acceptance. However, such programs generally have little potential to affect overall use of motor vehicles; each is expected to yield about 1–3% reductions in VMT (Apogee Research 1994; Howitt and Altshuler 1999).

## Controls on Transportation Infrastructure Planning and Investment

Another class of strategies that can be undertaken to reduce motor-vehicle emissions focuses on optimizing urban and regional transportation patterns and practices. Such strategies require careful integration of urban and regional plans for land use, transportation infrastructure, housing, and economic development. Because they are implemented on decadal time scales, their successful implementation requires a continuing commitment on the part of local and regional managers and politicians to long-term air quality improvement.

Within the general theme of transportation planning and AQM, highway capacity has become a highly contentious and complex issue. Because free-flowing traffic at moderate speeds produces less pollution per vehicle mile than highly congested traffic produces, construction of additional highway lanes and access roads would seem to improve air quality. However, highway expansion in a metropolitan area can encourage urban sprawl and low-density development (TRB 1995). Low-density development, which, in turn, increases the number and length of vehicle trips, decreases vehicle occupancy rates, and diminishes the practicality of pedestrian and transit trip making. Similarly, road building to alleviate congestion in densely developed corridors may induce additional travel, because a great deal of latent travel demand in such areas invariably has been suppressed by the existing congestion. The end result can be an increase in automobile travel and increased mobile-source emissions.[17]

## Linking Highway Capacity Expansion to Air Quality through the National Environmental Policy Act

In its initial forms, the SIP process lacked a formal procedure to ensure that automobile usage and VMT projections used by air quality planners in NAAQS attainment demonstration were consistent with regional plans for highway and road construction. Before the 1990 CAA Amendments, neither federal law nor the practices of metropolitan transportation planning linked air quality management with urban transportation investment policy. The National Environmental Policy Act (NEPA) of 1969 had attempted to address this problem by mandating that federally funded projects be broadly analyzed for their impacts on the environment. However, NEPA only required that environmental impacts be considered in evaluating projects; it did not provide substantive guidelines for permitting projects to proceed to construction, nor did it require consistency with the projections used in

---

[17]Others argue that the causal factors shaping metropolitan growth and development and associated vehicle usage are more complex than this viewpoint allows (TRB 1995).

SIPs. In addition, NEPA's project-by-project focus did not generally consider the cumulative environmental effects of multiple projects.

Further efforts to create links between air quality regulation and regional transportation planning and investment encountered significant institutional problems and resistance. Section 109(j) of the Federal-Aid Highway Act of 1970 required the secretary of transportation, in consultation with the EPA administrator, to issue regulations for the purpose of ensuring that federally assisted highway projects were consistent with the air quality plans for each pollution control area. However, the regulations were vague on the question of how consistency should be determined, and they had state transportation officials, rather than environmental regulators, making the consistency determinations. In most areas, EPA regional offices made little effort to activate Section 109(j). The 1977 CAA Amendments contained stronger language but were only marginally more effective (Howitt and Altshuler 1999).

### The Conformity Regulations

The CAA Amendments of 1990 and the Intermodal Surface Transportation Efficiency Act (ISTEA) of 1991 required much tighter integration of (or conformity between) clean air and transportation planning (Howitt and Moore 1999a). The conformity regulations most directly affect metropolitan planning organizations (MPOs), the public agencies that conduct transportation planning under federal transportation law (ISTEA from 1991 to 1997 and TEA-21 [Transportation Equity Act for the 21st Century] since then). Other agencies and stakeholders also play a role. A significant aspect of the conformity regulation is the incentive given to both air quality and transportation planners to maintain conformity. The penalty for nonconformity is a transportation funding cutoff. For some metropolitan areas, that can mean the loss of more than $100 million per year.

At the core of the conformity requirement is an EPA-mandated analytical procedure and regulatory test to ensure that transportation-related emissions in a nonattainment area stay within the limits used in the area's SIP. As described by Howitt and Moore (1999a), the process involves use of a computer simulation to make a 20-year forecast of emissions from the transportation system. The forecast takes into account changes in demographics, land use, economic development, and transportation infrastructure and services. The forecasted emission concentrations are compared with the maximum emissions permissible in certain milestone years under the state's SIP. If those concentrations exceed permissible levels in the SIP, an MPO must change its transportation plans and programs so that forecasted emissions would be within the emission budget constraints. Alternatively, the state must amend its SIP to reduce transportation-related emissions through additional mobile-source control measures or reduce emissions from stationary sources, such as industrial facilities or smaller area sources. In addition, an MPO must demonstrate timely implementation of transportation control measures in SIPs and fulfill the ISTEA "fiscal-constraint" requirement for transportation plans and programs by showing that it is likely to have sufficient financial resources to carry them out.

If an MPO cannot satisfy the conformity requirements within specified time periods, then penalties are imposed during a conformity "lapse" or "freeze." During this time, the MPO may not begin most new transportation projects, and the use of federal transportation funds is restricted. Conformity lapses or freezes can also result from certain shortcomings in a SIP, which may or may not involve transportation-related issues (Howitt and Moore 1999a).

To date, the most widespread effects of conformity have been procedural and cultural. Before the conformity regulations of the 1990 CAA Amendments, transportation planning and air quality regulation were effectively separate spheres of government activity, even within a given state. Conformity appears to have fostered greater interaction; as a result, it is likely that transportation and environmental agencies have gained more knowledge about and a greater appreciation for one another's missions, responsibilities, and procedures. Most transportation officials also seem to accept the legitimacy and high priority of environmental values in transportation decision-making (Howitt and Altshuler 1999).

Thus far, the conformity requirement has had the largest impact on NAAQS nonattainment areas experiencing rapid growth and, therefore, substantial economic and political pressure to expand transportation infrastructure. Atlanta, Charlotte, and Houston, for example, have had federal transportation funding interrupted because of conformity problems. As further emission reductions are required in the years ahead, nonattainment areas that are rapidly growing are likely to experience conformity as an increasingly salient constraint. It is uncertain how the conflicts between transportation and air quality goals in such areas will be resolved and whether the federal government will remain firm in enforcing the regulation.

The conformity requirement has had less impact on transportation planning in older, more slowly growing metropolitan areas. These areas have more mature highway networks and well-established transit systems, and most of their transportation projects, involving highway and transit reconstruction and modest expansions, are neutral or positive in terms of air quality. In addition, these areas often have more serious air pollution problems and thus more time to come into compliance with the CAA (see Table 3-1 in Chapter 3). Therefore, conformity has not required these areas to make major adaptations. However, these regions have not yet met their stiffest challenges, because they still must demonstrate future reductions in transportation-related emissions (Howitt and Moore 1999b).

## CRITICAL DISCUSSION OF MOBILE-SOURCE EMISSION-CONTROL PROGRAMS

The CAA has led to substantial successes in some aspects of its mobile-source programs. The emissions standards for individual LDVs and HDVs have promoted a host of new technologies, which have resulted in or are expected to result in emissions that are substantially lower than they were in 1970. These emission standards also resulted in eliminating the largest source of lead in the environment through the control of lead in fuels. Despite increasing travel, which has offset some of the gains, the number of communities having air that is not in attainment of the NAAQS for lead or CO (pollutants that are primarily from mobile sources) has dropped dramatically. As described in a recent NRC (2003b) report, that drop has resulted in substantial reductions in overall population exposures to high concentrations of ambient CO.

Mobile-source emission-control programs have also experienced and continue to experience challenges, including the persistence of high-emitting gasoline vehicles and older diesel vehicles, the relative lack of regulation of nonroad engine emissions, the contributions of mobile sources to HAP emissions (see discussion of HAPs in Chapter 5), the use of fuel policy to pursue other non-air-quality-related interests (for example, farm policy and reducing dependence on foreign oil), and the inability of AQM to affect individual travel behavior substantially.

The following section reviews and summarizes some of the lessons learned and the challenges ahead.

### Promotion of New Technologies Using Vehicle Emission Standards

The history of vehicle emission controls is one of long-term success, albeit success that came more slowly than policy-makers originally intended. Delays were due in part to opposition from the automotive industry facing the challenge of producing the required technologies in a cost-effective manner. Nonetheless, substantial improvements in technology have occurred: more efficient combustion resulting in fewer partially combusted hydrocarbons; emission-control systems with longer effective lifetimes; and the application of many new catalyst and sensor technologies, using increasingly sophisticated computer controls. Actual costs of the improvements have often proved to be less than anticipated (Anderson and Sherwood 2002; Cackette 1998). Lower costs were realized as production experience was acquired, economies of scale were achieved, and the desire to meet a mass market had fueled substantial competition. Some observers have suggested that such economies are more likely when policies target a relatively small number of large corporations and focus on inducing substantial (but not radical or technologically infeasible) changes in product technology (Howitt and Altshuler 1999).

Even so, the process can be marked by considerable controversy and even litigation. Delays are often unavoidable, and accommodations between legislative purposes and industry interests are a frequent part of the implementation process. Nevertheless, the technology promotion process appears to be able to provide the nation with a significant mitigation of air pollution in a cost-effective manner (Howitt and Altshuler 1999).

### High-Emitting Gasoline Vehicles

The emission-control program for mobile sources has been very successful in terms of its ability to reduce emissions from individual LDVs, operated in normal modes. The forecast emission reductions from recently promulgated regulations, if fully effective over the lifetime of an operational vehicle, will provide a further dramatic reduction in mobile-source emissions. However, analyses suggest that historically, a disproportionately large fraction of the mobile emissions come from a relatively small percentage of so-called high-emitting cars. Many of these high emitters are vehicles whose pollution control devices are not operating properly or whose engines are combusting combinations of fuel and lubricating oil. In addition to representing significant sources of VOC, $NO_x$, CO, and air toxics, high-emitting gasoline vehicles might also be significant contributors to ambient concentrations of $PM_{2.5}$ (Watson et al. 1998). That possibility suggests that attainment of the new NAAQS for $PM_{2.5}$ will require a much better understanding of the effect of engine operation and deterioration on the emissions of fine particles from gasoline vehicles.

The continued presence of high emitters, which were first described by Wayne and Horie (1983) and which continue to be observed in the contemporary vehicle fleet (NRC 2001c and references therein), will reduce the emission reductions that are forecast for programs to be implemented in the latter half of this decade. To date, the nation's AQM system has not come up with an effective and politically acceptable means to address this problem. One mechanism, which appears to have been at least partially successful, has been to require increasingly durable, warranted emission-control systems, resulting in substantially more robust emission-control systems. New vehicles are currently certified to 100,000 mile standards and major emissions control equipment, such as the catalyst and on-board computer, are warranted for 80,000 miles. Other mechanisms, such as inspection and maintenance and remote sensing, have been either politically controversial, technically ineffective, or both. A further opportunity, as yet unrealized, might lie in the improvement and monitoring of on-board diagnostic devices that immediately notify drivers of problems and in the long run provide officials and mechanics with accurate information on the performance of emission-control equipment.

### Reducing Emissions from Older and Nonroad Diesel Engines

Although there has been and probably will be substantial progress in reducing the emissions from new on-road diesel vehicles, three substantial challenges remain. First, the long life of diesel vehicles and the pattern in which older vehicles are used for shorter-haul routes in urban centers result in a large and continuing source of PM and $NO_x$ diesel emissions in urban settings with dense populations. Recent efforts by the states and EPA to promote buy-back, replacement, and retrofit programs have begun to address this issue, but these efforts have been modest. Second, the recent enforcement experience with on-road emissions being higher than those certified for the engines and the difficulty of inspection and maintenance for heavy-duty engines will require continued attention. Finally, as on-road LDV emissions continue to decline, emissions from various nonroad sources and on-road diesel vehicles will become increasingly important. Recent regulatory efforts by EPA (as detailed above) have begun to address this last issue. Although these recent efforts have begun to address the issue of in-use heavy-duty vehicle emissions, there is not yet a systematic, nationwide approach to the problem. Specific recommendations for addressing this problem are advanced in Chapter 7.

### Regulating the Content of Gasoline and Diesel Fuels

The past 15 years have seen a substantial increase in CAA programs regulating the composition of fuel. In some instances, these programs have been directly responsible for substantial decreases in exposure to important air pollutants—most notably, lead—and for a substantial reduction in benzene emissions and exposure with the advent of RFG. The elimination of sulfur is also likely to facilitate a new generation of gasoline and diesel control technologies, as occurred with the removal of lead in the 1980s. This potential for further technological innovation, however, depends on the successful implementation of the sulfur-reduction regulations on the time scale envisioned in the regulations.

The success of these new fuels in reducing key precursors for $O_3$ is less clear. Congressional requirements for oxygen in RFG that resulted in increased use of ethanol may have resulted in increasing evaporative emissions and, in some instances, a worsening of $O_3$ conditions (NRC 1999b). As indicated above, such actions were taken not only for air quality reasons but also for other reasons (for example, farm policy and energy security). The challenge of reducing $O_3$ precursors is made larger by the limitations of the current technical tools available for predicting motor vehicle emissions as a function of the fuel blend (for example, the Simple and Complex models used for RFG). For MTBE, the oxygen requirements also resulted in increasing the risk of groundwater and surface-water contamination (EPA 1999f).

One advantage of such programs has been to reduce emissions in older, less well-controlled vehicles. However, many of these older LDVs have now been replaced, but there is no provision for reviewing and, if appropriate, removing specific fuel requirements as the older LDVs are replaced. Finally, Tier II fuel requirements have included banking, averaging, and trading provisions to improve the cost-effectiveness of the rules. To date, however, with the exception of the lead phase-out, there has not been adequate experience to evaluate these programs and to determine their effectiveness. Such evaluation will be especially important as EPA implements the new sulfur-reduction requirements, and extends them to nonroad diesel fuel.

### Controls on Motorists' Behaviors

In contrast to the success of new motor vehicle standards, efforts to regulate citizen's personal travel behavior through restrictions or economic disincentives have typically provoked controversy and, in the end, have proved to be politically infeasible. Even where some effort has been made, the estimated pollution reductions have been modest. After 30 years of efforts to affect individual driving behavior, all involved—Congress, EPA, and state and local agencies—have opted to emphasize voluntary versions of such efforts (Howitt and Altshuler 1999).

### Conformity

In the related area of regulation of infrastructure investments, some modest progress appears to have been made. The 1970 and 1977 CAA Amendments were ineffective in ensuring consistency between state transportation investments and air quality improvement commitments. The 1990 CAA Amendments embodied a more realistic appreciation of how transportation decisions can affect air quality planning and backed the requirement for conformity with the tangible threat of federal fiscal penalties for failure to comply. The invigorated "conformity" requirement appears to have enhanced the attention paid to air quality objectives in metropolitan transportation planning, although it has not, to date, significantly affected state transportation investment decisions.

Perhaps the most important substantive effect of conformity may be a new way of looking at highway projects in the early stages of the transportation planning process. Proposals for major highway enhancements, which can be "emission budget busters," may now be less likely to move into the preliminary planning stage, particularly because planners must show that the financial resources needed to carry out the proposals are likely to be available.

Only a few rapidly growing areas have been motivated by conformity requirements to pursue major investments in mass transit, because such investments rarely produce significant air quality benefits. In the areas that already have extensive transit networks, however, the conformity requirements have reinforced the importance of modernization, service enhancements, and occasional extensions (Howitt and Moore 1999b).

The relationship between SIP development and the conformity process can create difficulties for both planning and regulatory purposes for several reasons: assumptions, data, and forecasts. These difficulties occur when the two planning processes are insufficiently interconnected. The federal conformity regulations mandate the use of the most up-to-date transportation-planning assumptions and data available (for example, for travel behavior, population, land use, and economic growth) and the most recent version of the emission forecasting models (that is, EPA's MOBILE model or California's EMFAC model). Because conformity analyses must be revised at least every 3 years in nonattainment areas and SIPs might not be updated for much longer periods, the underlying assumptions, data, and models used in the transportation-planning process may vary significantly from those in the air quality plan. The disjunction of modeling inputs and methods, depending on the circumstances, can create differences in estimates of future emissions that may only exist on paper in the planning process or mask real air quality problems. Thus, the disconnection between SIP development and the conformity process undermines the intended usefulness of conformity as a performance test allowing informed judgment of whether state transportation investments are consistent with air pollution reduction commitments in the SIP. Moreover, the planning horizons for air quality regulation and transportation planning may mesh poorly. Under current requirements, transportation plans are required to have a 20-year time horizon, and conformity is done on that basis. An attainment demonstration and associated maintenance plan, however, need only a 10-year time horizon. Transportation plans must therefore often use emission budgets that do not take account of future emission growth in transportation and other sectors (Harrington et al. 2003).

Both the possible disjunction of planning assumptions and forecasts between SIPs and conformity analyses and the poorly meshing time frames of the two processes can create important questions about regulatory fairness. The procedures of the CAA were designed so that, in setting emission budgets, decision makers would take forecasts of aggregate emissions and then explicitly allocate emission estimates across sectors in ways that created legitimacy and stakeholder commitment to the outcome. If transportation emission forecasts are updated while other sectors' are not, the validity and perceived fairness of the results can be questioned and stakeholder support for pollution reductions can be undermined.

Some proponents of conformity hoped that linking transportation and land use would encourage broader acceptance of land-use regulations to reduce emissions from mobile sources. However, the impact of conformity on land-use decision-making, which is in the hands of local governments that do not have a direct role in conformity, has been modest (Howitt and Moore 1999a).

## SUMMARY

### Strengths of the Mobile-Source Emission-Control Program

- Regulations on LDVs and LDTs have resulted in significant reductions in the emissions per mile traveled. In the case of CO, as shown in a recent NRC study (NRC 2003b), those reductions have resulted in significant reductions in overall population exposure. Further emission reductions are anticipated from the implementation of stricter emission regulations in the coming years.

- Emission regulations on LDVs and LDTs have promoted the development and application of new cleaner technologies for vehicles—technologies that are now used worldwide. Furthermore, the actual costs of these technologies were likely less than anticipated.

- Regulations on fuel properties, including content, have also resulted in air quality benefits, most notably is the phase-out of lead in gasoline that made the use of catalytic converters possible and reduced population exposure to lead. RFG resulted as well in reductions in population exposure to benzene. New regulations on sulfur content in fuels promise to further enhance the effectiveness of catalytic controls and reduce emissions of the on-road fleet.

### Limitations of the Mobile-Source Emission-Control Program[18]

- Gaps remain in the ability to monitor, predict, and regulate in-use vehicle emissions. The existence of high emitters is a major challenge, and

---

[18]Recommendations are provided in Chapter 7.

I/M programs have been less effective than expected in identifying high emitters and ensuring in-use compliance with emission standards. Other approaches, such as remote sensing and on-board diagnostics, show technical promise, but further testing of those techniques is needed. There is also a concern that remote sensing may be viewed as unacceptably invasive.

• As emissions from LDVs and LDTs decrease and the focus on attaining the NAAQS for $PM_{2.5}$ increases, emissions from nonroad engines and heavy-duty onroad vehicles are becoming increasingly important. New emission regulations for these sources are being implemented, but the long lifetime of these engines will slow the rate of penetration of the effects of these regulations into the fleet. Additional efforts in inspection and maintenance and in promoting retrofit and incentive buy-back programs are therefore needed.

• The inclusion of content-specific requirements in the fuel provisions of the 1990 CAA Amendments (for example, oxygen in the RFG program) can limit flexibility to meet standards in the most cost-effective way for areas required to implement the program. The standards also make it difficult to adjust the program in the face of new challenges (for example, in the case of MTBE contamination of groundwater and surface water) and have a small (sometimes even negative) impact on the control of some air pollutants (as might have been the case with the use of ethanol and potential increases in RVP).

• Growth in vehicle miles traveled, personal automobile usage, and popularity of fuel-inefficient vehicles (for example, SUVs) has offset a significant portion of the gains obtained from stricter emission standards on individual vehicles. With the exception of the conformity requirements of the 1990 CAA Amendments and subsequent related legislation, air quality managers remain unable to affect these societal and behavioral determinants of mobile-source emissions.

# 5

# Implementing Emission Controls on Stationary Sources

## INTRODUCTION

Stationary emission sources are divided into two categories in the Clean Air Act (CAA): major stationary sources (also called point sources) and area sources (see Box 5-1). Both contribute significantly to air pollution in the United States, and the CAA has contained provisions to regulate and control emissions from many of these sources for over three decades.

In principle, stationary sources can be controlled through the imposition of a design standard or a performance standard applied to individual facilities or through the imposition of an overall cap on a specific industry or segment of sources (see Box 5-2). The CAA applies these controls through a variety of programs that generally fall into five categories:

- Permits and standards for new sources or major modifications of existing sources (for example, the new-source review [NSR], New Source Performance Standards [NSPS], and prevention of significant deterioration [PSD]).
- Technology-based standards for emissions reduction in a class of existing facilities (for example, the reasonably available control technology [RACT] requirements for nitrogen oxides [$NO_x$], the acid rain $NO_x$ provisions, and the maximum achievable control technology [MACT] for hazardous air pollutants).
- Cap-and-trade provisions (for example, the acid rain sulfur dioxide [$SO_2$] program of the CAA Amendments of 1990).

---

**BOX 5-1   Major and Area Sources of Emissions**

The regulations and controls on stationary sources in the United States have generally been designed to focus on two types of stationary sources:

1. Major stationary sources whose emissions exceed a nominal threshold defined in the CAA or by a regulatory agency.
2. Area sources whose emissions fall below the threshold.

Major stationary sources (such as factories and electricity-generating facilities) are defined as single sources with emissions exceeding a threshold level that depends on the pollutant. This threshold is generally 100 tons/yr for criteria pollutants and their precursors; however, the threshold for Pb is 5 tons/yr. For the most severe $O_3$ nonattainment areas, the threshold for VOC can be as low as 10 tons/yr. For serious CO nonattainment areas, the CO threshold is 50 tons/yr. The reporting requirements for major stationary sources are more detailed than those for area sources (65 Fed. Reg. 33268 [2000]).

One important distinction between major stationary and area sources is the method used to estimate and control their emissions. Major stationary sources are inventoried individually and their emissions are generally controlled through a permitting process that requires specific regulatory review of each individual facility. Area sources, which are generally widely dispersed sources arising from relatively small industrial and business facilities (for example, agricultural fields and small copying and printing shops) or from application and use of consumer products (for example, architectural coatings), are inventoried and regulated collectively.

---

- Other trading and voluntary mechanisms (for example, pollution prevention programs and the U.S. Environmental Protection Agency's [EPA's] Project XL).
- Regulations on area sources (for example, consumer product specifications).

In addition to those programs, which either provide permits for, or mandate changes in, new and existing facilities and products, the CAA Amendments of 1990 also established the Title V operating permit program, which requires comprehensive operating permits for large stationary sources to record all operating requirements for a facility as a basis for tracking compliance.

Each of these programs is discussed below. This discussion is then followed by a summary of the strengths and weaknesses of the various programs and the lessons to be gleaned for future approaches to air quality management (AQM).

## BOX 5-2    Design Versus Performance Versus Cap and Trade

Traditionally, stationary sources have been regulated through the imposition of emission standards or limitations. The CAA defines such a standard or limitation "as a requirement established by the State or the Administrator which limits the quantity, rate, or concentration of emissions of a source to assure continuous emission reduction, and any design, equipment, work practice or operational standard promulgated under this Act." Although many specific programs regulate stationary sources in the CAA, the basic approaches that have been adopted to achieve emission reductions fall into three broad categories: technology specification standards (or design standards), performance standards, and the newer use of cap-and-trade requirements.

A design standard mandates that a set of design or technological options (for example, installation of particle traps in smoke stacks) must be adopted by the managers of the regulated facilities to meet emission targets. Although this approach has the potential to achieve substantial reductions in air emissions, it has been criticized as being overly inflexible and cost-ineffective. For example, even though there is often a substantial disparity in the marginal costs of emission reductions among facilities affected by the same technology standard, technology-specification standards do not allow market forces to use this disparity to minimize the overall costs of the desired level of emission reductions (Hahn and Stavins 1992; Stavins 2002). Moreover, because firms must use the technologies specified in the standard, the approach does not encourage individual firms to pursue ways to reduce emissions through potentially more effective alternative technologies and front-end process adjustments.

In contrast to a design standard, a performance standard simply specifies a maximum allowable rate of emission from a given type of source or facility, and the managers of the facility are free to choose any combination of technologies and operational practices to meet the standard. In principle, this approach provides an individual facility with greater flexibility to discover the most cost-effective way to meet the emission standard. Although the flexibility exists in theory, in practice the performance standard is normally set at the level that can only most readily be achieved by a known technology. Thus, unless readily available alternatives can meet the standards to the satisfaction of the regulators, there is likely to be a tendency for facilities to default to the known technology, thus also limiting the options for the affected industries. In that case, the necessary level of control is achieved by selection of parameters within that technology (for example, size of the control technology and flow rates of reactants).

The performance standard can set different degrees of control for different sources—low-emission sources may have a lower control requirement than high-emission sources. However, regulators faced with setting a performance standard often compromise, setting a standard at a lower level than the one that can be achieved by many facilities so that the facility with the largest uncontrolled emissions will not face an impossible task of control. As a result, marginal costs of emission reductions often continue to vary substantially among facilities. Further, once a performance standard is achieved at a facility, there is little incentive to discover more efficient ways of achieving the same or greater emission reduction,

**BOX 5-2** continued

and, most important, there is no mechanism for a company to profit from innovations that achieve emission reductions beyond the standard.

The choice between a design and a performance standard is often made as part of a rule-making process. In some cases, such as fugitive emissions, it is simpler to set a design standard because of the complexity of measuring the actual emissions. However, design standards do not provide the same limit on emissions that performance standards do; they simply provide a reduction over a set of baseline emissions. In either case, the standards are related to the output of the facility. Performance standards are usually expressed in such terms as pounds of pollutant emitted per million British thermal units of fuel heat input. In these cases, the actual amount of emissions is permitted to increase as the amount of fuel is increased.

In response to the limitations of both design and performance standards, a new approach (a market-based approach) based on cap and trade has emerged in the last decade. In this approach, each source category (or every source) in a given geographic area has its total emissions of a particular pollutant capped at a level below its current level, and each individual source is assigned an emissions allotment consistent in the aggregate with the overall emissions cap. The novel aspects of this total-emissions-based performance standard are (1) that it does not presume any particular technology or emissions standard for the sources, and (2) that it allows market forces to minimize costs and reward innovation. Each facility is allowed to achieve the required reductions in a variety of ways, including conventional pollution control, process change, and product substitution, as well as purchase of reductions at a more economical rate from other facilities that have exceeded their reduction target. Even with a cap-and-trade standard, an emission limit must be set that is based on feasible control technology or process operations. However, the ability to trade removes one of the problems faced by regulators when dealing with a range of existing sources. A greater control requirement can be set, and companies that cannot easily meet the requirement can trade emission-reduction credits to comply with the cap-and-trade requirement. There are challenges in applying this emission-control mechanism in every situation, as discussed later in this chapter. The mechanism does, at least in theory, offer the possibility of achieving substantial reductions while allowing individual sources to minimize costs and optimize efficiency.

## PERMITS AND STANDARDS FOR NEW OR MODIFIED MAJOR STATIONARY SOURCES

The CAA mandates that the states implement and EPA oversee permit programs to control and regulate pollutant emissions from major stationary sources in National Ambient Air Quality Standards (NAAQS) attainment and nonattainment areas. Under these programs, each new major stationary source of air pollutants must apply for a permit before beginning construction and, within the permit application, demonstrate that the new facility will meet appropriate emission-control standards. In recognition of the substantial costs of retrofitting, existing stationary sources are required by federal law to undergo the permitting process only when nonroutine modifications are planned that will result in a significant increase in pollutant emissions from that source.[1] The program for nonattainment areas is described in Part C of Title 1 of the CAA and that for attainment areas in Part D of Title 1. In the CAA, the program is the NSR in nonattainment areas and the PSD in attainment areas. Although NSR is often used generically for both types of programs, there are some differences, so the discussion below uses the CAA's terms.

## Background

Since 1970, the CAA has required EPA to promulgate NSPS for major and minor sources on a category-by-category basis. NSPS are national emission standards that are progressively tightened over time to achieve a steady rate of air quality improvement without unreasonable economic disruption. In recognition that the 1970 goals for attainment of air quality standards would not be met and that some attaining areas required measures to prevent conditions from worsening, NSR procedures became applicable to major stationary sources in nonattainment areas and in PSD attainment areas with the 1977 CAA Amendments.

The NSR provisions required that major new sources in nonattainment areas be constructed only if they created no net increase in emissions. That requirement was intended to ensure that major new or modified stationary sources of air pollution within nonattainment areas did not inadvertently undermine the state implementation plans (SIPs) that had been developed for those areas. To accomplish that, NSR requires that new and modified facilities use control equipment that has the lowest achievable emission rate (LAER) and that those facilities obtain an "emission offset" to offset the increase in emissions anticipated from the proposed construction or modification. As discussed in more detail below, the offset is actually required to be greater than the emissions increase from the proposed project to ensure progress toward attainment.

The PSD program for attainment areas was designed to do the following:

• Maintain public health protection in areas that already meet the NAAQS.

• Reduce the total amount of emissions entering the atmosphere and thereby provide general protection from public welfare damage and the impacts of pollutant transport.

---

[1]Throughout the remainder of this discussion, "modification" will be used as it is defined in the CAA to denote "any physical change in, or change in the method of operation of, a stationary source which increases the amount of any air pollutant emitted by such source or which results in the emission of any air pollutant not previously emitted."

• Protect visibility.

• Counteract the unintended incentive given to industries from the NSR program to relocate to less developed states and thereby avoid NSR permitting requirements.

PSD, like NSR, requires new and modified facilities to meet emission-control standards, although PSD standards need not be as restrictive as those for NSR. Moreover, PSD does not require emission offsets.

## NSR and PSD Requirements

### Applicability

NSR and PSD apply to major new stationary sources and major modifications of existing stationary sources. However, the definitions of these terms differ somewhat for the two programs:

• NSR applies only to criteria pollutant emissions and their precursors from major or modified sources. In most nonattainment areas, a major source for the purposes of NSR is a source that has the potential to emit 100 tons or more per year of any criteria pollutant. For ozone ($O_3$) nonattainment areas, the definition of "major stationary source" includes smaller sources in areas of more severe nonattainment and sources with the potential to emit as little as 10 tons per year of $NO_x$ and VOCs in extreme nonattainment areas.

• PSD generally defines a major source as one that produces 250 tons or more per year of any pollutant. However, 28 specific source categories are identified in the CAA for which the PSD definition of a major source is broadened to include sources that produce 100 tons per year or more of any pollutant.[2] A major modification is one that produces a "significant increase" in emissions as defined by PSD regulations (40 CFR 52.21(b)23).

---

[2]The 28 source categories are coal cleaning plants (with thermal dryers), coke oven batteries, Kraft pulp mills, sulfur recovery plants, Portland cement plants, carbon black plants (furnace process), primary zinc smelters, primary lead smelters, iron and steel mills, fuel conversion plants, primary aluminum ore reduction plants, sintering plants, primary copper smelters, secondary metal production plants, municipal incinerators capable of combusting more than 250 tons of refuse per day, chemical process plants, hydrofluoric acid plants, fossil-fuel boilers (or combination thereof) totaling more than 250 million British thermal units (Btu) per hour heat input, sulfuric acid plants, petroleum storage and transfer units with a total storage capacity exceeding 300,000 barrels, nitric acid plants, taconite ore processing plants, petroleum refineries, glass fiber processing plants, lime plants, charcoal production plants, phosphate rock processing plants, and fossil-fuel-fired steam electric plants of more than 250 million Btu per hour heat input.

The CAA indicates that existing facilities undergoing modifications that significantly increase pollutant emissions shall be subject to NSR or PSD review through the permitting process. The CAA also identifies the net emission increases for each of a series of pollutants that triggers such review. To implement this requirement, EPA developed rules to determine whether a proposed modification would result in a net increase in emissions. That determination involves prescriptions for estimating the present-day baseline emissions and the projected future emissions. In addition, EPA's rules exempt routine maintenance and repair activities from the definition of a modification that can trigger NSR or PSD review. EPA also exempts modifications that cause de minimis, or insignificant, emission increases. For a modification to trigger NSR or PSD review, an emission increase must exceed a "significant level" specified by EPA. That level varies by pollutant and attainment status of the area. As discussed later in this section, EPA's rules and regulations have been the subject of much debate and litigation and have been changed from time to time by EPA to reflect changes in policy priorities within the Executive Branch.

To implement the provisions, EPA usually delegates permitting authority to states or local districts. However, before such a delegation is made, the state or district must demonstrate that its permitting process is substantially the same as that used by EPA.

### Operation

The operation of NSR in nonattainment areas and PSD in attainment areas is conceptually similar. Major new or modified sources undergo a preconstruction review to qualify for a permit that allows the construction or modification to proceed. The new construction or modification must use stipulated control technology. However, the control technology requirements are different in attainment and nonattainment areas. Sources subject to NSR in nonattainment areas are required to use control equipment that provides the LAER and those subject to PSD in attainment areas are required to use best available control technology (BACT). Section 171(3) of the CAA defines LAER as the most stringent emission limitation based on either (1) the most stringent limitation in any SIP for that class or category of source or (2) the most stringent limitation achieved in practice for a certain class or category of source. Section 169(3) of the CAA defines BACT as an emission limitation based on the maximum degree of reduction of each pollutant subject to regulation under the CAA emitted from any major emitting facility, which the permitting authority, on a case-by-case basis, determines is achievable for such facility. The authority is required to take into account energy, environmental and economic impacts, and other costs. Thus, the emission limitations required under BACT can be less stringent than those under LAER.

In determining the specifics of a BACT requirement for a facility, EPA and the states must also ensure that the facility does not exceed the NSPS. In practice, NSPS, which are determined once and are changed only if the regulation is fully revised, serve as the minimum level for BACT and LAER determinations.

In nonattainment NSR, offsets are required to bring about a net decrease in emissions in the area. The size of the offset relative to the anticipated increase from the new or modified source varies from 1.1:1 for non-$O_3$ nonattainment areas and marginal-$O_3$ nonattainment areas to as much as 1.5:1 for extreme-$O_3$ nonattainment areas (see Box 3-3 in Chapter 3). The offsets must be reductions that would not otherwise occur (for example, as a result of other SIP activities) and can be obtained from new emission controls on, or shutdowns of, other facilities and sources. They must be real, permanent, and enforceable. In principle, offsets provide an incentive to try innovative control technologies. In practice, offsets are usually obtained by shutting down other facilities.

Offsets are not required in PSD permits. However, any emission increase after the use of BACT is limited to a fixed amount from all new facilities in the area. The amount of the increase allowed under PSD regulations, called the increment, is defined in terms of ambient air concentration and depends on the pollutant and the class of the area affected. National parks, wilderness areas, and similar sensitive areas are identified as Class I areas. Most areas are Class II. The regulations also define a Class III area, for which emission limits are more lenient than those for Class I and II areas; however, no areas have been designated Class III. The increments for various pollutants are shown in Table 5-1 for each area class.

PSD permit applications are also required to analyze the effects of the source on visibility, soils, and vegetation; the economic growth likely to result from the new source; and the effects of this new growth on the new emissions. When a Class I area is affected by the source, the federal land manager for that area must play a role in the permit process. According to PSD regulations, the manager can provide an analysis showing that there is an "adverse air quality impact" on the Class I area, even if all other permit requirements are met. In such cases, the manager can recommend that the permit not be issued, and the permit may be denied.

### Issues with the Application of NSR and PSD

NSR and PSD have been a positive force in AQM in the United States. NSR has provided a mechanism for economic development and new con-

TABLE 5-1   Allowable Concentration Increments (micrograms per cubic meter) for Prevention of Significant Deterioration (PSD)

| Pollutant[a] | Measurement | Increment in Areas That Are | | |
|---|---|---|---|---|
| | | Class I | Class II | Class III |
| Particulate matter | Annual arithmetic mean | 4 | 17 | 34 |
| ($PM_{10}$) | 24-hr maximum | 8 | 30 | 60 |
| Sulfur dioxide | Annual arithmetic mean | 2 | 20 | 40 |
| ($SO_2$) | 24-hr maximum | 5 | 91 | 182 |
| | 3-hr maximum | 25 | 512 | 700 |
| Nitrogen dioxide | Annual arithmetic mean | 2.5 | 25 | 50 |
| ($NO_2$) | | | | |

[a]$PM_{10}$ is particulate matter with an equivalent aerodynamic diameter of 10 micrometers or less.
SOURCE:  Clean Air Act, Section 163.

struction to proceed in nonattainment areas while maintaining progress toward NAAQS compliance.  Moreover, both NSR and PSD have mandated the use of modern, clean technologies and practices in new facilities and in modified existing facilities throughout the nation.  The application of BACT in conjunction with the requirements for NSPS has encouraged, indeed required, the continual development and application of new technologies that are more cost-effective, cleaner, or both.  However, NSR and PSD have some limitations as well.  Some of the more prominent aspects are discussed below.

*Complexity and Inefficiency*

The NSR- and PSD-permitting processes have become complex and time consuming, especially if there are disagreements between the permit seeker and the permitting agency.  The ever-growing nature of the process is illustrated by EPA's documentation describing NSR and PSD regulations, manuals, and guidance.  About 30 pages in length in the early stages of the program, the documentation now exceeds 1,000 pages and is contained in numerous documents with which permit applicants and writers must be familiar (EPA 2002k).  Representatives of industry complain that the process fosters inefficiencies and unduly discourages economic growth and innovation (NAM 2002).  The process as currently organized can lead to conflicts between the goal of implementing improved emission-control technology as quickly as possible on the basis of BACT or LAER requirements and the need for firms to know what control technology will be required for new construction or modifications to existing facilities.  Because of the lengthy time required to complete a permitting process and the rate at which control technologies are developing, the BACT or LAER requirements for a given project could change between the time that a project is proposed and the time that it is permitted.[3]

## Older, Dirtier Facilities Remain in Operation

When the NSR and PSD programs were enacted, Congress did not require that emission controls be placed on any existing facilities, in effect "grandfathering" these facilities, although the provisions left open the possibility that controls could be required on these older facilities as part of a SIP filed in a nonattainment area.[4] Such facilities were expected to reach the end of their operating lives about 30 years after their initial construction. Experience over the past 25 years, however, shows that older high-polluting facilities throughout the nation have continued to operate with minimal modernization or have undergone so-called lifetime-extension projects. These projects have been able to maintain the economic viability of the facilities well beyond their initial design lifetime without triggering NSR or PSD review and the addition of modern emission-control technology (see Box 5-3). In addition, placing controls on such facilities as part of a SIP has proved politically difficult. Facility owners and operators can demonstrate that the cost per ton of controls on such facilities, assuming a short remaining lifetime, is much greater than the equivalent cost for new facilities. Some maintain that the onerous nature of the NSR and PSD permitting process contributes to this state of affairs by raising bureaucratic hurdles against new construction. Others argue that industry keeps old facilities operating, sometimes unlawfully, to avoid the NSR and PSD requirements to use clean technologies, as discussed further below.

---

[3]Such requirement changes occurred in the South Coast Air Quality Management District during the 1980s. At that time, one of the BACT and LAER requirements for gas turbines was for the combustion devices to be designed and operated to produce low amounts of $NO_x$. Several companies were proposing to permit new gas turbines with low $NO_x$ combustors, as required. However, the district reviewed its BACT requirements and determined that selective catalytic reduction (SCR), not low $NO_x$ combustors, should be used as the BACT. That decision caused some of the projects with low $NO_x$ combustors to be dropped.

[4]One reason that new and modified emission sources were subjected to more stringent regulations than existing sources is that it was considered easier to develop an appropriate design standard for a facility that is being constructed or modified. Promulgating a design standard for existing facilities can represent a challenge to regulators because of the spectrum of operating conditions and preexisting emissions that are present in the field. Individual sources within a particular group, such as glass furnaces, might have significantly different operating temperatures and, consequently, different $NO_x$ emissions. If the sources were required to install the same technology or to obtain the same percentage of emission reduction, some sources would have higher emissions than others. Conversely, if a regulation were passed that required all sources to achieve the same level of emissions, sources that had low emissions would be able to comply with much less effort than other sources.

---

**BOX 5-3   The Grandfathering of Facilities**

When the CAA Amendments of 1970 were written, it was assumed that electric power plants would be decommissioned when they completed their expected functional lifetime. Instead, many older power plants have received extensive maintenance to extend their operation and remain in use. Although no specific exemption was created for these facilities, their extended operation amounts to a de facto grandfathering, and that term is commonly used to describe facilities with little or no emissions control that continue in operation well beyond their original expected lives. Table 5-2 provides a breakdown by vintage of $NO_x$ emission rates for coal-fired power plants operating in 1999 in pounds per megawatt hours (the amount of $NO_x$ emitted per unit of electricity generated) and the percent of total $NO_x$ from coal-fired boilers emitted by each vintage. The table indicates that some facilities have remained operational long after their expected 30-year lifetime and that these older facilities emit $NO_x$ at a much higher rate than newer facilities. They also emit a disproportionately large fraction of $NO_x$ relative to the power they produce, as indicated in the far right column of the table. Substantial reductions in emissions from coal-fired power plants could be obtained by retrofitting plants or by replacing them with newer coal-fired or gas power plants.

Although there are reasons for focusing regulation primarily on new facilities or major modifications—most notably the relatively lower cost to incorporate control technology in the planned construction—the net effect of de facto grandfathering of facilities can be a substantial source of emissions. The use of a total cap on emissions with emissions trading can provide a monetary incentive for older facilities to reduce emissions. (See discussion on advantages and challenges of such programs later in this chapter.)

---

TABLE 5-2   $NO_x$ Emissions from Coal-Fired Boilers in 1999 by Vintage

| Power Plant Vintage | Avg. $NO_x$ Emission Factors (lb/MW-hr) | % of Coal-Fired Electricity Capacity | % of Coal-Fired Electricity Generation | % of Total $NO_x$ Emitted | % of $NO_x$ Emitted per % of Electricity Generated |
|---|---|---|---|---|---|
| Pre-1950 | 7.44 | 1 | 0.4 | 0.6 | 1.50 |
| 1950–1959 | 5.97 | 15 | 12.9 | 14.9 | 1.16 |
| 1960–1969 | 5.95 | 20.6 | 18.8 | 21.6 | 1.15 |
| 1970–1979 | 5.37 | 37.6 | 38.6 | 40.1 | 1.04 |
| 1980–1989 | 4.09 | 23.5 | 26.6 | 21.1 | 0.79 |
| 1990–1999 | 3.55 | 2.4 | 2.6 | 1.8 | 0.69 |

Abbreviation: lb/MW-hr, pound per megawatt-hour.
SOURCE: Data from Burtraw and Evans 2003.

## Definition of Significant, Nonroutine Modification

Probably the most controversial aspect of the NSR and PSD programs is the set of rules, regulations, and guidance that EPA developed to determine whether a proposed modification to an existing facility would result in a significant net emissions increase of any regulated pollutant under the Clean Air Act and thus trigger NSR or PSD review. These requirements have been subject to considerable debate and litigation. EPA has always used a number of factors in determining whether physical changes in a plant constitute routine maintenance. Owners of sources have complained that the agency has not consistently applied those factors over time as Administrations have changed. Environmental groups have desired stricter rules and have claimed that owners of older high-polluting plants have thwarted Congress's intention by improperly labeling lifetime-extension projects as routine, thereby avoiding the requirement to meet PSD or NSR emission standards. This viewpoint was supported by the U.S. Court of Appeals of the Seventh Circuit in its landmark 1990 decision in the case of Wisconsin Electric Power Company (WEPCO) versus EPA. The court ruled that the "massive" overhaul of an existing WEPCO facility could not be exempted as routine maintenance. However, the WEPCO decision was viewed by EPA as an isolated case and did not substantially change its policy concerning lifetime-extension projects. Such a change did appear to occur in the late 1990s when EPA asked the Justice Department to file lawsuits against seven utilities that had completed lifetime-extension projects on numerous power-generating plants without submitting NSR or PSD permit applications.

## Reforming NSR

Since the early 1990s, EPA has discussed potential reforms for the NSR permitting process. The goal of these reforms is to make permit applications more straightforward and certain for the applicants, while continuing to provide the environmental protection required by the CAA. In 1996 and 1998, EPA published formal notices of proposed changes in NSR and PSD (61 Fed. Reg. 38250 [1996]; 63 Fed. Reg. 39857 [1998]). However, these proposals were never finalized.

The national energy policy enunciated by the Bush administration in 2001 called for another review of NSR and PSD, and in June 2002, the review recommended several changes to NSR that were substantially different from the NSR reforms proposed (by the Clinton administration) in 1996 and 1998. These new rules modify the requirements for NSR on an existing facility and will generally make it easier for projects to be undertaken at existing facilities without triggering NSR. A description of the new

NSR rules is available in the *Federal Register* (67 Fed. Reg. 80186 [2002]; 68 Fed. Reg. 61248 [2003]).

The new rules promulgated by the Bush administration have been controversial. Some states, local governments, and environmental groups have argued that the new rules do not adequately address the problem posed by older, grandfathered facilities (see Box 5-3). On the other hand, affected industries generally welcome the rule changes as improving the flexibility for permit reviews, providing certainty for decisions concerning routine repairs and maintenance, and ultimately making it easier to improve the efficiency of existing facilities and limit their pollutant emissions. The committee did not assess the potential impacts of these new rules. In the Consolidated Appropriations Resolution for federal fiscal year 2003 (Public Law 108-7), signed by the President in February 2003, Congress called for an independent evaluation of the impacts of the revisions to the NSR program by the National Research Council. The evaluation will be carried out by a different committee.

## OTHER TECHNOLOGY-BASED STANDARDS IMPOSED ON MAJOR FACILITIES

In addition to the requirements for new and modified emission sources, the CAA contains explicit requirements for the imposition of emission-based standards and emission reductions on specific types of sources regardless of whether they are new or being modified. Examples of these requirements are reasonably available control technology (RACT) within nonattainment areas, and maximum achievable control technology (MACT) for major sources of hazardous air pollutants (HAPs). The Acid Rain Program in the 1990 CAA Amendments focused primarily on the $SO_2$ cap-and-trade program but also included more modest and more traditional controls for reduction of nitrogen oxide ($NO_x$) emissions from existing facilities.

### Reasonably Available Control Technology

Beginning with the CAA Amendments of 1977, nonattainment areas have had to apply RACT to all major sources in their areas as part of their attainment-demonstration SIPs (see Chapter 3). RACT is determined by EPA through a process of developing control technique guidelines (CTGs) that take into account cost as well as other factors for each of a number of industrial facilities. To date, EPA has promulgated over 60 CTGs. RACT is generally implemented in each state in accordance with the CTGs. In some cases, states have gone beyond the CTGs in their RACT rules.

## The Acid Rain NO$_x$ Provisions

To address the problem of acid rain, Congress included provisions in the CAA Amendments of 1990 that were designed to reduce emissions of SO$_2$ and NO$_x$ (see Chapter 2). The SO$_2$ provisions, which are described in detail in the following section, set a goal for reduction of emissions by 10 million tons and then set specific caps on allowable emissions from individual facilities to achieve that goal. Unlike the SO$_2$ provisions, the 1990 CAA Amendments set a target reduction of 2 million tons of annual NO$_x$ emissions and relied on the promulgation of emission standards on specific facilities to attain the reductions. The requirements were implemented in two phases. Starting in January 1996, Phase I required use of low NO$_x$ burner technology in tangentially and dry-bottom-wall coal-fired power plants, producing reductions in emission rates of about 40%. Phase II, which began in 2000, extended emission standards to more types of coal-fired power plants. Though generally prescriptive, the program did allow some flexibility. Companies were able to average emission rates for commonly held facilities, but intercompany trading was not allowed. Facilities that did not achieve the standard after making specified investments could appeal for alternative emission limits. The imposition of Phase I of the Acid Rain Program's NO$_x$ performance-oriented emission standards has substantially decreased the emission rates of the affected facilities.

## Maximum Achievable Control Technology

As discussed in Chapter 2, the CAA Amendments of 1990 required EPA to establish emission standards that "require the maximum degree of reduction in emissions of the hazardous air pollutants . . . that the Administrator determines is achievable" for sources that emit more than 10 tons per year of any listed HAP or 25 tons per year of a mixture of HAPs. In response to that mandate, EPA established standards of maximum achievable control technology (MACT) for relevant HAP sources. In principle, the standards are performance standards, because each facility is only required to match the emission rate obtained using the MACT standard, and it may use any technology. However, the manner in which MACT standards are promulgated provides an incentive for companies to opt for the sanctioned technology and thus may ultimately have the affect of a design standard (see Box 5-2).

The 1990 CAA Amendments specified that the HAP standards covering 25% of the identified categories of emission sources be issued by 1994; another 50% by 1997, and the remaining 25% by 2000. However, this schedule has not been met. The promulgation of these standards has been delayed considerably from its original time frame, proving to be a source of criticism (Williams 2003) and litigation (Sierra Club v. U.S. Environmental Protection Agency, No. 02-1135 [DC Circuit]). As of February 2003, EPA had promulgated 79 MACT standards, affecting 123 source categories (T. Clemons, EPA, Washington, DC, personal commun., March 31, 2003). In May 2003, EPA indicated that it fully expects to finalize all MACT standards by the time states would be required to set MACT limits on a facility-by-facility basis according to Section 112(j), as amended (EPA 2003b). This section, referred to as the "MACT hammer," requires states to set MACT standards by facility-specific permit limits if EPA fails to set standards within 18 months of the statutory deadline.

Implementation of MACT standards developed so far is estimated to have already reduced HAP emissions by about 25% (EPA 2001a). Presumably, even larger reductions will be achieved when the program is fully implemented. The use of standards that directly (through a design standard) or indirectly (through a traditional performance standard) mandate the installation of specific pollution control technologies appears to have resulted in substantial reductions in pollutant emission rates (see Chapter 6). However, these technology-specific approaches have some deficiencies (see Box 5-2). They are not set up to minimize costs across all companies who must comply, nor do they provide an incentive for creative problem solving within companies that can result in the development of new, more efficient approaches to reducing emissions.

## EVALUATION OF TRADITIONAL CONTROL PROGRAMS FOR MAJOR STATIONARY SOURCES

Although the technology-specific control programs have considered a range of pollutants from the start, consideration has been segmented by the programs, such as NSPS, MACT, RACT, NSR, and PSD, at federal and state levels operating on different time frames and under different levels of stringency. The result has been to make it difficult for any one facility to implement multipollutant controls in a systematic and cost-effective fashion.

Perhaps most important, older plants have not had to make emission reductions in many cases, and the emissions from these older facilities provide a substantial contribution to the emission inventories of some pollutants (see Box 5-3). This situation has been caused by (1) a complex system of requirements for new, modified, and existing facilities that has provided incentives for not retiring or modifying facilities; (2) RACT and other rules that have either been less stringent than originally intended or not been energetically enforced by some states; and (3) the relatively high cost of retrofitting, thereby providing an incentive for operators facing cap and trade to purchase reduction allowances from others who can make the reductions less expensively (see cap-and-trade discussion below).

In addition, there are several concerns regarding HAP emission controls. First, there is a concern that MACT standards stifle the development of pollution prevention technologies (for example, product substitution and reformulation) that are more cost-effective than the predetermined control technologies that are the basis for the standards. Such development is encouraged in the 1990 CAA Amendments. Second, there is a question of whether the MACT hammer requirement of Section 112(j) of the Clean Air Act will need to be imposed. This requirement has been especially challenging for mercury emissions (see Box 5-4). Third, there is the challenge of compliance monitoring and enforcement for HAPs. Continuous emission monitors do not exist for most of them, and substantial contributions from fugitive emissions and temporary excursions from normal operating conditions ("upsets") are likely. Finally, questions remain concerning HAP con-

---

**BOX 5-4   Electric Steam-Generating Units Regulation for HAPs—Focus on Mercury**

The CAA treats HAPs from electric utility steam-generating units (EGUs) differently from HAPs emitted by other sources. Instead of subjecting EGUs to the generic regime for HAPs regulation, the CAA requires EPA to conduct a study of the potential hazards from utility air toxics and to determine whether regulation is "appropriate and necessary." In December 2000, EPA completed the study and made the required determination, concluding that regulation of HAPs from coal- and oil-fired EGUs is necessary and that mercury is an air toxic of most serious concern. A subsequent report by the National Research Council indicates that an estimated 60,000 newborns are at risk each year for adverse neurodevelopmental effects from in utero methyl mercury exposures (NRC 2000a).

EPA established a subcommittee of the CAA Advisory Committee to advise it on setting the standards for HAP emissions from EGUs. The advisory group met for well over a year but was unable to reach consensus on a number of key issues. Notwithstanding the difficulties, however, a settlement agreement in a court case set the schedule for EPA to promulgate the HAP regulations for EGUs. The schedule called for EPA to propose a rule by December 2003 and finalize the rule by December 2004. On December 15, 2003, EPA proposed two alternatives for controlling emissions of mercury from EGUs.[a] One alternative would require utilities to install controls known as maximum achievable control technologies (MACT) under section 112 of the CAA. The other alternative is to create a market based cap-and-trade program, which would require a revision of EPA's December 2000 finding that it is "appropriate and necessary" to regulate HAPs emissions from EGUs using the MACT standards. Given this schedule for rule promulgation, the CAA requires compliance by December 2007, although the Bush administration has indicated that it may extend the compliance date by 1 year.

---

[a]EPA proposed its options for reducing mercury emissions from EGUs after the committee completed its deliberations.

trol following MACT.  As discussed in Chapter 2, the 1990 CAA Amendments require EPA to establish additional emission standards following an assessment of residual risk.  Questions also remain about the science behind residual risk assessments:  the adequacy of emission inventories and dispersion and exposure models, the need for more timely updates to EPA's Integrated Risk Information System (IRIS), the uncertainty in dose-response relationships, and the lack of information on some health outcomes.

## COMPLIANCE ASSURANCE FOR TRADITIONAL CONTROL PROGRAMS

EPA and the states rely on two primary means to ensure compliance with the emission-control requirements described in the previous sections: on-site inspection and compliance monitoring.  To facilitate these efforts, the 1990 CAA Amendments enacted provisions for the Title V operating permit program, which was designed to record all relevant requirements and conditions in one document and to provide a fee structure for the states to provide resources to support the compliance efforts. The following sections discuss the Title V program, the means to ensure compliance (on-site inspections and compliance monitoring), and the challenge of off-normal emissions.

### Title V Operating Permit Program

Operating permits for major stationary sources are required under Title V of the CAA.  The Title V permit lists all the control requirements for a particular source to provide the necessary information for inspectors to verify compliance.  In addition, the Title V permits are public documents allowing public review of source-emission data and control requirements. The 1990 Amendments of the CAA set forth a schedule that called for all Title V permits to be issued by November 1997.

The Title V permits can play an important role in limiting emissions. Emission standards are usually written in terms of some unit related to production (for example, pounds of emission per million British thermal units [Btu] of heat input).  The actual emissions depend on the capacity of the unit and the hours of operation.  Operating permits can place limits on those factors, thus providing an effective limit on the emissions rate from a source.

Before the Title V provisions, permit programs in different states and local districts varied in their requirements.  State and local air pollution control officers were concerned about the Title V program during enactment and during promulgation of the rules, although they ultimately supported the program.  Some permitting agencies thought that the final EPA regulations implementing the Title V provisions were overly concerned with details. Agencies with already strong permit programs thought that the Title V regulations increased the complexity of the permitting process without improving the overall results. These agencies believed that the regulations were not sufficiently flexible to allow for effective local options.

One important element of Title V amendments was the provision that permitting agencies could collect fees for their costs associated with the permitting system. These fees have provided a useful source of income to agencies that allows them to develop a more effective permit program.

EPA data show that 36,953 stationary sources were potentially subject to the Title V requirements (EPA 2003j). Of these, 17,998 received revised permits that limited their operating hours or emissions below the level that classified them as stationary sources. Such sources, known as synthetic minors, became exempt from the Title V requirements, because they were no longer major stationary sources. Of the remaining 18,955 major stationary sources, only 14,247, or 75%, have received a permit as of September 30, 2002. The EPA inspector general issued a report in 2002 that examined the reasons for delays in the issuance of these permits (EPA 2002f). These reasons included delays in the promulgation of the initial EPA regulations for state and local Title V permit programs, delays in the EPA approval of proposed permit programs, and longer times required for review and approval of permits. There was also some confusion about the monitoring requirements for sources that were included in Title V. The delay in issuing guidance for specific monitoring requirements in the Title V regulations was also cited as a reason for the delay in the implementation of the Title V program.

The complexity of the effort required to issue these permits was underestimated at the time the 1990 CAA Amendments were passed. Sources and state agencies had to pay more attention to emission data, which were not accurately recorded for many sources, to ensure that sources were properly classified as major or synthetic minors. In some cases, the sources and the permitting agencies had to determine which of several contemporaneous requirements to include in the permit.

Part of the complexity is due to the nature of large stationary sources. Such sources have many emitting units that must be considered separately in the permit for the source. An example of a complex permit is one granted to an oil production company in California. This Title V permit, available on the ARB web site, contains 1,460 pages to describe the conditions applied to all the permit units in the source (CARB 2003b).

Overall, despite these difficulties, the Title V program has improved the application of operating permits nationally and given enforcement staff a better understanding of the specific requirements for individual permit units at a source.

## Government On-Site Inspections of Stationary Sources

Inspections by government officials (or government contractors) are a fundamental component of compliance assurance for stationary sources (EPA 1992). EPA and state inspectors typically conduct two kinds of inspections: "routine" inspections, which are not based on any suspicion of violation, and "for-cause" inspections, which target a particular facility because of suspicions that it is noncompliant. Inspectors conducting routine inspections observe visible emissions; examine data on control devices and operating conditions for comparison with those specified in the facility's permit; and review records and log books on the facility's operations (GAO 2001a). During a for-cause inspection, the inspector typically focuses on identifying possible sources of the emissions causing the inspection and specifying appropriate corrective action. The inspector is free to investigate other possible violations that he or she observes.

Federal and state inspectors perform over 17,000 routine inspections per year (GAO 2001a). In 1999, 88% of the 17,812 facilities routinely inspected by EPA and state officials were reported to be in compliance with permit requirements (GAO 2001a). However, a General Accounting Office report (GAO 2001a) concluded that routine inspections at large air-pollution-emitting facilities fail to detect significant noncompliance with permit requirements that can be revealed by more intensive but rarely undertaken investigations. In intensive investigations of three industries conducted by EPA, 75 of the 96 facilities investigated were not in compliance with their permit requirements (GAO 2001a).

## Compliance Monitoring of Stationary Sources

In addition to government inspections, compliance can be assessed through monitoring. Because of the limited enforcement resources available to state and federal agencies, most compliance monitoring is done by the individual companies themselves pursuant to state and federal legal requirements and various permit conditions. A variety of approaches for compliance monitoring are accepted by the nation's AQM system. These include (1) emissions estimation models, which do not actually monitor emissions directly; (2) periodic source testing; (3) parametric emissions monitoring (PEM); (4) continuous emissions monitoring (CEM); and (5) remote sensing.

### Emissions Estimation Models

This approach makes use of the same method as that used in the development of emission inventories (discussed in Chapter 3), in which emissions are estimated as the product of an empirically derived emission factor and a facility-specific activity factor. Because this method does not involve any measurements of actual emissions from the facility, it is not strictly speaking a form of monitoring. Nevertheless, given the high cost and unavailability of other monitoring methods, many of the regulated facilities in the United States rely on this method to calculate permit fees and report facilitywide emissions to agencies that develop the areawide and nationwide emission inventories. As discussed in more detail in Chapter 6, because this method assumes that emission controls placed on a facility are effective, it is not a scientifically robust way to document independently the success of AQM in reducing pollutant emissions.

### Periodic Source Testing

This method uses periodic emission measurements to assess emissions from a facility. Such an approach certainly provides a more accurate assessment of emissions at the time of the test than that from an emissions estimation model, but it is still limited by the need to extrapolate from the measured emissions (made under specific conditions) to the total emissions. Thus, the validity of periodic source tests depends on the accuracy of the data from the tests, the representativeness of the conditions during the tests, and the validity of the model used to extrapolate. Because such testing has to be prescheduled, periodic emission measurements are likely to be conducted during optimal conditions that are not necessarily reflective of conditions over the long run (GAO 2001a).

### Parametric Emissions Monitoring

Parametric emissions monitoring (PEM) applies computer models to data gathered from the source on various operating parameters—such as temperature, pollutant flow rates, and oxygen levels—to estimate emissions continuously, thereby providing a continuous record of a facility's compliance status (GAO 2001b). Two categories of PEMs have been developed: (1) those that apply first principles (for example, thermodynamics or the ideal gas law) to calculate emissions from operating parameters; and (2) those that predict emissions based on a profile of a source's pattern of past emissions (developed using neural networks or regression) as related to the operating parameters. The first approach is limited by assumptions needed to solve the complex equations involved, and the second depends on the facility and the sensors behaving in a way consistent with the historical data (EPA 2000c). EPA enforcement officials consider parametric procedures to be generally less accurate than either CEM systems discussed below or periodic source tests. Nevertheless, source facilities are allowed to use PEM to demonstrate compliance with emission standards if the monitoring of the process and the pollution-control-device operating parameters indicate that the facility is operating within a permitted range (40 CFR § 64.3). A common approach to determining compliance is a combination of PEM and periodic testing.

### Continuous Emissions Monitoring

A CEM system includes all equipment necessary to determine the rates of pollutant emissions, either by taking measurements within the stack or by extracting a sample from the stack for analysis, and to record the emissions on a continuous basis (Schwartz et al. 1994). CEM is potentially the most accurate method for measuring emissions and ensuring compliance. It is also the most expensive and the most technologically and operationally demanding. CEM systems that meet the requirements for regulatory purposes are available for carbon monoxide (CO), hydrogen chloride (HCl), $NO_x$, $SO_2$, carbon dioxide ($CO_2$), and oxygen ($O_2$), opacity, total reactive sulfur, hydrogen sulfide ($H_2S$), and volatile organic compounds (VOCs) (NRC 2000c; GAO 2001b; EPA 2003k). Instruments to monitor emissions of mercury, multiple metals, particulate matter (PM), and total nonmethane hydrocarbons continuously are being developed (Hunter et al. 2000; NRC 2000c; Seltzer 2000; GAO 2001b; Mitra et al. 2001). Extractive methods to measure several pollutants at once with Fourier transform infrared (FTIR) spectroscopy and gas chromatography (GC) are also available but not yet used in a regulatory setting, despite the availability of reference spectra for 100 of the 189 HAPs listed in Title III of the 1990 CAA Amendments (EPA 2003l).

CEM systems provide only a few percent of all determinations of stationary-source emissions for compliance purposes. Such systems are required by EPA for only a limited number of source categories; these include new waste incinerators, some chemical plants and petroleum refineries, utilities and large boilers subject to the Title IV acid rain emissions-trading program, and as a condition to consent decrees in enforcement actions (NRC 2000c; GAO 2001b). In all, about 6,750 CEM systems are in use in the United States (GAO 2001a). About 80% of those measure CO, opacity, $NO_x$, and $SO_2$, and less than 1% measure hydrocarbons and air toxics (GAO 2001b).

EPA enforcement officials in general find CEM systems preferable to parametric monitoring, and somewhat preferable to periodic source monitoring (GAO 2001b). EPA has, however, been reluctant to require regulated sources to install CEM technologies "because of time and resource constraints in issuing new rules, as well as the perception that advanced monitoring technologies are too expensive" (GAO 2001b). Companies have been reluctant to use CEM technologies because they are expensive and their use might reveal additional violations (GAO 2001b). The vendors of CEM technologies are likewise reluctant to put resources into more research unless they have some assurance that EPA will require companies to install them (GAO 2001b). At the same time, EPA must be confident that potentially expensive CEM technologies will work over the long haul before it prescribes their use. Thus, the relevant actors find themselves in a cycle in which promising monitoring technologies are not developed or implemented (see discussion on advances in environmental instrumentation in Box 7-5 in Chapter 7).

### Remote Sensing

Remote-sensing techniques quantify the concentrations of gases and particles emanating from a source by measuring the spectral properties of light waves that have interacted with these gases and particles. Remote sensors can be deployed at surface sites in the vicinity of a source, on aircraft, or even on satellites. They are particularly useful for detecting leaks or "fugitive" emissions, because they allow a large area to be rapidly sampled. Remote sensing has not been used widely for routine compliance monitoring (EPA 1992), although a recent report prepared for EPA identified several remote-detection technologies (see Chapter 7, Box 7-5) that, if used instead of portable monitors, could drastically reduce the costs of such programs (ICF 1999).

### Off-Normal Emissions

A basic assumption underlying technology-based standards (which include EPA's NSPS, LAER, BACT, RACT, MACT, and most SIP requirements) is that a well-operated and maintained source can achieve a specified emission standard or limit under all expected operating conditions by using control equipment that has been shown through a performance test to be capable of achieving that limit (62 Fed. Reg. 54900 [1997]). Emission-control technologies, however, are not perfect, and even the most advanced technology may experience off-normal conditions that cause emissions to spike upward. An agency seeking to regulate the emissions from a facility through the application of emission-control technologies has two alternatives: (1) set a lenient emission standard so that a facility can be expected to remain in compliance 100% of the time, including during off-normal conditions; or (2) set a stringent standard and allow for "upsets" or "excursions" in emissions during off-normal conditions identified in the relevant permits and compliance monitoring regulations. For the most part, EPA has elected the second option, often permitting an operating-emission range for a source. Determination of whether a source is in compliance then depends on the number and severity of emission excursions and "the particular circumstances at the source" that resulted in the excursions (62 Fed. Reg. 54900 [1997]). Although operationally efficient, this approach has some negative aspects. If a facility is allowed an unknown number of emission excursions of unknown magnitude, how can the citizens living and working in the vicinity of the plant be assured that their health is being adequately protected? Likewise, how can the operators of the plant determine when an excursion is sufficiently severe to warrant radical corrective action?

## CAP-AND-TRADE PROVISIONS FOR MAJOR STATIONARY SOURCES

Starting in the 1980s, air quality managers began exploring the use of market-based approaches to pollution control. This shift was motivated in part by the belief that market-based approaches would help level out the marginal costs of emission reductions among the affected facilities, dramatically reducing the overall costs of control. It was also believed that market-based approaches would provide greater incentive to innovate, which could in turn contribute to further cost savings and perhaps even greater emission reductions. Another distinct advantage of many market-based approaches is that they place a definite limit, or cap, on the aggregate emissions from a particular type of source or even from all sources of a pollutant. As a result, target emission levels are maintained even while economic and population growth continues.

The most commonly known market-based approach is a cap-and-trade program in which discrete emission quantities, such as a ton of a pollutant, are traded among sources. Trading takes place under an aggregate emission cap set by the regulating agency, presumably based on a determination of the maximum level of emissions consistent with protecting human health and welfare. Cap-and-trade programs have been shown to be effective at achieving emission reductions at much less cost to the regulated facilities than traditional technology-based or performance-oriented standards. However, there are potential disadvantages, some of which are discussed in more detail at the end of this section.

### The Acid Rain $SO_2$ Emissions Trading Program

Emissions trading programs were first used in the United States during the 1980s (see Box 5-5); however, the best-known program is that involving the trading of $SO_2$ emissions from electric utilities in the Acid Rain Program (Title IV) of the 1990 CAA Amendments. As discussed in Chapter 2, the goal of this program was to cap nationwide $SO_2$ emissions from electric utilities at 8.9 million tons beginning in 2000.

To accomplish the emissions reduction, Congress established a national cap on $SO_2$ emissions from electric utilities and authorized EPA to allocate an annual emissions allowance to each of the affected electricity-generating facilities. The allowance assigned to each utility is based on its historical rate of resource consumption (expressed in units of equivalent million British thermal units of energy contained in the fuel used annually) and entitles the holder to emit a specified amount of $SO_2$ in the year the allowance was issued or in any later year. If the initial allocation does not cover the desired emissions for a given utility, that utility has a variety of options. It can reduce its emissions to the allowable amount internally through the use of technologies to capture emissions before they escape to the atmosphere (for example, scrubbers), the use of renewable or alternative clean technologies, and conservation and pollution prevention (for example, the use

---

### BOX 5-5 Early Trading Programs Implemented in the United States

#### Emission-Reduction Credit Programs

Emission-reduction credit (ERC) programs implemented in the 1980s were the first emission-trading programs in the United States. These programs generally were organized around measuring and trading of emission rates rather than discrete units of emissions and shared some of four characteristics: (1) offsets, or emission reductions at existing facilities that must be obtained by new facilities in areas that fail to attain NAAQS; typically, the emission reductions at existing sources must be greater (usually by a factor of 1.3:1or more) than the new emissions; (2) bubbles, or average emission rates over several facilities in a region; (3) netting, which applies to expansion within a facility at existing sources; and (4) banking, which refers to intertemporal trading; banked credits are typically lost if a plant shuts down. The first-generation ERC program yielded about $10 billion of savings in capital costs but still yielded far less than the expectations of economists (Tietenberg 1990; Hahn and Hester 1989a,b).

One problem with the ERC program was the relatively high transaction costs associated with trades. State regulators also wavered in their commitment to the program. That undermined the willingness of parties to engage in trading (Liroff 1986). A final failing of the ERC program was that it translated an existing prescriptive program into a trading program without any associated gains for environmental quality. Hence, the program was perceived as solely in the interest of business, resulting in widespread mistrust on the part of environmentalists and some regulators (Hahn and Stavins 1991).

#### Lead Phase-Out

The phase-out of lead (Pb) in gasoline affected emissions from mobile sources (Nussbaum 1992). It was implemented at refineries, however, and thus

*(continued on next page)*

---

**BOX 5-5**   continued

affected stationary sources. Unleaded fuel was mandated to be available beginning in 1974, but implementation was eventually postponed until 1979. The
maximum allowable Pb was reduced to a maximum of 1.1 gram (g) per gallon
(gal) by 1982, to 0.5 g/gal by 1985, and to 0.1 g/gal by 1986. To achieve this
phase-out, a program of "interrefinery averaging" was initiated on a quarterly
basis from 1983 to 1986. Banking across quarters was allowed as the average
allowable Pb content was reduced. Several hundred refineries, representing
about half of all refineries affected by the phase-out, participated in the trading
program. Savings amounted to about 20% of program costs (EPA 1985). The
trading program enabled political support for the Pb phase-out to take shape.
There is also compelling evidence that the trading program accelerated the process of technological change that lowered costs, an important lesson for other
regulatory programs (Kerr and Newell 2001).

### Chlorofluorocarbon Phase-Out

The Montreal protocol and international agreement to protect stratospheric $O_3$
required national reductions in use and eventual phase-out of chlorofluorocarbons
(CFCs) and halons, gases that cause depletion of stratospheric $O_3$. The trading
programs to achieve these reductions were implemented nationally (with the exception of trading within the European Union). The trading program implemented
in the United States was innovative because, under Title VI of the 1990 CAA
Amendments, it recognized that different chemicals had different $O_3$-depleting
potentials (ODPs), and the chemicals were assigned different weights, depending
on their ODPs. Subsequently, this program was supplemented by a tax on virgin
CFCs. The tax was introduced principally to capture the profits that accrued to
holders of permits that were created by the scarcity of permits. Ultimately, the tax
may have been the most effective instrument in the CFC phase-out in the United
States.

---

of low-sulfur fuels). The utility can also elect to meet its emission allocation
in part or in whole by purchasing additional allowances from others with a
surplus. In exchange for the flexibility given to emission sources to meet
their emission allocations, each source was required to install CEM systems
and report emissions to EPA on a quarterly basis. EPA is responsible for
managing an allowance tracking system and for ensuring compliance with
the program. Failure to comply results in a $2,000 per ton fine and a
requirement to reduce emissions the next year by an additional ton for each
excess ton of $SO_2$ emitted.

Congress designed the $SO_2$ trading program to be implemented in two
phases. Phase I began in 1995 and affected electricity-generating facilities
with a capacity of 100 megawatts (MW) or greater and with an emission
rate of 2.5 pounds or more of $SO_2$ per million Btu of heat input (GAO

2000). Phase II began in January 2000 and expanded to include all electricity-generating facilities over 25 MW. At the beginning of the Phase II program, annual electric utility emissions of $SO_2$ were capped at 9.2 million tons. By 2010, the cap is scheduled to drop to 8.95 million tons, which will be equivalent to a 10-million-ton or a 50% reduction in emissions from electricity generation when compared with those in 1980.

### Economic and Emissions Performance of the $SO_2$ Trading Program

The $SO_2$ emissions trading program provided the large reductions in $SO_2$ emissions from stationary sources and at much less estimated cost than would have been obtained from traditional technology-based control strategies (GAO 2000). EPA reported that all companies maintained compliance from the inception of the program through 2000; some noncompliance occurred in 2001 apparently as a result of simple and small accounting errors (EPA 2001d). A decrease of 3 million tons of $SO_2$ was seen in the beginning of the program (EPA 2001d), and the total achieved reductions from Phase I was approximately 22% over the base allocations (see Figure 5-1). The beginning of Phase II saw an additional 1-million-ton decrease in $SO_2$, bringing the nationwide total emissions to 11.2 million tons. This

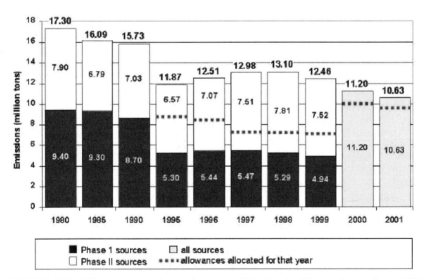

FIGURE 5-1  $SO_2$ emissions from electric utilities in the United States from 1980 to 2001. The emission reductions between 1990 and 1995 are attributable to actions taken to implement the acid rain section of the CAA 1990 Amendments and have been documented by continuous emission monitors. SOURCE: EPA 2002l.

total was slightly higher than the cap for the year 2000, as sources began drawing on their banks of unused emission allowances accumulated from overcontrol during Phase I, but the total was still lower than that during any year of Phase I. A detailed discussion of the spatial pattern of these reductions is included later in this chapter.

Overall, EPA believes that the $SO_2$ cap-and-trade program has produced more reductions more rapidly and at a lower cost than anticipated when the legislation was passed. At the time of its passage, it was estimated that Phase I and Phase II would cost $4.6 billion per year. However, current estimates project total costs at only $1 billion (see Box 5-6). The compliance flexibility allows facilities with high-cost controls to purchase

---

### BOX 5-6   Savings from the $SO_2$ Emissions Trading Program

The savings obtained in the cap-and-trade program to reduce $SO_2$ emissions from electric utilities appear to have been substantial. At the time of enactment of the program mandated in Title IV of the 1990 CAA Amendments, the costs for the program were projected at $4.6 billion per year. However, as has often been the case for the emission-control programs mandated in the CAA, the actual costs have been substantially less than first projected. It is now estimated that when the bank of emission allowances achieves equilibrium in about 2010, the cost of the program will be just over $1 billion per year.

Is that cost a significant savings over the costs of more conventional approaches? The answer appears to be yes, although estimates of the cost savings depend on the baseline against which the program is compared. Compared with programs using performance standards, the emission-allowance trading program is estimated to have reduced implementation costs by 30–50% (Carlson et al. 2000; Ellerman et al. 2000). Compared with a prescriptive technology approach, such as the requirement of scrubbers at a certain class of facilities, the trading-program savings have been estimated to be substantially higher (Carlson et al. 2000).

There has been a tendency to overstate the savings of the trading program. That stems from a failure to recognize a trend in the electricity, coal, and railroad industries toward greater use of low-sulfur coal at existing facilities, thus reducing $SO_2$ emissions without new regulations in 1990. The $SO_2$ trading program enabled the industry to capitalize on this trend, however, by making use of the flexibility inherent in permit trading. A program based on the imposition of performance standards also could have taken advantage of the fortuitous trend in the availability of low-sulfur coal (although probably not as thoroughly as a cap-and-trade approach).[a] A prescriptive program (for example, one that requires the installation of scrubbers) would not have gained a similar benefit.

---

[a]The innovation of blending coals with different sulfur content provided savings for many companies, but it would not have been a fully useful strategy under a performance standard, because it did not achieve a low enough emission rate at any one facility, thus requiring a facility to also install control equipment. Trading enabled emission-control decisions to be made in the most cost-effective manner.

allowances from others with low-cost compliance options. A portion of the cost decrease can be attributed to the financial incentive to improve scrubber technology. About 40% of the reductions came from the use of scrubbers. Another important low-cost strategy estimated to be responsible for about 60% of the reductions came from switching to low-sulfur coal—a switch that was greatly assisted by the deregulation of railroad transport, which made western low-sulfur coal cheaper (Ellerman et al. 2000). This particular option for low-cost control may not be available in the trading of every pollutant; however, the potential financial incentives under a cap-and-trade program can be expected to result in the reduction of costs in unanticipated ways.

The features that appear to have contributed to the successful design of the $SO_2$ trading program include large emission reductions, simplicity, effective monitoring, transparency, certain penalties, and the opportunity for banking emission allowances.

- *Substantial emission reductions:* The overall cap set by Congress for $SO_2$ emissions represented a 50% reduction in emissions nationwide. Such a large emission-reduction mandate was most likely key to ensuring that trading between regions did not produce spatial pockets of emission increases.
- *Simplicity:* The $SO_2$ trading program set a clear aggregate cap on the total annual allocation of emission allowances for all large fossil-fuel-fired electricity-generating facilities and allowed nearly unfettered opportunities to trade or bank emission allowances.
- *Availability of CEM systems:* CEM systems have come at a substantial cost and might not have been required technically, because engineering formulas could reliably predict $SO_2$ emission rates from the sulfur content of coal. However, mistrust between environmental advocates and the regulated community led to uncertainty about when and whether post-combustion controls, such as scrubbers, would operate. Compliance assurance under trading required measurement of actual emissions. The technology to do so, as well as the computing capability to process large amounts of emission data, emerged just in time to facilitate the $SO_2$ trading program.
- *Transparency:* The electronic allowance tracking system maintained by EPA posts allowance holdings and transactions and emission data in a prompt fashion. The resulting transparency facilitated the simplicity of the $SO_2$ trading program, which in turn contributed to its overall success.
- *Certainty of penalties:* Noncompliance triggers a prespecified financial penalty plus the surrender of a number of allowances for a subsequent year.
- *Opportunity for banking emission allowances:* This opportunity has contributed greatly to the cost savings of the $SO_2$ trading program (Ellerman et al. 2000). Banking provided a mechanism for firms to insure

against adverse conditions caused by fuel markets or their own compliance activities. In addition, it facilitated "buy-in" by some of the regulated parties into the program design, because banking endows them with an asset that is only of value when the program is successful and stable.

## $NO_x$ Emissions Trading Programs

### Regional Clean Air Management

The South Coast Air Quality Management District (SCAQMD) implemented the Regional Clean Air Management Program (RECLAIM) in 1994 to stem the tide of rising industrial emissions in the Los Angeles Basin. RECLAIM was initially designed to control stationary-source emissions of $NO_x$, $SO_2$, and certain reactive organic gases by using cap-and-trade provisions. However, reactive organic gases trading did not function because of the difficulty in establishing a baseline for allocating allowances, and $SO_2$ trading was problematic because of emissions being concentrated at a small number of power-generating facilities. As a result, RECLAIM focused on trading $NO_x$ emissions. The program operates under a total emissions cap, which declines at a rate of 5–8% per year. When fully implemented in 2003, overall emissions are expected to be reduced by almost 80%.

The $NO_x$ program encountered difficulties because of two unforeseen eventualities:

• The initial allocation was based on a period of high economic activity. When the economy slowed, companies found themselves endowed with substantial excess allowances. Interyear banking of emission allowances was therefore prohibited.

• The year 2000 was pivotal for the program because it was the first year when aggregate annual allocations would fall below aggregate emissions, thereby requiring actual sectorwide emission reductions. Unfortunately, most companies had not elected to make investments in control technologies before 2000. At the same time and for reasons unrelated to RECLAIM (for example, deregulation of the electricity market and alleged market manipulation by suppliers), there were widespread problems in electricity distribution in California in the summer of 2000. The summer of 2000 was also unusually hot in California, and that led to a high demand for electricity. The confluence of the factors, along with the absence of a banking mechanism in RECLAIM, created difficult challenges for the program. Electricity demand spiked, and to prevent widespread power disruptions, the participation of power generators in the RECLAIM program was suspended so that these facilities could operate at rates that caused them to exceed their emission allocations.

In addition, EPA recently reviewed the RECLAIM program (EPA 2002m). The lessons EPA derived from RECLAIM for other market-based programs include the following:

- Market-based programs require substantial planning, preparation, and management during development and throughout the program's life.
- Market information is a key factor affecting decision-making at individual facilities.
- Regulators should strive to create confidence and trust in the market by making a full commitment to the program and ensuring consistency between the market and their policies.
- Unforeseen external circumstances and policies (such as energy deregulation) can have dramatic impacts on market-based programs. Therefore, these programs must be designed to react quickly and effectively to unforeseen external factors.
- Periodic evaluation, revisiting of program design assumptions, and contingency strategies are crucial for keeping programs operating effectively.
- RECLAIM's experience seems to demonstrate that cap-and-trade programs can work with new-source review. That finding may be a function of the types of sources included or the controls in place at many facilities.
- Regulators need to have a strong understanding of the regulated facilities and the factors affecting their decision-making.

Since 2000, the SCAQMD has substantially amended and imposed the RECLAIM rules (SCAQMD 2001), and trading prices have returned to pre-2000 levels. At this writing, SCAQMD is actively considering seeking additional $NO_x$ reductions from RECLAIM (SCAQMD 2003).

## Northeast $O_3$ Transport Region $NO_x$ Budget Trading Program

This program was developed by the states of the Ozone Transport Region (OTR) to supplement the $NO_x$ emission reductions mandated under CAA provisions for RACT in Title V and the Acid Rain Program of Title IV, which did not adequately address the regional transport of $NO_x$ and $O_3$ and the resultant nonattainment of the $O_3$ NAAQS in the region. The OTR mandates the use of a cap-and-trade system involving large stationary sources to reduce $NO_x$ emissions during the May-to-September $O_3$ season. It aims to reduce total emissions by 55–65% in the 1999–2002 interval and by 65–75% starting in 2003. The program was initially designed to have two trading zones to avoid possible regional disbenefits of trading but was finally implemented without constraints to maximize cost savings. Interyear banking of allowances is limited in the program through a provision known as "progressive flow control," which limits the size of the aggregate bank. (Banking is discussed later in this chapter.)

## NO$_x$ SIP Call Trading Program

The next bellwether in emissions trading is expected to be the initial implementation of the NO$_x$ SIP call trading program in 22 states and the District of Columbia during the 2004 summer O$_3$ season. The program will target large stationary sources of NO$_x$ emissions, mostly electricity-generating facilities, by requiring that the affected states revise their SIPs to achieve NO$_x$ emission-reduction targets assigned by EPA. EPA will allow, but not require, states to participate in an interstate trading program. This program will subsume the current OTR trading program and is expected to meet or exceed the annual cost of the SO$_2$ Acid Rain Program. The program imitates the SO$_2$ trading program in many respects, but like the OTC program, it has limited opportunity for inter-annual banking of emission allowances.

## Cap and Trade in Proposed Multipollutant Legislation

At this writing, new programs are under consideration that would mandate further multipollutant emission reductions from the electric utility industry in a cap-and-trade program. Among others, the Bush administration has proposed the Clear Skies Act, which would set caps and provide a trading program for SO$_2$, NO$_x$, and mercury emissions. Senator James Jeffords, Senator Tom Carper, and other senators and congressmen have proposed programs that would extend the emission controls to include CO$_2$ and would mandate shorter times and steeper goals for compliance. For the first time, these proposals offer the opportunity to apply multipollutant control measures by using cap-and-trade techniques in an entire electricity-generating industry sector. If properly structured, the programs have the potential to make a valuable and relatively cost-effective contribution to the nation's AQM system.

Key elements of the proposed legislation are highly controversial, and the different versions may have substantially different effects on emissions, on compliance strategies used by the regulated facilities, and on regions of the country. Some important differences in these proposals are whether to regulate CO$_2$, whether to include mercury in the trading system or mandate reductions at each facility, and the method for allocating permits. This committee was not charged to evaluate these proposals in depth and has not done so. In designing the proposals, the challenges and opportunities discussed in the next section may prove useful in ensuring that any proposal is optimally designed and implemented.

### Evaluation of Cap-and-Trade Approaches to Air Regulations

The advent of cap-and-trade systems over the past decade has provided an opportunity to achieve substantial reductions in emissions at a cost that appears to be substantially below that required in a traditional emission-control program. At the same time, these programs have met with varying success. Even in the case of the acid rain $SO_2$ emissions-trading program, which has been the most successful application, there are issues to be addressed and improvements that could be made. The following identifies the major issues and opportunities for ensuring that cap-and-trade programs can be an effective and growing part of the AQM system.

### Spatial Redistribution of Emissions

One of the major reservations expressed about cap-and-trade programs is that they may produce spatially and temporally heterogeneous patterns in emission reductions with undesirable environmental consequences. Although more conventional prescriptive approaches can address regional and seasonal issues by defining technology or performance standards that are more restrictive in areas where or, at times, when environmental problems are more critical, the least costly trading programs allow trading across regions and banking of emission allowances without regard to the possible environmental consequences. Such trading might worsen ecological hot spots or increase the number of persons exposed to pollution. That possibility is especially of concern for toxic air pollutants. Spatially heterogeneous emission reductions are also of concern for a pollutant like $NO_x$, whose emissions can have little or no effect on $O_3$ pollution in one region of the nation and a dominant effect in another. On the other hand, heterogeneous emission reductions are not of concern for nonreactive, long-lived gases, such as $CO_2$ and methane ($CH_4$).

Some analysts of cap and trade point out that there is little possibility that any given area will have negative impacts from the program, provided the cap is set low enough to reduce emissions by a large percentage (Burtraw and Mansur 1999). They also point out that emissions trading cannot be used to avoid meeting NAAQS (or other health and safety regulations) and that states retain the authority to set tighter emission limits to meet NAAQS. Both Massachusetts and Connecticut have adopted more stringent $SO_2$ emission levels for electricity generators for that reason, and New York is in the process of developing them. These arguments appear to be largely borne out by the results of the acid rain $SO_2$ emissions trading program, which set a stringent national emissions cap on a criteria pollutant. As illustrated in Figure 5-2, the acid rain emissions trading program has achieved $SO_2$ emission reductions in each of five major multistate regions of the contiguous

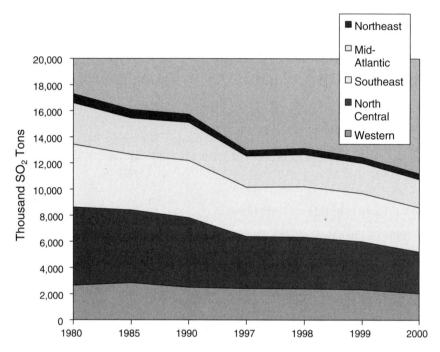

FIGURE 5-2    Regional $SO_2$ emission from electric utilities. Northeast Region: CT, ME, MA, NJ, NY, NH, RI, and VT; Mid-Atlantic: DE, MD, PA, VA, WV, and DC; Southeast: AL, FL, GA, KY, MS, NC, SC, and TN; North Central: IL, IN, MI, MN, OH, and WI; and Western: AZ, CA, LA, NM, OK, TX, IA, KS, MO, NE, NV, CO, MT, ND, SD, UT, and WY.  SOURCE: EPA 2001d.

United States. Even more significant, regions with the highest emissions, such as the north-central region, have had the largest reductions.[5] On the other hand, some emission increases have occurred on smaller geographic scales (for example, some states have had emission increases) (NAPAP 1998).

To guard against even the possibility of regional disbenefits, a modified cap-and-trade program with "zones" can be adopted.  For example, the RECLAIM program in California allows trading between two zones but only in one geographic direction.  This approach is guaranteed to prevent an increase in emissions in the zone that has the greatest potential to affect

---

[5]The pattern of trading also appears to have added slightly to the aggregate benefits to ecosystem health, but by far the most significant benefits have been realized as a result of the dramatic overall reduction in emissions (Burtraw and Mansur 1999; Swift 2000).

ambient air quality. However, the existence of multiple zones serves to reduce the potential number of trades and to raise the cost of environmental regulations. Indeed, the original design for the acid rain $SO_2$ emissions-trading program would have divided the nation into two trading zones to prevent trades that might worsen ecological problems in northeastern states. However, policy-makers recognized that a more constrained market would increase the cost, and they opted for the uniform national market approach currently contained in the statute (Hausker 1992). As discussed above, that decision does not appear to have caused substantial problems.

### Banking Emission Allowances for the Future

Another major issue concerning the environmental performance of market-based programs is the role of banking emission allowances. Though not as common, prescriptive approaches can explicitly account for temporal issues, for example, by placing restrictions on open burning or implementing seasonal $O_3$ controls. The opportunity to bank emission allowances from one time period to the next can contribute substantially to cost savings, because it can facilitate a rational investment plan across facilities. Some facilities may be retrofitted early if they are utilized heavily. The emission allowances that are not used at that facility may be used to keep another with a shorter remaining operating life running until it is retired, thus avoiding expensive retrofits. Banking can provide an important element of economic stability to a permit market (for example, the absence of banking for the RECLAIM program was one of the factors contributing to its instability). However, the accrual of banked emission allowances can distort the perception of environmental performance under a trading program. In the early years, emissions appear to be below their allowable level, suggesting unexpected environmental success. In later years, when the banked allowances are drawn down, emissions appear to be above the allowable level, suggesting that the program is unsuccessful.

Banking within a single emissions season can also have beneficial or adverse effects. For example, seasonal limits on $NO_x$ emissions that allow emissions to vary in a manner that might enhance $O_3$ concentrations over the summer $O_3$ season can be temporally mismatched with peak $O_3$ concentrations that result from the reaction of $NO_x$ and VOCs during specific episodic meteorological events (for example, warm stagnant conditions). More recent cap-and-trade programs, such as the Ozone Transport Commission's $NO_x$ budget, contain flow-control provisions that attempt to limit increases in emissions during meteorological conditions conducive to high $O_3$ concentrations by discounting the value of allowances when a high number of allowances are banked.

## Fairness in Allocating Emission Allowances

For a market-based trading program to be effective, it must be per-ceived as being fair. Key to a fairness perception is the method used to assign emission allowances. In most previous programs, including the acid rain $SO_2$ emissions-trading program, fairly simple principles served to guide the regulatory process. The general principle was to allocate emission allowances on the basis of historical generation (heat input at the plant). This grandfathering approach provides cost-free allowances to larger emit-ting sources that will be most affected by the program. That can result in giving larger allowances to those facilities that have not taken action to reduce—in effect penalizing those utilities that have taken earlier actions.

A second method is an allowance allocation program based on output. Fixed numbers of allowances are allocated on the basis of the percentage of electricity (or other product) generated by each source. Typically, the allocations are updated as the generation mix changes. Many stakeholders argue that this method encourages greater efficiency than the grandfathering method (EPA 1999g).

A third method is to hold an auction where all affected sources must go to purchase allowances to cover their emissions. An analysis of allocating $CO_2$ allowances under the three methods found that the auction is the least costly to society (Burtraw et al. 2001b). A series of important general equilibrium studies over the last decade also focused on the virtues of an auction for allocating emission permits (Goulder et al. 1997, 1999). The social cost of an allowance auction is expected to be dramatically less than allocation at zero cost. The reason is that the regulatory program (regard-less of how allowances are allocated) raises costs similar to a new tax and thereby serves to exacerbate distortions from preexisting taxes. However, an auction provides a source of revenue that can be used to reduce preexist-ing distortionary taxes. The main criticisms of auctions are that they increase the cost of a program from the viewpoint of companies, which have to pay for emission allowances, as they pay for other inputs to produc-tive activity, such as fuel and labor.

## Setting and Revising the Emissions Cap

The concept of placing an overall limit on total emissions was an important innovation in the cap-and-trade approach in environmental policy. A shortcoming of the approach as currently practiced is the absence of a mechanism for changing the cap in response to new scientific and economic information. During the 1990s, for example, the cost of control-ling $SO_2$ emissions from coal-fired power plants fell to less than half of the amount predicted at the time of the 1990 CAA Amendments (Ellerman et al. 2000). Moreover, as will be discussed in Chapter 6, there is growing scientific evidence that $NO_x$ emissions contribute more to the deleterious ecological effects of acid rain than had been recognized at the time of enactment of the 1990 Amendments, an understanding that came after the relatively modest target set for reducing $NO_x$ emissions in the acid rain rule. Despite these new understandings, regulators have not been able to adjust the $SO_2$ emission cap and $NO_x$ emission targets assigned by Congress in Title IV (Zuckerman and Weiner 1998), and any further adjustment now awaits the passage of the multipollutant legislation discussed above or some other legislation.

An alternative to a firm cap would be one that was adjusted in response to new information. One example is the cap on specific $O_3$ depleting substances. The cap was dramatically revised downward when new scientific evidence was discovered, because the initial cap definition did not prevent change in the cap.[6] Others have suggested similar trigger mechanisms on emission caps to provide economic relief if costs are greater than expected (Pizer 2002), but such an approach might better be coupled with a mechanism that provides environmental improvement when costs are less than expected. One possibility is a cap that is set to decline over time in anticipation of new technologies and less expensive ways to achieve emission reductions. However, if the cost of reductions exceeded a ceiling, the cap would be frozen until costs again fell below the ceiling. The value of a tradable emissions allowance could serve as an index for such a mechanism.

## Implicit Emission Increases Following Transition to a Trading Program

Another concern about market-based programs, but one that can be addressed in their design, is that an important source of cost savings has to do with implicit emission increases at some facilities under market-based trading compared with those under prescriptive regulation (Oates et al. 1989). Under a prescriptive regulation, each facility must meet or exceed an emission-reduction target. Because of the nature of emission-control investments and the liabilities incurred from noncompliance, a facility will generally exceed the target reduction, meaning that emissions will be less than allowable. An emissions-trading program provides the flexibility to overcomply at one facility and to apply the unused emission allowances to increase emissions at another facility. Even if facilities have identical costs, when faced with uncertainty about production and emissions, a group of facilities can come closer to meeting allowable aggregate emissions with the

---

[6]Another example is fisheries, which define a tradable right to be a share of the cap (not a specific number of tons). In this way the cap could change, and the allowances could be changed automatically in response (see Tietenberg 2002).

same level of collective risk of noncompliance by managing their operations as a portfolio. The slack in a regulatory program, defined as the difference between the allowable emissions and the actual emissions, inevitably will be lessened by the move from a prescriptive to a market-based approach. This may be of environmental concern; in any case, it may provide justification for lowering allowable aggregate emissions under a market-based program.

## Compliance Assurance and CEM

Because the market ultimately determines the locations and times at which emission reductions take place, market-based approaches, such as cap and trade, require a greater investment in monitoring than other regulatory approaches (EPA 1992). In the case of the acid rain $SO_2$ emissions-trading program, Congress recognized this concern and mandated that all affected facilities install and operate continuous emissions monitoring (CEM) systems to document compliance. In that program, monitoring is required to collect data at 15-minute intervals and report consolidated data on an hourly basis (CAA § 412(b),(c); 42 USC § 7651k(b)(c); GAO 2001b; Swift 2001). CEM systems also played a critical role in the South Coast Air Quality Management District's RECLAIM trading program for $NO_x$ and $SO_2$ emissions (GAO 2001b). The importance of CEM systems raises a challenge for extending the cap-and-trade approach to other pollutants for which cost-effective CEM technology is not available.

## OTHER TRADING AND VOLUNTARY STATIONARY-SOURCE PROGRAMS

In response to a desire to achieve emission reductions in the most cost-effective manner, a number of other approaches have been proposed, tested, and in some cases implemented. These approaches include open-market and other noncapped forms of trading (described briefly in Table 5-3) and voluntary programs, such as Project XL. These efforts are reviewed briefly below.

### Open-Market and Other Forms of Trading

Open-market trading represents an alternative to cap-and-trade emissions trading. In this approach, there is no aggregate emissions cap, usually because there is not a meaningful inventory of emission sources or a mechanism to monitor their emissions. The program is intended to provide an incentive for companies and facilities to reduce emissions voluntarily by participating in the program. Several states, including Michigan, Texas, and New Jersey, have implemented open-market trading programs for VOCs (NAPA 2000). To participate, a company must first establish its

TABLE 5-3   Open-Market and Other Noncapped Forms of Trading

| Advantages | Disadvantages |
|---|---|
| **Open-Market Emissions Trading**[a] | |
| • Helps to identify the emission inventory | • Ill-defined market because emissions are not capped |
| | • The prospect of new entrants increases economic risk to those participating in the program[b] |
| | • The veracity of emission baselines and reductions may be doubted |
| **Pollution Offset Trading**[c] | |
| • Allows economic growth without emission increases | • Administratively complex to implement |
| | • Generally unable to achieve the environmental goals at the least cost |

[a]An emissions trading program without an aggregate emissions cap and without mandatory enrollment. Each source must estimate its baseline emissions and implement an emissions monitoring program.

[b]The economic risk to those currently participating in the program is that their investments in pollution reduction equipment could be rendered uneconomic due to an infusion of low-price permits. Such a concern could undermine the incentive to make investments in emission reductions.

[c]When a source can increase its emissions only by obtaining tradable permits sufficient to offset its impact on ambient pollution, offset trading might facilitate the permitting of the source.

individual emissions inventory baseline. The company must then establish a mechanism for monitoring its emissions on an ongoing basis. If the managers of a facility believe that they can reduce emissions at a cost that is below the marginal cost in an established trading program, they have the incentive to join the program and attempt to sell some of the emission reductions achieved.

The primary argument for open-market trading is that it provides a way for expanding market-based emission-control programs to reach sources beyond the large stationary sources that are amenable to cap and trade. However, open-market trading has been widely criticized by economists and environmentalists. Economists have criticized the lack of well-defined markets in such programs because emissions are not capped. The possibility of new facilities entering open-market trading and bringing substantial emission permits at low cost is a deterrent to investments by facilities that already participate in the program. Environmentalists have criticized open-market trading because it provides an opportunity for a trading program to collect emission permits for supposed reductions in emissions that might have occurred anyway. Furthermore, the establishment of an emission baseline is problematic, and many doubt the veracity of baselines that are established under existing programs. Indeed, recent problems identified with the open-market trading in New Jersey (EPA, 2002n) suggest that greater attention must be paid to documentation of baselines and to compliance monitoring if such programs are to succeed in reducing emissions and gain the trust and support of all parties.

## Voluntary Programs to Improve Permitting Processes

In response to concerns that the current stationary-source permitting process inhibits new and cost-effective emissions control, several innovative permitting approaches have been attempted or proposed over the past decade. These generally have been directed toward (1) improving the administrative efficiency in the permit development and review process; (2) increasing flexibility and responsiveness for the regulated sources, recognizing the need for rapid modification of production processes and products in response to technology and market developments; and (3) improving the cost efficiency and the environmental benefits expected to result from the prescribed emission controls. Most of the proposed and experimental approaches have not been widely adopted, but they demonstrate a continuous search for improved efficiency in the permitting process. Some of these approaches are embodied in the NSR reforms discussed previously. In addition, EPA launched Project XL as part of the Reinventing Government Program initiated in 1995. Project XL encouraged demonstrations of flexible, multimedia (not limited to ambient air emission control), facilitywide permitting. Although several worthwhile demonstrations were initiated under the project, it was discontinued within 5 years and had little national impact on permitting practice. Nevertheless, the Project XL experience is interesting because it demonstrates the difficulties inherent in any attempt to generalize flexible permitting approaches, frequently because the operating details of single sources and the interests of the various local stakeholders are unique to each situation.

Although EPA has authority in many cases to make innovative revisions to its regulations to enhance their performance, the limited progress achieved by Project XL, as well as several other flexible permitting initiatives by EPA, states, and other stakeholder groups, was partly due to provisions in existing legislation that make such experiments difficult. In some cases, legislative guidance rather than regulatory experiments may be needed to achieve nationwide improvements in permitting efficiency.

## AREA-SOURCE REGULATIONS

Although emissions from one area source are generally small, the total emissions from all area sources can be significant. Area sources appear to be especially important contributors to $O_3$ pollution (via the emissions of VOC) and PM. In recognition of their importance, states first began working together to develop rules to control VOC emissions from some types of area sources in the late 1980s. For example, California and other states began adopting product formulation requirements for low-VOC architectural coatings. These were eventually translated into national requirements in the 1990 CAA Amendments, which contained several provisions requiring EPA to identify and regulate area sources of such products. A major impediment to making progress on area-source emissions arises from the large number of uncertainties associated with emission inventories for these sources. Specific challenges include the many sources in any given category and the wide variation in the conditions and operating practices under which the emissions can occur. To address the problem, the Emission Inventory Improvement Program (EIIP) includes the development of inventories for area sources (EPA 2001b), but it is unclear, barring a more ambitious program of measuring and testing, whether such an inventory-development process will result in a more accurate understanding of the contribution of area sources to air pollution in the United States.

Section 183(e) of the 1990 CAA Amendments required EPA to conduct a study of consumer and commercial products by 1993 to identify products that needed regulation and then to proceed with developing control technology guidelines (CTGs) for each priority category. EPA completed the study in 1995, identifying four groups of consumer and commercial products, including such items as aerosol spray paints, architectural coatings, and industrial cleaning solvents. EPA proceeded with developing CTGs for the categories of products within those groups, albeit slowly. Regulations for the first group were scheduled to be in place by 1997, and regulations for all four groups were to be completed by 2003 (64 Fed. Reg. 13422 [1997]). Currently, however, only three of the six product categories within the first group have been regulated; no regulations for the other three groups (scheduled for completion 1999, 2001, and 2003) have been completed. EPA noted in its 1999 *Federal Register* notice (64 Fed. Reg. 13422) that it "intends to exercise discretion in scheduling its actions under section 183(e) in order to achieve an effective regulatory program." The slow pace appears to be due to the perception that such rules are not central to efforts to attain the NAAQS for $O_3$. However, in the absence of high-quality inventories and the contributions of these sources to pollution, it is difficult to verify whether the perception is accurate.

Title III of the 1990 CAA Amendments also requires EPA to list and regulate area sources of HAPs. In some cases, MACT standards have been developed for several categories of area sources, including commercial dry cleaning, hazardous waste incineration, secondary aluminum production, and secondary lead smelting. The amendments also required EPA to address area sources in an integrated urban air toxics strategy. The 1999 strategy document identifies 32 high-priority HAPs (plus diesel exhaust) and lists 29 area-source categories that have been or will be targeted for regulation, including 13 new categories for which standards are to be issued by 2004. For area sources, the agency has the discretion to issue either MACT standards or less stringent generally available control technology (GACT) standards. EPA is now proceeding with implementing these regulations.

### Status of Area-Source Controls

To date, the efforts to control area sources have been relatively scattered and have slipped far behind mandated implementation schedules. Recent efforts, such as the integrated air toxics strategy, have begun to bring some strategic vision to the control of such sources. However, in the absence of a high-quality inventory of such sources, it is nearly impossible to quantify their emission contributions and to set priorities. Yet, those few analyses that have been done (for example, by EPA in the HAPs inventory) suggest that area-source emissions are significant and will be even more important after the imposition of MACT has reduced emissions from the major industrial sources. A renewed focus on improving area-source inventories and controlling key sources at an accelerated pace will be important in crafting effective strategies to reduce emissions and exposure to many important pollutants.

## SUMMARY OF KEY EXPERIENCES AND CHALLENGES FOR STATIONARY-SOURCE CONTROL

### Strengths of Stationary-Source Control Programs

• CAA programs have achieved substantial reductions of pollutants from existing stationary sources, for example, $SO_2$ (through the Acid Rain Program), VOCs (for example, through RACT), and HAPs (through MACT). The Ozone Transport Commission, and the upcoming $NO_x$ SIP call trading program have the potential to achieve substantial reductions in $NO_x$ emissions.

• NSR and PSD requirements have encouraged the continuous development and application of cleaner technologies and emission controls for major new stationary sources and have resulted in reduced emissions from those sources.

• The development of emission cap-and-trade programs that use market forces to limit the cost of pollution control has provided the AQM system with a mechanism that is capable of achieving substantial emission reductions at reduced costs.

## Limitations of Stationary-Source Control Programs[7]

- Although the NSR and PSD programs appear to have been effective for new facilities, such programs and related economic factors (1) provide an incentive for industries to extend the life of higher-emission (grand-fathered) facilities and (2) lead to litigation alleging facility modifications (primarily to extend their useful lifetime) without prior approval.

- The NSR and PSD programs do not affect a large fraction of the existing facilities that have not undergone major modifications but have remained in operation.

- With the exception of CEM, there is limited ability to quantify stationary-source emissions.

- Facility-specific emission standards have lowered emissions, but because they are often production-based standards, the potential remains for emissions to increase as economic activity or product demand increases.

- The next phase of control on HAP emissions is predicated on the conclusions of a residual risk analysis that is fraught with scientific uncertainty.

- Achieving the full potential of cap and trade will require applying the technique to a broader range of pollutants, implementing a less cumbersome process for revising caps and targets, developing enhanced and more cost-effective CEM and other monitoring technologies, and guarding against deleterious geographical and temporal distribution of emission reductions.

- Controls on area sources lack focus and are hampered by a large number of uncertainties in the magnitude of the emissions from these sources.

- To date, many emission-control programs for stationary sources have addressed pollutants separately, resulting in different time lines and different requirements for reductions of various pollutants at the same facility. That approach can raise the cost of emission control without adding any appreciable benefit in emission reductions. Recent legislative proposals for multipollutant reductions at electricity-generating facilities offer an opportunity to merge one of the most successful techniques—cap and trade—with a multipollutant approach. This merger would enable these facilities to develop long-term plans for capital improvements that minimize costs and reduce all relevant pollutant emissions at the same time.

---

[7]Recommendations are provided in Chapter 7.

# 6

# Measuring the Progress and Assessing the Benefits of AQM

## INTRODUCTION

Implementation of the Clean Air Act (CAA) via the procedures and methods described in Chapters 2 through 5 has resulted in significant improvements in air quality in the United States, according to the U.S. Environmental Protection Agency (EPA), and these improvements have had a positive impact upon public health and welfare. EPA estimated that emissions of the six criteria pollutants have decreased by about 30% over the past three decades, despite sizable increases in population, energy use, and gross domestic product (see Figure 1-4 in Chapter 1). The estimated benefits of these reductions are substantial; they include an estimated 100,000 to 300,000 fewer premature deaths per year and 30,000 to 60,000 fewer children each year with IQs below 70 (EPA 1997). In economic terms, these benefits have been estimated to amount to trillions of dollars. In this chapter, the committee discusses how such estimates of the progress and benefits of AQM in the United States are made, the uncertainty of the estimates, and what can be done to reduce the uncertainty.

## MONITORING POLLUTANT EMISSIONS

### Direct Measurement

The most direct way to confirm that specific emission-control technologies are working effectively is to measure the rate at which pollutants are released from relevant sources. However, with a few exceptions, pollutant emissions are not routinely monitored in the United States. One notable exception is the congressional requirement for continuous emissions monitoring (CEM) of sulfur dioxide ($SO_2$), nitrogen oxides ($NO_x$), and particulate matter (PM) from any source regulated under the acid rain provisions of the 1990 Amendments to CAA. As discussed in Chapter 5, the inclusion of such monitoring is viewed as being essential to ensuring the success of the cap-and-trade mechanism incorporated into the legislation. Moreover, the application of CEM has provided direct evidence of substantial reductions in $SO_2$ emissions from utilities since the implementation of the acid rain controls (see Figure 5-1 in Chapter 5). Inspection and maintenance (I/M) programs for motor vehicles, which were mandated in the CAA, could serve, in principle, as a check on the effectiveness of mobile-source emission controls. However, as discussed in Chapter 4, the effectiveness of I/M programs has been limited because of shortcomings in program design and effectiveness, and public resistance to such programs in some areas of the country.

There are a number of reasons why emissions are not routinely monitored. There are myriad stationary and area sources that contribute to pollution, and technologies are not available to monitor their emissions routinely and reliably. Given the resources and measurement technologies available to the AQM system, a program that attempted to monitor emissions comprehensively through direct measurement would be unrealistically expensive and complex. In addition, efforts by the government to monitor certain types of emissions on a continuous basis (for example, mobile emissions) might be viewed by some as an unacceptable invasion of privacy. On the other hand, the application of new technologies and creative measurement strategies could help to make the task more tractable and less invasive. For example, a number of emerging technologies and methods could be deployed to augment I/M for mobile emissions. Remote sensors have been used to identify high-emitting vehicles without inconveniencing motorists or interfering with traffic (Stedman et al. 1997; Bishop et al. 2000); on-board diagnostic systems are being developed to automatically monitor and document problems that lead to increased emissions from individual motor vehicles; and standard air quality monitors could be deployed inside tunnels and along roadways to help characterize in-use emissions from fleets of vehicles (Kean et al. 2001). As discussed in Chapter 5, CEM technologies are very valuable in tracking stationary-source emissions and could be used more widely, but the development of a broader range of CEM systems has been slow.

### Using Ambient Concentrations to Confirm Emission Trends

EPA estimates that nationwide emissions of volatile organic compounds (VOCs), $SO_2$, PM, carbon monoxide (CO), and lead (Pb) have decreased substantially since the early 1980s; the decrease in $NO_x$ emissions is estimated to be more modest (see Table 6-1, part A). These emission decreases can be reasonably ascribed to the promulgation of emission controls related to the nation's implementation of the CAA requirements. Because the importance of $NO_x$ for ozone ($O_3$) control was not recognized until the late 1980s or early 1990s, the slower pace of $NO_x$ emission decreases, compared with other pollutants, is probably due to the later implementation of $NO_x$ controls.

The decline of pollutant emissions during a period of substantial growth in population, energy consumption, and gross domestic product in the United States is cited by EPA and others as evidence of the substantial progress of the AQM system. However, the trends listed in Table 6-1, part A, have been developed from inherently uncertain emission inventories (see Chapter 3), so significant uncertainties must also be attached to the emission trends portrayed in EPA's reports. Because of such uncertainties, a technically robust AQM system should have mechanisms in place that could

TABLE 6-1  Summary of EPA's Trends in Estimated Nationwide Pollutant Emissions and Average Measured Concentrations

| Pollutant | 1983–2002 | 1993–2002 |
|---|---|---|
| **A. Changes in Estimated Pollutant Emissions, %[a]** | | |
| $NO_x$ | −15 | −12 |
| VOC | −40 | −25 |
| $SO_2$ | −33 | −31 |
| $PM_{10}$[b] | −34[c] | −22 |
| $PM_{2.5}$[b] | No trend data available | −17 |
| CO | −41 | −21 |
| Pb | −93 | −5 |
| **B. Changes in Measured Ambient Pollutant Concentrations, %** | | |
| $NO_2$[d] | −21 | −11 |
| $O_3$ 1-hr | −22 | −2 |
| $O_3$ 8-hr | −14 | +4 |
| $SO_2$ | −54 | −39 |
| $PM_{10}$ | No trend data available | −13 |
| $PM_{2.5}$ | No trend data available | −8[e] |
| CO | −65 | −42 |
| Pb | −94 | −57 |

[a]Negative numbers indicate reductions in emissions and improvements in air quality.
[b]Includes only directly emitted particles.
[c]Based on percentage change from 1985.
[d]The trends in $NO_2$ should be viewed cautiously because of potential artifacts from the instrumentation. Also, because NO is readily converted to $NO_2$ in the atmosphere, ambient monitoring data are reported as $NO_2$.
[e]Based on percentage change from 1999.
SOURCE: Adapted from EPA 2003.

independently establish the validity and accuracy of emission trends derived from emission inventories. Given the current inability to monitor all emissions comprehensively, trends cannot be verified directly. However, in principle, verification can be done indirectly by using long-term measurements of primary pollutant concentrations in the ambient air. The underlying assumption of this approach is that, all things being equal, there should be an approximate 1:1 relationship between a change in the total emissions of a primary pollutant (for example, CO and $SO_2$) and a change in the pollutant's average atmospheric concentration. Although this approach is straightforward in concept, it can be difficult to implement without an appropriately designed network of pollutant monitors.

Since the 1980s, the United States has had an extensive air quality monitoring network that routinely measures the concentrations of selected air pollutants in some locations. (The objectives and features of this network are discussed in some detail in the next section.) Trends in the concentrations of relevant air pollutants derived by EPA from data obtained from this network are listed in Table 6-1, part B. Initial inspection of these trends indicates qualitative consistency with the estimated pollutant emission trends discussed above—that is, the trends of both emissions and concentrations are downward. However, a more detailed quantitative comparison of the trends indicates significant inconsistencies. As illustrated in Figure 6-1, the downward trends in the average pollutant concentrations tend to be significantly greater than those of the emissions.[1] That result could be interpreted to mean that pollutant emissions have decreased more than estimated from emission inventories. However, there are other viable explanations. Significant uncertainties can exist in the concentration trends derived from the nation's air quality monitoring network (see later discussion on trends analysis). More important, the nation's air quality monitoring network was not designed to track nationwide emission trends or evaluate emission inventories; instead, it was largely designed to monitor urban pollution levels and compliance with National Ambient Air Quality Standards (NAAQS). As a result, most of the sites in the air quality monitoring network are urban; thus, the trends derived from them are more representative of urban pollution trends than national trends. Because many emission controls on stationary sources between 1970 and 1990 were aimed at urban emissions, urban areas might be expected to have larger decreases in pollutant concentrations than those seen overall. In any event, it would appear that air quality monitoring data provide qualitative but not quantitative confirmation that pollutant emission trends are downward (especially in urban areas) in the United States.

---

[1]$PM_{10}$ is an exception; however, note that the emissions change plotted in Figure 6-1 for $PM_{10}$ is based on the estimated change from 1985 to 2002 and not 1983 to 2002.

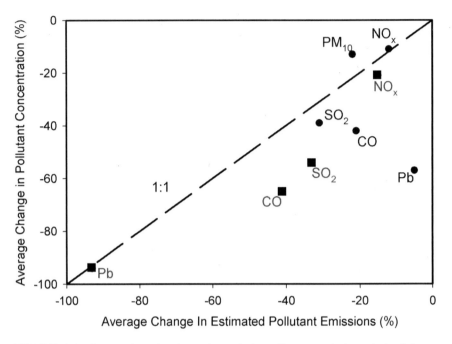

FIGURE 6-1  Scatterplot of estimated trends in pollutant emissions derived from emission inventories and changes in average pollutant concentrations derived from air quality monitoring networks. The squares indicate the trends for the period 1983–2002 (except for $PM_{10}$ emissions, which are for the trend period 1985–2002) and the circles for 1993–2002. SOURCE: Data from EPA 2003.

## MONITORING AIR QUALITY

The 1977 Amendments of the CAA state that

the Administrator shall promulgate regulations establishing an air quality monitoring system throughout the United States which (1) utilizes uniform air quality monitoring criteria and methodology and measures such air quality according to a uniform air quality index, (2) provides for air quaity monitoring stations in major urban areas and other appropriate areas throughout the United States to provide monitoring such as will supplement (but not duplicate) air quality monitoring carried out by the States required under any applicable implementation plan, (3) provides for daily analysis and reporting of air quality based upon such uniform air quality index, and (4) provides for record keeping with respect to such monitoring data and for periodic analysis and reporting to the general public by the Administrator with respect to air quality based upon such data.

In response to this and subsequent congressional mandates, EPA has overseen the development and operation of several national monitoring networks. These networks generally fall into one of two categories: ambient air quality monitoring networks that measure the atmospheric concentrations of pollutants at various locations, and deposition networks that measure the rate at which pollutants are deposited on the earth's surface. Collectively, these networks provide the best and most detailed (although by no means comprehensive) data available today for assessing the progress of the AQM system.

Because air quality and atmospheric deposition can exhibit large daily, seasonal, and interannual variations independent of any changes in pollutant emission rates, long-term monitoring is needed to detect and interpret air quality trends and thereby determine the effectiveness of regulations. The requirement for complete, precise, and accurate long-term monitoring of air quality presents significant technical challenges that require substantial investments in financial and human resources. In the United States, more than $200 million is spent annually to support air monitoring (EPA 2002p). In addition to providing an essential performance measure of the effectiveness of air quality regulations, air quality monitoring networks provide critically important information to scientists attempting to advance the understanding of the causes and remedies of air pollution. Thus, these networks represent a valuable and significant national resource.

The major federally supported monitoring networks for atmospheric composition and deposition operating in the United States are summarized below and in Table 6-2. An example of the spatial distribution of one of these networks (used to monitor $O_3$) is presented in Figure 6-2.

## Atmospheric Composition Monitoring Networks

### National, State, and Local Air Monitoring Stations

The CAA requires every state to establish a network of air monitoring stations for the criteria air pollutants. These networks are called the state and local air monitoring stations (SLAMS). SLAMS currently consist of approximately 4,000 monitoring stations whose size and distribution is largely determined by the needs of state and local air pollution control agencies to meet their respective state implementation plan (SIP) requirements (for example, to assess their NAAQS attainment status). A subset of the SLAMS monitoring sites (1,080 stations) comprises the national air monitoring stations (NAMS). They are located in urban and multisource areas to provide air quality data in areas where the pollutant concentrations and population exposures are expected to be the highest.

TABLE 6-2  Summary of Major U.S. Monitoring Networks

| Network | Start Year | Lead Agency | No. of Sites | Measurements |
|---|---|---|---|---|
| National air monitoring stations and state and local air monitoring stations (NAMS/SLAMS) | 1980 | EPA | ~4,000 | Continuous $O_3$, $NO_2$, CO, and $SO_2$ measurements; $PM_{10}$ and $PM_{2.5}$ and total suspended particulates (TSP) at least once every sixth day; other measurements taken at selected sites |
| Photochemical assessment monitoring stations (PAMS) | 1994 | EPA | 85 | $O_3$, NO, $NO_2$, surface meteorological conditions ($NO_y$ at some sites); 56 VOCs; some sites include 3 carbonyl compounds. |
| $PM_{2.5}$ networks | 1999 | EPA | 1,100 | 1,100 federal reference method: mass (every third day, 5% daily); 300 continuous mass (hourly); 200 speciation (every third day) |
| Interagency Monitoring of Protected Visual Environments (IMPROVE) | 1987 | NPS EPA | Increasing with time, ~160 | All sites:  every third day elemental and organic C, $SO_4^{2-}$, $NO_3^-$, Cl⁻, elements between Na and Pb, and $PM_{2.5}$ and $PM_{10}$ mass Selected sites:  hourly light-scattering and/or extinction coefficient, humidity, temperature; photographic scene monitoring |
| Clean Air Status and Trends Network (CASTNet) | 1990 | EPA | 84 (additional sites scheduled to begin operation) | Weekly particulate $SO_4^{2-}$, $NO_3^-$, $NH_4^+$, gaseous $HNO_3$ and $SO_2$; and continuous $O_3$; meteorological conditions for calculating dry deposition rates |

| Network | Year | Operator | Sites | Measurements |
|---|---|---|---|---|
| National Atmospheric Deposition Program and National Trends Network (NADP/NTN) | 1978 | Cooperators[a] | ~250 | Weekly measurements of wet deposition: pH, $SO_4^{2-}$, $NO_3^-$, $NH_4^+$, $Cl^-$, $Ca^{2+}$, $Mg^{2+}$, $Na^+$, and $K^+$, precipitation amounts |
| NADP/Mercury Deposition Network (NADP/MDN) | 1996 | — | ~80 | Total mercury and sampling period precipitation; some sites report methylmercury |
| Atmospheric Integrated Research Monitoring Network (AIRMoN) | 1984 | NOAA | Wet: 9 Dry: 5 | Research and measurements of wet deposition, including $SO_4^{2-}$, $NO_3^-$, $NH_4^+$, $PO_4^{3-}$, $Cl^-$, $Ca^{2+}$, $Mg^{2+}$, $Na^+$, and $K^+$; dry deposition, including $HNO_3$ and $SO_2$ and particulate $SO_4^{2-}$; $NO_3^-$; $NH_4^+$; measurements designed to provide information at greater temporal resolution |
| Gaseous Pollutant Monitoring Network (GPMN) | 1986 | NPS | ~30 national parks (multiple stations in some) | Continuous $O_3$ and meteorological data; some sites measure continuous $SO_2$, CO, $NO/NO_y$, and periodically VOCs |

[a]NADP has several hundred cooperators that provide monetary and in-kind support, including several federal, state, local, tribal, and nongovernmental programs.
SOURCES: NAMS/SLAMS 2002; PAMS 2002; GPMN 2002; IMPROVE 2002; AIRMoN 2002; CASTNet 2002; NADP/MDN 2002; NADP/NTN 2002.

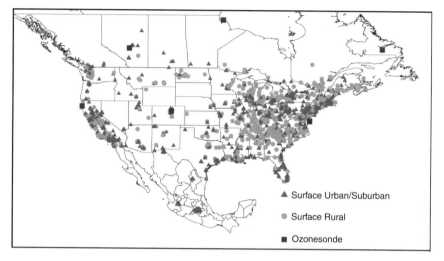

FIGURE 6-2   Locations of surface $O_3$ monitoring sites and ozonesonde sites in North America.  SOURCE: NARSTO 2000.  Reprinted with permission; copyright 2000, EPRI, Palo Alto, CA.

## Photochemical Assessment Monitoring Stations

The CAA Amendments of 1990 required EPA, in partnership with state and local agencies, to carry out more extensive monitoring of $O_3$ and its precursors in areas with persistent exceedances of the $O_3$ NAAQS (those $O_3$ nonattainment areas that are classified as severe or worse).  In response, EPA established a network of photochemical assessment monitoring stations (PAMS) in 24 urban areas (see Figure 6-3) to collect detailed data for VOCs, $NO_x$, $O_3$, and local meteorology.  The chief objective of PAMS data collection is to provide an air quality database that will assist air pollution control agencies to assess and refine $O_3$ control strategies and specifically to evaluate the trends in and effectiveness of controls implemented on VOC and $NO_x$ emissions in an area and to evaluate photochemical models being used by state and local agencies to carry out the attainment demonstrations required for their SIPs (see Chapter 3).

In principle, the data from the PAMS network could be extremely useful for the regulatory and scientific communities.  However, it appears that the full potential of the data has yet to be realized (NARSTO 2000).  Questions concerning the accuracy and specificity of the VOC and $NO_x$ concentrations obtained from the PAMS instrumentation have limited the ability of researchers to use the data to empirically assess the relationships between ambient VOC and $NO_x$ concentrations and $O_3$ formation (Parrish et al. 1998; Cardelino and Chameides 2000).  In spite of those limitations,

Operating PAMS Sites, 1998

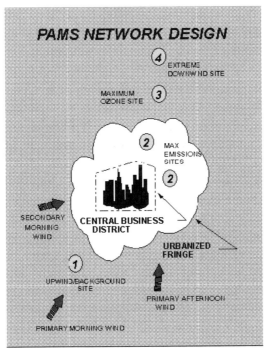

FIGURE 6-3    The PAMS network. Top Panel: Map showing the locations of PAMS sites in 1998. Lower Panel: A schematic showing the general design of the network within an urban area with upwind and downwind sites as well as sites near the largest sources of precursor emissions and sites where $O_3$ concentrations are typically at a maximum. SOURCE: PAMS 2002.

the PAMS data sets can probably still be used to evaluate trends, but this type of evaluation generally requires a record of measurements of a decade or more, and the PAMS network is just now reaching that level of maturity.

### Gaseous Pollutant Monitoring Program

Complementing the aforementioned urban-focused networks is the Gaseous Pollutant Monitoring Program (GPMP) operated by the U.S. National Park Service. The goal of this network is to provide data on baseline and trend concentrations of $O_3$ and other pollutants in national parks, data that can be used to assist in the development of policies to protect park resources.

### Interagency Monitoring of Protected Visual Environments

The Interagency Monitoring of Protected Visual Environments (IM-PROVE) is a monitoring system established to assess visibility levels and trends and to identify sources of visibility impairment primarily in national parks and wilderness areas. Through the IMPROVE program, annual and seasonal spatial patterns in PM and light extinction can be assessed (Box 6-1).

---

**BOX 6-1   Monitoring Visibility**

Visibility impairment from air pollution arises primarily from the scattering and absorption of light by suspended particulate matter (PM) with an aerometric diameter less than 2.5 μm. The contribution of human activities to visibility impairment in wilderness areas and national parks is assessed by monitoring the concentration and composition of PM in these areas and deriving so-called light extinction coefficients and visibility ranges from these measurements. Visibility trends thus derived vary by region and are not fully intercomparable because of the infrequent PM sampling schedules typically used. In the eastern United States, visibility has shown some improvement in the last decade but remains seriously degraded. The mean visual range is about 24 kilometers (km), compared with visibility in a "pristine" atmosphere in the range of 75–150 km. In the western United States, visibility levels do not appear to have changed significantly in the past decade; the mean visual range is 80 km compared with natural visibility of 200–300 km. (Visibility is less in the eastern United States even in the absence of human activities because of the higher humidity, which enhances the ability of particles to scatter light.) There is evidence that over the period 1988–1998, visibility declined in some national parks because of area increases in sulfur emissions (Sisler and Malm 2000). As discussed in the preceding chapters, EPA adopted a new regional haze program in 1999 to help address this problem.

## Enhanced PM$_{2.5}$ Monitoring Networks

Many of the monitoring networks described above have been collecting routine observations of airborne total suspended particulates (TSP) and PM$_{10}$. However, with the promulgation of a new NAAQS for PM$_{2.5}$[2] in 1997 came the need for a national monitoring program for PM$_{2.5}$. The resulting network could be used to help meet four objectives: (1) establish locations of nonattainment of the PM$_{2.5}$ NAAQS; (2) aid in the design of SIPs and then track their effectiveness; (3) assess regional haze and visibility (Box 6-1); and (4) provide data for health effects studies and other ambient aerosol research. Two types of measurements are undertaken in the network: (1) monitoring of the total mass concentration of PM$_{2.5}$, and (2) determination of its chemical composition. The network, as currently designed, uses a filter-based technique to measure and characterize PM$_{2.5}$. Particles from the atmosphere are first collected on filters by passing ambient air through them over a period of time (typically 24 hr), and the filters are then analyzed gravimetrically to determine PM$_{2.5}$ mass concentration and chemically to determine PM$_{2.5}$ composition. Although the filter-based technique is relatively straightforward for state and local agencies to implement, it is labor-intensive and subject to a variety of positive and negative artifacts, especially for the determination of chemical composition (Pierson et al. 1980; EPA 2002b). Although considerable progress has been made in eliminating these artifacts (for example, through the use of annular denuders, filter packs, and backup sorption beds), a number of problems remain unresolved. To enhance the long-term ability of the PM$_{2.5}$ network to meet the needs of the regulatory, health effects, and air pollution research communities, those remaining problems will need to be addressed adequately, or other techniques less susceptible to artifacts will need to be deployed. A new class of semicontinuous PM monitors shows promise in this regard.

To address the problems with the filter-based techniques and to respond to the recommendations of the National Research Council (NRC 1998b, 1999a, 2001d), EPA has implemented a number of PM "supersites," where more detailed research is addressing the scientific uncertainties associated with fine PM. Program goals focus on fine particulate characterization, methods testing, and support of studies of health effects and exposure. Locations of the initial PM supersites are shown in Table 6-3.

## Hazardous Air Pollutants

There is not as extensive a nationwide monitoring network for hazardous air pollutants (HAPs) as there is for criteria pollutants. A large number

---

[2]PM$_{2.5}$ and PM$_{10}$ designate particulate matter with aerodynamic equivalent diameters of 2.5 micrometers ($\mu$m) or 10 $\mu$m or less, respectively.

TABLE 6-3   Locations of Initial $PM_{2.5}$ Supersites

| Location | Oversite Institution |
| --- | --- |
| Baltimore, MD | University of Maryland at College Park |
| Fresno, CA | Desert Research Institute and California Air Resources Board |
| Houston, TX | University of Texas at Austin |
| Los Angeles, CA | University of California at Los Angeles |
| New York, NY | State University of New York at Albany |
| Pittsburgh, PA | Carnegie Mellon University |
| St. Louis, MO | Washington University |
| Atlanta, GA | Georgia Institute of Technology |

SOURCE: Adapted from EPA 2002o.

of HAPs are monitored at PAMS sites, and a number of local and state HAP monitoring programs are in place. These programs vary considerably from one location to the next in terms of the species measured, the types of localities monitored, and the frequency and quality of measurements (EPA 2000e; Kyle et al. 2001). Although the data needed to assess trends in HAPs are sparse, EPA has reported trends for a small subset of the most critical HAPs (Box 6-2). EPA is working to create a more consistent, comprehensive national monitoring network for a small set of HAPs. In 2001, it began a pilot monitoring project that includes 10 locations and an initial trends network that includes 11 cities. These trial efforts are aimed at helping to assess sampling and analysis precision requirements, sources of variability, and minimal detection levels.

## Deposition Monitoring Networks

### National Atmospheric Deposition Program and National Trends Network

Wet deposition of major solutes is monitored throughout the country by the interagency-supported National Atmospheric Deposition Program and National Trends Network (NADP/NTN) (see Figure 6-4). The primary aim of the network is to determine the exposure of natural and managed ecosystems to nutrient and toxic elements and acidity arising from anthropogenic activities, such as fossil fuel burning and agriculture. The spatial pattern of the wet deposition is estimated with a hybrid data-assimilation and modeling system that incorporates NADP/NTN data with information on spatial patterns of topography and precipitation. Application of this system appears to confirm that the aforementioned reductions in $SO_2$ emissions from power plants achieved as a result of the Acid Rain Program of the 1990 CAA Amendments have resulted in a decrease in sulfate deposition in the eastern United States (Box 6-3). Moni-

BOX 6-2    Preliminary Indications of Progress on HAP Control

EPA (2000d) estimates that the 1996 level of HAP emissions decreased by 23% compared with the 1990–1993 baseline period. It is difficult to confirm the accuracy of these emission-trend estimates using ambient data because of the short time span being considered and the limited coverage of monitoring sites. Nevertheless, EPA reported trends for 33 urban HAPs for the period 1993–1998 (EPA 2000d) based on existing data and found evidence of downward trends in the majority of HAPs being monitored. The greatest reductions were seen for benzene (Figure 6-5) and suspended lead, both due primarily to changes in fuel formulation that resulted in either substantial reduction in emissions (in the case of benzene) or their elimination (in the case of lead). (See Chapter 4 for a more detailed discussion of these efforts.)

toring for current and future toxic air pollutants of concern in a consistent manner across the nation and in a way that provides data to support exposure and risk characterization is an important consideration, and research into methods to accurately estimate ambient concentrations and exposure to HAPs is needed.

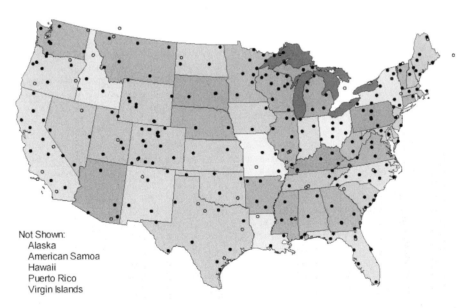

Not Shown:
  Alaska
  American Samoa
  Hawaii
  Puerto Rico
  Virgin Islands

FIGURE 6-4   Locations of the National Atmospheric Deposition Program and National Trends Network (NADP/NTN) monitoring sites in the contiguous 48 United States. SOURCE: NADP/NTN 2002.

> **BOX 6-3  Deposition Networks Confirm Decrease in SO$_2$
> Since Implementation of the Cap-and-Trade
> Emission-Control Program for SO$_2$**
>
> As a result of the application of continuous emissions monitoring (CEM) to utili-
> ties regulated under the acid rain provisions of the CAA, it is reliably estimated that
> annual SO$_2$ emissions from utilities have declined by about 5 million tons during the
> 1990s (see Figure 5-1 in Chapter 5). The next step in an AQM assessment of
> progress is to ensure that these reductions in emissions produced the appropriate
> atmospheric response—a reduction in sulfate deposition rates. That reduction has
> occurred, as illustrated in Figure 6-6 and according to an analysis of data from the
> National Atmospheric Deposition Program and the National Trends Network (Stod-
> dard et al. 2003). The final step is to determine if these reductions have led to the
> appropriate ecosystem response. This issue is discussed in Box 6-6.

### National Atmospheric Deposition Program and Mercury Deposition Network

The Mercury Deposition Network (MDN) was established as part of the NADP to measure wet deposition of total mercury (as well as methyl-mercury at some sites). The MDN is located mostly in the northeastern

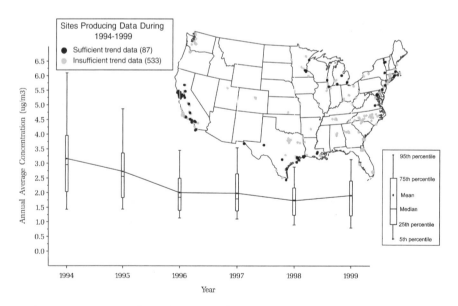

FIGURE 6-5  National trend in annual benzene concentrations in metropolitan areas, 1994–1999. SOURCE: EPA 2001a.

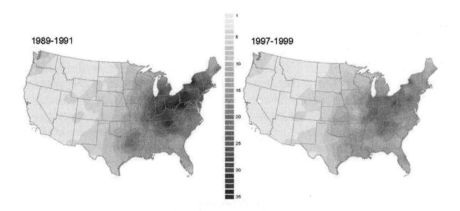

FIGURE 6-6    Trends in wet sulfate deposition in the United States using data from the Clean Air Status and Trends Network (CASTNet) and the National Atmospheric Deposition Program/National Trends Network (NADP/NTN) (1989–1991 vs. 1997–1999).    Data are in kilograms per hectare.    The largest decreases are indicated across the mid-Appalachians, the Ohio River Valley, and the Northeast regions of the country. SOURCE: EPA 2003o.

United States and is largely funded by state agencies. The long-term objective of the MDN is to assess national-scale spatial patterns and temporal trends in wet deposition of mercury. However, because the network has operated for only a few years on a small number of sites, it can only provide information on site-specific deposition.

## Clean Air Status and Trends Network

The Clean Air Status and Trends Network (CASTNet) measures the components of atmospheric deposition that enter the environment in dry form, such as particles and gases. Monitoring dry deposition is critical in determining the total pollution load across the United States. For example, dry deposition in some areas contributes as much as 60% of the total sulfur deposition. Two-thirds of CASTNet sites are located east of the Mississippi River. Dry deposition rates are calculated using atmospheric concentrations, meteorological measurements, vegetation characteristics, land use, and surface conditions. Such factors impart a high amount of uncertainty to estimates of the spatial patterns of dry deposition. Therefore, regional patterns and long-term trends have not been well characterized.

**Atmospheric Integrated Research Monitoring Network**

The Atmospheric Integrated Research Monitoring Network (AIRMoN) provides highly resolved information over time on precipitation and dry deposition by using daily sampling methods. AIRMoN is designed to provide an intensive research-based foundation to support the routine operations of NADP and CASTNet. Whereas NADP and CASTNet are designed to characterize long-term trends, AIRMoN uses a small number of sites at selected locations to more quickly detect a temporal trend in deposition that might arise (for example, from emission controls or from new sources) and to evaluate the effects of these trends on sensitive ecosystems. There are currently two components of AIRMoN—7 sites to detect changes in wet deposition and 13 sites to evaluate changes in dry deposition. The flat funding for this program for 10 years has resulted in the closing of three dry deposition sites.

## Air Quality Monitoring Discussion

The nation's air quality monitoring system has evolved considerably over the past 30 years and will need to continue evolving to meet future air quality challenges. In this section, particular strengths, weaknesses, and other aspects of the air quality monitoring system are considered. EPA is developing a new national ambient air monitoring strategy, which includes a new network design called NCore, as well as a continuous monitoring implementation plan. The report outlining these new plans has been released in draft form and is undergoing review by state and local agencies and the academic and technical communities (EPA 2002p). The committee did not assess the extent to which these plans will help address the concerns discussed below.

### Monitoring Objectives

Monitoring networks are an essential part of any air quality management system and can help to meet one or more of the following objectives (Chow et al. 2002a,b; Demerjian 2000; NARSTO 2003):

- Determine the specific air pollutants or their mixtures that are associated with health and welfare effects, including visibility impairment.
- Estimate exposure of human populations or sensitive ecosystems.
- Measure compliance with ambient air quality standards.
- Develop information about the sources contributing to the pollution.
- Develop information on processes controlling the formation, removal, and transport of the various pollutants and their precursors.

- Measure the effectiveness of emission-control programs.
- Serve as a warning system to alert individuals that are sensitive to poor air quality and to aid in the development of air quality forecasts.

Each one of the objectives places different demands on a network in terms of the species measured, instrument sensitivity, time resolution and frequency of the measurements, and location of the monitoring stations. Even though more than $200 million per year is spent on routine air quality monitoring in the United States (EPA 2002p), monitoring networks have limited resources. As a result, they are able to address only some of the above objectives adequately. At least two major issues arise as a result of resource limitations (NARSTO 2000):

- *Identifying the problem versus finding the solution and assessing program effectiveness.* Most of the existing networks have been designed only to measure compliance with the existing NAAQS and reveal little about the appropriate management strategies needed to solve the problems or measure the success of various emission-control strategies (see earlier discussion on emission trends). Network design should be evaluated and expanded to make air quality networks in the United States more relevant to other important objectives of monitoring. Some specific changes are discussed below.
- *Measuring critical species in a regular monitoring mode.* Because of the variety of criteria pollutants, their precursors, and HAPs and their large range of concentrations, monitoring is technically challenging as well as expensive. In some cases, these challenges are beyond the capabilities of state and local regulatory agencies. Creative mechanisms for fostering collaboration and technology transfer among regulatory agencies, research and academic institutions, and small businesses need to be devised to meet the challenges (see discussion in Box 7-5 in Chapter 7).

## Siting of Air Quality Monitoring Stations

Following enactment of the CAA, the AQM system in the United States emphasized urban-scale air quality problems and the use of controls on local emissions to solve air pollution. Moreover, major urban areas were the only areas specifically identified in the congressional mandate that initially directed the EPA administrator to develop a national monitoring program. As a result, the nation's ambient air quality monitoring networks have been dominated by urban sites. The intervening years have seen a growing concern for large-scale air quality problems that extend over multistate airsheds and are affected by long-range transport. However, a concomitant evolution of the nation's air quality networks has not occurred for gaseous pollutants. For example, NAMS and SLAMS for $O_3$ contain about 1,600 sites. Those that are urban and suburban sites outnumber rural sites by a factor of 2 (see Table 6-4).

The largely urban-based networks in use in the United States have proved quite useful for characterizing urban air quality for the specified pollutants; however, there are several limitations to the information that can be gained from these networks. In particular, they are inadequate for characterizing air quality on a regional scale; they provide little information to address processes related to the production of secondary pollutants, such as $O_3$; and the spatial coverage of these data is relatively small and not well suited for use in air quality model performance evaluation. Moreover, to assess urban emission impacts, many of the so-called rural sites are located within or directly downwind of urban areas and are therefore often affected by urban pollution plumes (NARSTO 2000). The sites in PAMS, the other major network for gaseous pollutants, are also by congressional mandate exclusively located in and around major metropolitan areas (see Figure 6-3). The current lack of rural monitoring sites severely hampers the AQM system's ability to address multistate airshed pollution problems or to assess and mitigate pollutant impacts on natural and managed ecosystems, such as agriculture. It also may hamper the ability to identify the true extent of NAAQS nonattainment. For example, analysis of monitoring data from rural areas suggests that a large fraction of rural counties in the eastern United States might be in violation of the new 8-hr $O_3$ NAAQS (Chameides et al. 1997). It is not possible, however, to assess the extent of this rural nonattainment without more extensive rural monitoring.

Another shortcoming of the current monitoring networks for gaseous and aerosol pollutants is their strong emphasis on measuring ambient concentrations rather than concentrations in specific microenvironments that

TABLE 6-4    Ozone Monitoring Sites in the United States[a]

| Number of Sites Designated Urban or Suburban[b] in AIRS[c] | Number of Sites Designated Rural[b] in AIRS[c] | Number of Sites Measuring Ozone Vertical Profiles Using Ozonesondes |
|---|---|---|
| 1,071 | 504 | 4 |

[a]For the 48 contiguous states.
[b]As noted in text, many of the sites designated rural are not truly rural in character.
[c]EPA's Aerometric Information Retrieval System, now called AirData 2003.
SOURCE: Adapted from NARSTO 2000.

may sometimes be hot spots.[3] In the absence of data on pollutant concentrations in hot spots, characterization of pollutant exposures of persons who work in, reside in, or travel through them is problematic (see Box 7-5 in Chapter 7).

## Air Quality Measurement Techniques

Standard operating procedures, measurement methods, and quality assurance and quality control (QA/QC) procedures are critical to ensure that the data sets from monitoring networks can be directly compared and integrated for use in trends analysis, for investigations of atmospheric processes, and for improvement of predictive models. In accordance with the *Code of Federal Regulations* (40 CFR Part 53 [2002]), EPA established federal reference methods (FRM) and federal equivalent methods (FEM) to be used in measurements of criteria pollutants (EPA 2002q). Further development of monitoring techniques is needed, however, to address some of the following concerns:

- The instrumentation used to measure NO and $NO_2$ in the FRM does not have the sensitivity or the specificity to measure nitrogen oxides accurately over the full range of conditions typically encountered by such instrumentation (McClenny et al. 2002).
- $PM_{10}$ and $PM_{2.5}$ are generally measured once every 3 or 6 days, a sampling rate that is too infrequent to capture the true variability of PM concentrations.
- New analysis systems were developed in response to the PAMS initiative and its requirements for measurements of speciated hydrocarbons and carbonyl compounds. However, routine monitoring of hydrocarbons and carbonyls is a challenging task, even for research institutions, and thus far, delivery and analysis of quality-ensured data from PAMS have been limited (see NARSTO 2000).
- $PM_{2.5}$ measurements will continue to pose a technical challenge because of the complex multiphase mix of constituents in ambient aerosols. As noted earlier, current sampling methods are subject to various artifacts.

---

[3]Hot spots are locales where pollutant concentrations are substantially higher than concentrations indicated by ambient outdoor monitors located in adjacent or surrounding areas. Hot spots can occur in indoor areas (for example, public buildings, schools, homes, and factories), inside vehicles (for example, cars, buses, and airplanes), and outdoor microenvironments (for example, a busy intersection, a tunnel, a depressed roadway canyon, toll plazas, truck terminals, airport aprons, or nearby one or many stationary sources). The pollutant concentrations within hot spots can vary over time depending on various factors including the emission rates, activity levels of contributing sources, and meteorological conditions.

For example, nitric acid and organic vapors may adsorb on the filter used for the collection of atmospheric particles (positive artifacts), and ammonium nitrate or organic PM may evaporate from the filter during sampling (negative artifacts). Many studies have been completed or are being conducted to develop and test more reliable monitoring methods and analytical procedures to determine the chemical composition of PM (see NARSTO [2003] for a review). A promising new technology based on automated semicontinuous sampling technology has been developed in the supersites programs and could be used in the routine Speciation Monitoring Network.

• There is an emerging interest in bioaerosols—fine PM of biological origin, which may include allergens, viruses, bacteria, or fungi. They can enter the atmosphere inadvertently as a result of animal-feeding operations (NRC 2003c) or intentionally as a result of bioterrorism (NRC 2002d). Identification of these materials will be more complex than simple chemical or elemental analysis and may involve microbiological tests.

• Because the primary focus of monitoring strategies has been on documenting NAAQS attainment, there is little motivation to develop methods to measure ambient concentrations that are substantially below the standards. Such measurements are needed to provide critical data for understanding atmospheric processes and providing input for air quality models.

### Air Quality Trend Analysis Techniques

A variety of techniques can be used to determine long-term trends in ambient pollutant concentrations. The method most commonly used by EPA is to compute yearly averages of concentrations at all stations within a metropolitan statistical area (MSA). Values for missing yearly averages are linearly interpolated if in a middle point; missing end points are replaced with the nearest year of valid data. Linear regression analysis is done, and a method known as the Theil test (Hollander and Wolfe 1973) is applied to detect the significance of any trend. This approach is a fairly simple, straightforward method of trend detection, but it has several important limitations. As noted earlier, the monitoring stations may not be, and often are not, spatially distributed to document the average air quality over the region of interest, and thus the trends derived might not be representative of the region. The results of a linear regression analysis can be highly sensitive to such factors as the time interval chosen for analysis and the frequency of observations. In some cases, sparse data make it difficult to calculate trends. For instance, samples of PM are generally collected at a frequency that is longer than most PM episodes. Consequently, several years of data collection will be needed to characterize seasonal and spatial patterns accurately and an even longer period of sampling will be needed to detect long-term temporal trends.

One of the greatest challenges in concentration trend analysis—carried out with the specific purpose of determining if pollution control efforts have been successful—is accounting for meteorological influences, which can cause a great deal of variability in concentrations and are often large enough to obscure changes in anthropogenic emissions. A variety of statistical approaches (such as regression-based modeling and extreme value approaches) can be used to filter out meteorological influences from a trend; however, the results often vary, depending on the method and the specific concentration metric used.[4] For all these reasons, most pollutant concentration trends cited by EPA and other organizations (including the trends noted in this report) should be used with caution (Box 6-4).

## Data Availability

EPA issues a yearly *National Air Quality and Emission Trends Report* (EPA 2000e), which provides a general summary of the nation's air quality. The summary includes the results of an analysis of air pollutant trends based on data obtained from the monitoring networks described above. EPA also issues a yearly Toxics Release Inventory (TRI) (EPA 2003m), which contains information about the types and amounts of approximately 650 chemicals that are released from certain industries and federal facilities. The reported information includes annual estimates of emissions as well as releases to water and land and quantities of chemicals sent to other facilities for waste management. However, the trends reports and TRI reports are intended for a general audience and do not contain enough information to facilitate more extensive analyses by the scientific and technical community. For such purposes, EPA has developed the Aerometric Information Retrieval System (AIRS), a framework that provides access to monitoring data from different sources. Although EPA has made efforts to update AIRS, problems remain with the system that limit its utility. All monitoring data are not stored in AIRS, and data stored in AIRS have not been "standardized"; for example, the data have not been converted to uniform measuring units or to universal standard time. Most important, the data are not available in real time (as are, for instance, data from the National Weather Service). Although EPA maintains the AIRNow web site where individuals can access information on present and past air quality conditions at localities throughout the United States, the information is qualitative and thus not suitable for quantitative analyses. Many states have begun to make their air quality monitoring data available in real time over the

---

[4]A trend analysis for CO in Fairbanks and an approach by Reddy (2000) for estimating the probabilities of future CO exceedances in Denver are described in earlier NRC reports (NRC 2002b, 2003b).

---

**BOX 6-4   Case Study on Pollutant Trend Analysis: Ozone**

EPA analyses show that over the past 20 years, national 1-hr average $O_3$ concentrations have decreased by about 18%; most major metropolitan areas showed downward tends, and the Northeast and West showed the greatest improvements. In contrast, the South and North-central regions of the country and many national parks show increased $O_3$ concentrations over the past 10 years. However, numerous uncertainties are associated with those $O_3$ trend estimates. The formation and accumulation of $O_3$ are strongly affected by prevailing meteorological conditions. Therefore, $O_3$ trends, driven by emission-control policies, can be easily masked by year-to-year meteorological variability, and trend analyses can reach very different conclusions based on the methods used to factor out meteorological influences. $O_3$ trends have been routinely examined using the second highest daily maximum 1-hr average concentration in a given year, because that value is used to establish compliance with the 1-hr $O_3$ NAAQS; however, that value tends to fluctuate with a larger amplitude than a more robust statistic, such as the 95th percentile. As a result, subtle trends can be masked by the fluctuations of the extreme values.

Several research groups have questioned EPA's reported $O_3$ trends and carried out independent studies using data from national monitoring networks. A comprehensive review of such studies is presented by Wolff et al. (2001). Some locations, such as Los Angeles, have shown consistently decreasing $O_3$ trends, but national-scale studies paint a more complicated picture. Milanchus et al. (1998) examined daily maximum 1-hr $O_3$ concentrations over the period 1984–1995 and found that most locations exhibited downward trends during that time, and a few isolated sites showed upward trends. Fiore et al. (1998) examined median summer afternoon $O_3$ concentrations for the years 1980–1995 and found that trends were insignificant over most of the continental United States, although $O_3$ concentrations decreased in New York, Los Angeles, and Chicago. A follow-up study examined trends in the number of exceedances of the $O_3$ standards (Lin et al. 2001). Over the period 1980 to 1998, strong downward trends in the number of days in exceedance of either $O_3$ standard were found along the northeastern and southwestern coasts, and significant upward trends were found in various locations elsewhere in the country. In most locations, the $O_3$ air quality improvements seen in the 1980s leveled off in the 1990s. The results differ among those and other studies because they are based on different yearly ranges, time periods within each year, pollution metrics, and methods to account for meteorological effects. Thus, it is very difficult to compare and evaluate the validity of reported $O_3$ trends systematically based on past analyses.

---

internet. That is an important development that should be encouraged at the national level by EPA. The availability of air quality data in real time will open up significant entrepreneurial and research opportunities. For example, real-time air quality data could greatly facilitate the development of predictive (4-dimensional data assimilation) air quality modeling systems that could be used to assist air pollution mitigation efforts (Box 6-5).

**BOX 6-5   Air Quality Forecasting**

The role of air pollution forecasts is growing as an AQM tool. Most forecasts produce 1–3-day $O_3$ forecasts that predict exceedances of specific concentration thresholds. These forecasts can be used as a warning system for individuals sensitive to poor air quality and as an AQM tool. The techniques used to make these forecasts usually fall into one of four general types (Dye et al. 2000; NOAA 2001):

- Phenomenological forecasts depend most heavily on the skill of the forecaster, who subjectively processes both air quality and meteorological information and, on the basis of past experience, formulates a prediction of pollution levels in the future. Despite its subjective nature, this approach can useful when used in conjunction with one or more of the objective methods discussed below, which have their own limitations.
- Climatological methods rely on the association observed between increased pollution levels and specific meteorological conditions to predict future pollution episodes. The predictions can be based on a variety of approaches ranging from simple assumptions of persistence (for example, pollution levels remaining high the day after high levels occur) to more complex weather typing schemes (for example, identifying recurring weather patterns that are accompanied by high pollution levels).
- Statistical methods rely on more quantitative relationships between pollution concentrations and meteorological parameters (for example, temperature and humidity) derived from past observations and weather forecasts. The statistical relationships can be as simple as linear regressions to more complex neural networks. These statistical relationships are then used in conjunction with weather forecasts to predict future pollution levels.
- Mechanistic methods use chemical transport models (CTMs) in a so-called four-dimensional data assimilation mode to predict future pollutant concentrations. This approach typically makes use of data from monitoring networks to specify current air pollutant concentrations and meteorological fields and detailed meteorological forecasts from the National Weather Service to calculate pollutant concentrations in the future (McHenry et al. 2000; Chang and Cardelino 2002).

Although air pollution forecasting for secondary pollutants, such as $O_3$, is a relatively new scientific endeavor, and the accuracy and predictive skill of the methods described above have yet to be evaluated comprehensively (NRC 2001a), such forecasting has begun to be used in an operational mode.

In some instances, predictions of future air pollution episodes are released to the public as an advisory. Individuals, especially those most sensitive to air pollution, can, if they choose, use the information from these advisories to alter their behavior in a way that would minimize their exposure to the pollution. For example, in AIRNow, probably the largest national air pollution forecasting effort, EPA provides the public with same-day and next-day $O_3$ forecasts as well as $PM_{2.5}$ forecasts for many cities. The AIRNow program developed an infrastructure to collect daily air quality forecasts from state and local agencies for over 250 cities and provide real-time and forecasted air quality data to the public through media outlets (Wayland et al. 2002). EPA compiles air quality forecasts and posts them to the AIRNow website for public utilization (www.epa.gov/airnow).

*(continued on next page)*

**BOX 6-5**  continued

Similarly, the National Weather Service (NWS) intends to develop and implement a nationwide system for distributing air quality forecasts in conjunction with NOAA's Office of Oceanic and Atmospheric Research, EPA, and the air quality modeling community to provide to the public information such as current ozone conditions from live weather broadcasts (Stockwell et al. 2002). NOAA has also developed several experimental forecast products (including models to predict smoke dispersion from forest fires) and made them available on the internet, and the National Park Service has begun to issue $O_3$ advisories for some national parks to protect its workforce and to alert park visitors.

In more ambitious operational applications, air pollution predictions are used by air quality managers to trigger specific emission reductions—that is, temporary measures to lower pollutant emissions to avert or at least limit the severity of the episode. A growing number of state and local air quality agencies are using summertime $O_3$ forecasts to take steps to reduce precursor emissions. Many cities now use forecasts to declare "ozone action days," when the public is encouraged on a voluntary basis to modify their activities to reduce $O_3$-producing emissions. To date, their effectiveness has not been independently evaluated.

From a technical point of view, a major concern of the growing use of air pollution forecasting as an operational tool is the apparent lack of a concomitant effort to design and implement tools to assess the accuracy of the forecasts and the impact their use has on averting adverse health effects and reducing pollutant emissions. If air pollution forecasting is to reach a level of sophistication and use comparable to that of meteorological forecasting, a serious program of assessment must be undertaken.

## Monitoring Vertical Profiles of Air Pollutants

Vertical profiles of pollutants are only available from intensive, research-oriented field campaigns and thus are very limited in time and space. Results from several observational campaigns in the United States and abroad have demonstrated the importance of making vertical measurements of pollutants (see Kleinman and Daum 1991; Berkowitz et al. 1996; Hamonou et al. 1999; Simo and Pedros-Alio 1999; Ansmann et al. 2000). A variety of techniques can be used routinely to measure vertical profiles of atmospheric composition and related meteorological variables, including use of aircraft, which can provide a valuable platform for studying atmospheric composition over a wide range of altitudes. Kleinman and Daum (1991) studied the vertical distributions of airborne particles, water vapor, and insoluble gases in measurements taken during aircraft flights conducted near Columbus, Ohio, and found plumelike features at high altitude where pollutant concentrations increased by 50% or more for distances of several kilometers. These features were attributed to air near the earth's surface lofted by convective storms. These types of studies are important in quantifying the role of long-range transport (see next section), because pollutants that have been lofted above the boundary layer will travel farther from their sources, and in better elucidating the relationship between pollution and meteorological conditions.

High-resolution vertical profiles of pollution plumes aloft can also be obtained with balloon-borne instruments called ozonesondes that measure concentrations with an electrochemical cell. For example, the National Oceanic and Atmospheric Administration's (NOAA) Climate Monitoring and Diagnostics Laboratory (CMDL) has a network of eight sites that make weekly $O_3$ vertical profile observations from the surface to about 35 kilometers (km) (NOAA 2003). In addition, differential absorption lidars (a laser-based instrument that measures the light backscattered by particles along the beam path) can be used to obtain vertical profiles of atmospheric constituents.

### Monitoring Long-Distance Transport of Air Pollutants

As state and local emission controls have successfully reduced local pollution sources, long-range transport has become an increasingly important source of NAAQS violations, especially for $O_3$ and $PM_{2.5}$. Yet, very few air quality monitoring sites allow meaningful examination of long-range pollution transport. Additional monitoring sites located outside urban areas provide important information for developing effective regional control strategies and would be an important element in helping to implement recommendations made in Chapter 7. However, current ground-based networks are inadequate for a quantitative determination of long-range pollutant transport, as much of it occurs above the boundary layer. Some additional information on long-range transport can be gleaned from high-elevation rural monitoring sites. Over the longer term, satellite remote-sensing techniques have the potential for comprehensively monitoring pollutant transport on regional and global scales. For example, Fishman et al. (2003) have used satellite-based instruments to provide detailed maps of tropospheric $O_3$ that are capable of defining the regional extent of $O_3$ pollution in areas around the world. Current satellite instruments can provide a qualitative picture of the atmospheric distribution of some aerosol types, but quantitative information on aerosol type and mass flux is difficult to obtain because of the poorly characterized radiative properties of most aerosols (NRC 2001a).

## ASSESSING HEALTH BENEFITS FROM IMPROVED AIR QUALITY

A primary goal of the nation's AQM system is the protection of human health. Therefore, an important aspect of measuring progress in AQM is demonstrating that reductions in pollutant emissions and improvements in air quality have resulted in measurable health benefits or that preventative measures have averted a worsening of public health. Such a demonstration not only helps to confirm that the appropriate air pollutants are being regulated but also helps to establish that the societal expenditures made to improve air quality have benefited society directly and concretely. Although the process of setting goals and standards discussed in Chapter 2 has much in common with the process of assessing health benefits from improved air quality, they are two distinctly different and independent activities. The former identifies adverse effects and risks and sets goals and standards to avoid or decrease these effects and risks; the latter establishes retrospectively that specific improvements in air quality have resulted in a reduction in adverse health effects.

Ideally, an analysis of the benefits of air pollution mitigation for public health would rely on independent toxicological and epidemiological data that linked specific long-term air pollution trends with specific trends in public health. However such linking has proved to be particularly difficult (EPA 2003n). Although there is a wealth of overlapping data on temporal patterns of disease within specific populations and on patterns of air pollution in areas where the population resides and/or works, air pollution is believed to account for a small proportion of the current incidence of disease in the United States (HEI 2000). Changes in other risk factors, such as age, socioeconomic status, amount of smoking, occupation, and weather, whose distribution and effect in the population can vary over time, can easily mask any benefits related to air quality improvements. For example, the potential for reduced exacerbation of asthma due to reductions in air pollution might pose an opportunity to track actual health benefits. Associations of PM with exacerbations of asthma have been repeatedly demonstrated epidemiologically in time-series analyses of emergency-room visits and hospital admissions and in panel studies examining associations with peak flow, medication use, and symptoms (Ostro et al. 1998). However, the underlying prevalence of asthma has increased significantly, especially among children, complicating efforts to track changes in asthma exacerbation in relation to changes in air pollution (for example, by looking at the number of asthma hospitalizations). Even if air pollution is declining, the size of the potentially vulnerable population is increasing, and the absolute number of hospitalizations might rise even as levels of pollution fall. This challenge is made even more daunting when one tries to track changes in multipollutant, multipathway exposures and the resulting changes in health.

Because of the current inability to directly link observed population health trends with contemporaneous ambient air quality benefits, less direct approaches have been adopted to assess the health benefits of reductions in air pollution. Aspects of three such approaches are discussed below.

### Assessments Based on Data from Short-Term Air Pollution Events

Extreme air pollution events, such as those experienced in London, England, in the 1950s and 1960s, provide distinctive data sets linking the effects of such events with population health. For example, one of the most infamous air pollution events occurred in December 1952, in which air pollution dramatically increased over a 10-day period with a coincident large increase in cardiorespiratory disease, manifested by increases in hospital admissions and a doubling of the daily mortality. The number of daily deaths declined when air quality improved, albeit in a time-lagged manner (U.K. Ministry of Health 1954).

A related approach uses data from specific events that caused a temporary cessation of air pollutant emissions to establish a link between improvements in air quality and human health outcomes. For example, Friedman et al. (2001) found a significant statistical correlation between closure of inner city Atlanta to traffic during the 1996 Olympics and reduced visits to emergency departments for asthma in children. Similarly, Ransom and Pope (1995) used data from Utah Valley during two intervals in the late 1990s, when a steel mill was shut down due to a labor dispute, to characterize the relationship between $PM_{10}$ concentrations and the incidence of bronchitis and asthma.

Data from extreme and short-term events can demonstrate the link between air pollution and adverse health effects as well as the ameliorative power of improved air quality. However, the data are only marginally relevant to the assessment of health benefits that accrue from gradual changes in pollutant concentrations expected under a broad, incremental program to control air pollution.

### Assessments Using Risk Functions and Exposure Estimates

A second type of assessment method used by EPA in its examination of the health benefits of the CAA combines "risk functions" derived from the health effects literature and data from air quality monitors to estimate public health benefits (EPA 1997). The risk function is an estimate of the incremental change in a health indicator, such as daily mortality, hospital admissions, emergency-room visits, restricted-activity days, respiratory-symptom days, and asthma attacks, that results from an incremental change in the concentration of an air pollutant (or mix of air pollutants). This risk function is then multiplied by the observed change in ambient air pollution over some span of time using data from air quality monitoring (or estimates of emission reductions) to estimate the resulting change in the number of adverse health events.

The risk-function approach is widely used by the regulatory and public policy communities. For example, it has formed the basis for most of EPA's estimates of the health benefits of AQM in the United States (EPA 1996a). However, this approach has at least two potential pitfalls: (1) the trends in ambient air pollution derived from air quality monitors may not be representative of the actual trend in population exposure to pollution; and (2) the cause-and-effect link between improved air quality and improved public health is not directly measured. With respect to the second point, because the studies used to derive the risk functions can be the same as, or similar to, those used in the goal and standard setting process, this type of assessment does not provide an independent verification that the air quality improvements attained by an AQM system had the intended health benefits. Despite these limitations, the risk-function approach has the advantage of providing a relatively simple and straightforward method for translating observed changes in air quality into an associated health outcome and as such will likely continue to be used. Some specific measures that could be taken to improve the estimates derived using this approach are discussed below.

### Assessments Based on Tracking Public Health Status and Criteria Pollutant Risk over Time

There have been a number of epidemiological studies throughout the world directly linking daily variations in PM with either mortality or hospitalization for heart and lung problems. The results of these studies provide a direct assessment of some of the population health impacts of ambient PM for specific locations and time periods. To date, however, there have not been consistent efforts to track health changes over time and to relate those changes to changes in air quality.

There are some indicators of pollution levels in the United States that have been and are being tracked over long periods (for example, EPA's AIRS database for ambient pollutants). However, except for the National Death Index and Cancer Registry, there is no national tracking system for the full range of acute and chronic disease and associated environmental risk factors. Requirements for environmental health tracking are not consistent among the various state and local registries that are in use for chronic disease (EHTPT 2000), making it difficult to draw robust conclusions linking chronic disease clusters to environmental factors beyond local or state boundaries. Several programs are in place that can potentially be expanded or modified to provide a coordinated approach that identifies hazards and clusters of disease occurrences, evaluates exposure and environmental factors, and tracks the overall population health. These programs are the following:

• The Pew Environmental Health Commission examined a number of national health outcome databases for information on chronic diseases linked to environmental factors.  These databases included the National Hospital Discharge Survey, the National Ambulatory Medical Center Survey, and the National Hospital Ambulatory Medical Care Survey.  All provide information on patient demographics and health-care services administered, but they are not designed to describe the communities in which the illness occurred or the environmentally related health outcomes (EHTPT 2000).

• The Centers for Disease Control and Prevention (CDC) Environmental Health Laboratory has an ongoing study to evaluate human exposure to environmental contaminants.  Blood and urine samples are collected from a subset of participants from the larger NHANES[5] effort and are tested for a range of environmental contaminants.

• The National Children's Study (NCS 2002), now in development, is designed to track the health of 100,000 children, starting at a prenatal stage and continuing into the adult years.  The study includes tracking environmental factors, such as chemical exposures and nutrition, but does not monitor chronic diseases that may develop in the later years of life.

In addition, Congress provided CDC with funding in fiscal year 2002 of over $17 million to begin developing a nationwide environmental public health tracking network and to further develop capabilities to track environmentally related health problems within state and local health departments.  CDC's (2003a) goal is to develop a tracking system that integrates data on environmental hazards and exposures with data on diseases that are possibly linked to the environment to do the following:

• Monitor and distribute information about environmental hazards and disease trends.

• Advance research on possible linkages between environmental hazards and disease.

• Develop, implement, and evaluate regulatory and public health actions to prevent or control environmentally related diseases.

With this appropriation, CDC funded 17 states, 3 local health departments, and 3 schools of public health to begin developing such a tracking network and state and local capabilities.  Congress also is considering a broader bill entitled the Nationwide Health Tracking Act to develop a comprehensive system for identifying and monitoring chronic diseases and potentially correlating their causes with environmental, behavioral,

---

[5]NHANES (National Health and Nutrition Examination Survey) is a series of surveys conducted by CDC designed to collect data on the health and nutritional status of the U.S. population.

socioeconomic, and demographic risk factors. The information gathered from such a program could be used as a basis for taking steps to alleviate the sources of the disease affecting a particular community or the population as a whole.

There are a number of ways that an improved national health database, when coupled with ambient air quality data, could be used to derive a metric for the effects of AQM on health outcomes. For example, periodic epidemiological studies could be carried out that independently test for the association between ambient pollutant concentrations and the incidence of adverse health effects (Burnett et al. in press). If, over the course of the studies, the association between health outcomes and PM concentrations remained constant while PM concentrations declined, the association between the two would likely be robust and could be used along with the ambient data to directly estimate the premature deaths or hospitalizations avoided as a result of the air quality improvements. If the association changed—suggesting either a stronger or weaker association between air pollution and health—the indication would be that AQM actions taken changed the pollution mix to make it less or more toxic and that more analysis was needed before the actual health benefits accrued from AQM could be assessed. Although this approach is relatively new and has not been fully tested, it is encouraging to note that a similar approach was recently applied to analyzing health trends following an Irish government action to ban the use of coal in homes in Dublin, and a measurable improvement in premature mortality was observed (Clancy et al. 2002).

The development of these and other new approaches is indicative of the need perceived by the health effects research community to measure health changes more directly as air quality improves. This information could be used to hold air pollution regulations "accountable" for their health performance and to confirm that the improvement projected by new regulations actually occur. Several new initiatives are under way to develop these approaches further—for example, EPA is developing scientifically rigorous environmental indicators that can be tracked over time (Whitman 2001), and the Health Effects Institute is funding the development and testing of a variety of such techniques (HEI 2002).

### Efforts to Track the Effects of HAP Emission Reductions

The challenges for assessing progress in health improvement as a result of reductions in emissions of HAPs are daunting, probably more so than those associated with criteria pollutants. As discussed above, the absence of a comprehensive air monitoring network for even a subset of HAPs makes direct measurement of long-term changes in exposure impossible (although recent efforts are attempting to fill that void). In addition, unlike the criteria pollutants for which much of the health effects data provide evidence of relatively short-term, measurable changes in health, a major focus of HAP effects is cancer, which has long latency periods, making short-term tracking less fruitful. For example, recent reductions in benzene exposure might not result in measurable changes in cancers for decades. As a result of these limitations, EPA declined to provide an analysis on HAPs in its 1997 report to Congress on the benefits and costs of the CAA (EPA 1997).

Despite these challenges, there have been a number of efforts to develop baseline cancer risk estimates and hazard index calculations using dose-response information and exposure estimates derived from EPA's 1990 cumulative exposure project (Axelrad et al. 1999) and similar projects. Cancer risk is estimated as the potency of the substance multiplied by an average daily dose. For substances assumed to have a threshold dose (below which health effects are not expected), exposure measures are compared with a reference concentration. Cumulative hazard index is the sum of ratios of pollutant exposure estimates to reference levels. These efforts are analogous in some ways to the baseline estimates for criteria pollutants used to justify regulatory action under the NAAQS and at least provide a start for future assessments as changes in HAP emissions are implemented.

The largest of these efforts is EPA's National Air Toxics Assessment (NATA) (EPA 2002e), which characterizes inhalation risks from exposures to HAPs identified in the urban air toxics strategy (EPA 2000b) and from exposures to particles in diesel exhaust. NATA has provided a tool for exploring control priorities and has served as a preliminary attempt to establish a baseline for tracking progress in reducing HAP emissions. NATA provided estimates for the year 1996 and attempted systematically to evaluate and link emissions, ambient measurements, atmospheric transport, and human exposure. The main components of the assessment were the list of urban air toxics (EPA 2000b), the 1996 national toxics inventory of emissions, the Assessment System for Population Exposure Nationwide (ASPEN) dispersion model for developing estimates of ambient concentrations of HAPs from stationary emissions, the hazardous pollutant exposure model for developing exposure estimates for vehicular HAPs, microenvironment factors to establish the relationship between ambient concentrations and microenvironments of interest, toxicity metrics relating exposure to risk, and risk assessment guidelines providing further guidance on estimating risks from exposures.

A number of preliminary findings were of interest. Twenty-three of the 33 HAPs analyzed were estimated to present risks greater than 1 excess cancer incident in 1 million people, a risk threshold noted in several portions of Section 112 of the CAA. Major and area stationary-source emissions and on-road and nonroad mobile emissions were each associated with risks of concern. A number of caveats were noted by EPA, indicating undercharacterization of risk and identifying major future research and data needs. For example, higher localized risks were not captured, and comparison of a national impact assessment with local impact assessments suggests underestimation for certain microenvironments in both urban and rural areas by factors of 30 and 100, respectively. Further, NATA's attempts to compare modeling results with ambient data found the model underpredicted ambient exposures to some metals by a factor of 5. The tendency toward underestimation has been suggested to be due to gaps in the emissions inventory, but reductions in pollution releases since 1996 have effectively reduced the significance of these gaps, increasing the consistency between NATA estimates and current monitoring data (Environmental Defense 2002). Observed ambient concentrations of benzene were slightly underpredicted by the models, reflecting a lack of understanding of the emissions and environmental behavior of this compound. The assessment did not address dioxins, indoor air exposures, or noninhalation pathways. It also made no adjustments for other HAPs not listed in EPA's urban air toxics strategy.

EPA (2002r) and the EPA Science Advisory Board (EPA/SAB 2001) have acknowledged that the current NATA effort has many limitations. In the absence of systematic monitoring of HAPs and the large gaps in personal HAPs exposure data, no comprehensive validation of the NATA models has been done. Although the health risk data for the compounds vary in quality, the estimates of cancer and other risks continue to be limited by the lack of a robust human database that avoids some of the challenges of extrapolation from higher concentrations. In its urban air toxics strategy, EPA is funding work to improve both the health database and the NATA. Pilot monitoring projects have been initiated to better characterize trends in ambient concentrations and to generate information on the temporal and spatial variability of ambient HAP concentrations (STAPPA/ALAPCO/EPA 2001). These efforts will likely address some of the pressing problems associated with tracking health impacts from HAPs; however, the remaining data gaps are significant.

## Other HAP Assessments

In addition to NATA, there have been a number of conceptually similar assessments of HAPs, although none has evaluated trends at the national level. These assessments have included the following:

• Cancer and noncancer HAP risks calculated as part of EPA's comparative risk project were considered sufficiently serious (relative to risks addressed by other EPA programs) to rate HAPs as high risk (EPA 1987).

• The California South Coast Air Quality Management District completed an assessment of trends in cancer risk from the mid-1980s to 1997 on the basis of a monitoring and modeling effort for 20 HAPs. The multiple air toxics exposure study (MATES II) used traditional techniques to study urban HAPs in California's South Coast Air Basin (encompassing Los Angeles, Riverside, Orange, and San Bernardino counties) (SCAQMD 2000).

• An assessment for Minnesota estimated exposure to 148 HAPs from EPA's 1990 cumulative exposure project. Relatively high concentrations were found in the St. Paul metropolitan area and many smaller cities throughout the state.

• The California Comparative Risk Project (CCRP) established risks of high concern for a number of HAPs, especially when more highly exposed populations were addressed (CCRP 1994). A more recent attempt to use monitoring data to conduct a comprehensive evaluation of the impact of HAPs noted the paucity of toxicity measures for many HAPs and the limited monitoring data for HAPs in California, a state with one of the most extensive monitoring networks. It was concluded that the information necessary to assess fully the health significance of HAPs was not available (Kyle et al. 2001).

### Tracking Progress in Reducing HAPs-Related Health Effects for the Future

In summary, the ability to assess changes in health related to reductions in HAP emissions has been limited by the absence of high-quality data on ambient concentrations, the limitations in the health database, and the challenges of tracking health progress for diseases, such as cancer, that have long latency periods. Over the past decade, EPA and others have made strides to improve the techniques available for such assessments, most notably through the recent development of NATA. Both that assessment and the urban air toxics strategy noted limitations in the current analyses and the steps needed to be taken to improve them. Over time and with better air quality data, improved validation of exposure models, and enhanced health understandings, these techniques will track progress much more directly than those used today. Implementation of these efforts will require sustained attention and resources over the coming decades.

### Monitoring Actual Human Exposure

AQM progress can be also be evaluated by directly tracking human exposure to air pollutants over time. Because concentrations measured by ambient monitors may not reflect human exposures, personal monitors worn by study participants have been used, and sampling of biomarkers in, for example, blood and breast milk has been done. Corresponding environmental samples provide information on source contribution (for example, ambient air, indoor air, and food) (EPA 2002s). EPA has initiated the National Human Exposure Assessment Survey (NHEXAS) to document status and trends of national distributions of human exposure, and to that end, EPA has begun a series of pilot projects. A stated objective is to evaluate the efficiency of EPA's regulations to reduce exposure. NHEXAS attempts to evaluate total human exposure to chemicals of concern and from contributions from different sources (see, for example, Clayton et al. 2002). Personal air monitors measure air breathed by individuals, diet and other samples are taken to address other routes of exposure, and blood and urine samples are obtained to characterize total exposure. An exploratory study under NHEXAS also evaluated selected subpopulations that might exhibit higher exposures (Pellizzari et al. 1999).

Methods using biomarkers as a surrogate for exposure to air pollution and other sources of environmental stress are developing rapidly and have great promise. Human biomarkers of air pollution are specific compounds or cellular changes that appear within the body, indicating the occurrence of exposure and a stress-related response to a specific pollutant or set of pollutants. These biomarkers can be the pollutant itself, one of its metabolites, or an enzyme produced in response to the pollutant or its metabolites. They can also be specific for modification of DNA (for example, DNA adducts). Biomarkers can potentially appear in blood, urine, or a specific organ where a pollutant might accumulate or do damage.

Because exposure biomarkers are a specific indication of exposure and a primary step in the causal chain of a disease outcome, their appearance provides stronger evidence of an actual health outcome from exposure to a pollutant compared with measurements of ambient pollutant concentrations. As such, tracking the appearance and abundance of specific biomarkers within a population over time may be a useful and convincing way to document a public health benefit (or lack thereof) from air quality improvements. Risk functions based on the presence and concentration of a given biomarker can then be used to estimate specific health changes more reliably. A classic example of the utility of this approach is the study linking decreases in Pb emissions from mobile sources to declining concentrations of Pb in ambient air and human blood (see Figure 6-7A-C). A recent CDC report illustrated another example: a measurable decline in the levels of cotinine—a marker for exposure to environmental tobacco smoke —in a random sample of the U.S. population found in tracking the increasing restrictions on smoking in public spaces (CDC 2001).

However, the development of compound-specific markers of actual effects or early stages of effects has been much more difficult. To date, several compound-specific markers of exposure have been developed (for

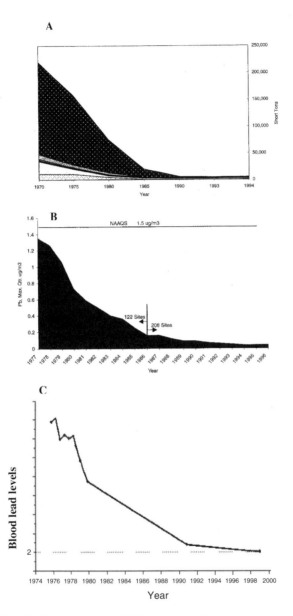

FIGURE 6-7 (A) Total estimated U.S. lead emissions by major source category from 1970 to 1994. (B) Maximum quarterly observed lead particulate concentrations (in micrograms per cubic meter) at U.S. monitoring sites from 1977 to 1996. (C) Blood lead levels (in micrograms per deciliter) in the U.S. population from 1976 to 1999. SOURCES: A and B, from NARSTO 2000. Reprinted with permission; copyright 2000, EPRI, Palo Alto, CA. C, from CDC 2002.

benzene and 1,3-butadiene in addition to Pb and cotinine), but markers of actual effects (for example, DNA changes) have in many cases been linked to more than one compound, perhaps tracking the effects of the mixture but making it more difficult to understand how changes in components of the mixture affect health. In addition, science has been unable to develop an adequate marker of either exposure or effects for some compounds (for example, diesel exhaust) (HEI 1995).

Efforts to develop such markers—either of compounds in the body or, ideally, of actual effects—could potentially be very useful in tracking the presence and effects of biopersistent chemicals. For example, there has been some success in population-based monitoring of breast milk for persistent organic pollutants (POPs) (Solomon and Weiss 2002). Impacts via noninhalation exposure routes (for example, bioaccumulation of ingested substances such as dioxins) are typically undercharacterized, and a population-based biological monitoring program could significantly improve the capacity to track progress in reductions of bioaccumulating HAPs. In the long run, the development of a battery of sensitive and relatively easy to apply markers that could be measured in blood and urine samples of the U.S. population would be invaluable in tracking the current status of exposure to air pollution and its effects. The effort of the CDC to extend the NHANES program to investigate population exposures to a number of such chemicals (CDC 2003b) is especially valuable and merits continued support and enhancement, as does the EPA NHEXAS effort.

## ASSESSING ECOSYSTEM BENEFITS FROM IMPROVED AIR QUALITY

In light of the CAA's clearly stated goal of protecting public welfare, assessment of benefits to ecosystems should be viewed as an integral part of the AQM system. Measuring such benefits relies on a multitiered approach that includes the following:

• Long-term monitoring of ecosystem condition simultaneously with air pollutant exposure and meteorology to elucidate patterns and trends in exposure and response in time and space.
• An enhanced network of meteorology and exposure measurements, including more measurement locations and variables (for example, incident diffuse and direct radiation) and further development of meteorology and exposure models for landscape and regional estimates of exposure and effects.
• Intensive ecosystem studies at a subset of representative locations to understand mechanisms of response to multiple factors, including air quality, climate, and topography.
• Process model development and application to investigate causes and mechanisms of observed responses. Also, to assess current and future

exposures and responses of ecosystems to multiple stressors at multiple temporal and spatial scales.

• Risk assessment research that combines the above approaches to develop methods for quantifying susceptibility of ecosystems to multiple stressors at multiple levels (Linthurst et al. 2000).

The greatest challenge to understanding responses of ecosystems to air pollution is that natural systems are exposed to multiple stressors, and the integrated response can be different from expectations based on pollutant exposures in controlled environments. For example, the response of a forest to a pollutant can be exacerbated or minimized by climatic conditions, topography, forest community dynamics, and other factors, such as increased atmospheric $CO_2$ (Drake and Leadley 1991). Moreover, the effects of air pollutants on forest ecosystems are often indirect, involving the alteration in the physical or chemical environment of the vegetation or increasing the susceptibility of affected vegetation to attack by pests (see Johnson and Lindberg 1992).

Unfortunately, the nation's AQM system has not been able to build a cohesive program capable of reliably reporting the status and trends in exposure and ecosystem conditions across regions and the nation (Farrell and Keating 1998). Some aspects of the current program and its deficiencies are discussed below.

### Tracking Ecosystem Exposure

Evaluation of ecosystem exposure is conducted using data from the air quality and deposition monitoring networks. A major deficiency of the current network design for air pollutants is the scarcity of air quality monitors in rural and forested areas. A more general and more basic deficiency is that the monitors can make measurements only at fixed locations. As a result, aerial coverage is relatively small. Satellite-based instruments, such as the total $O_3$ mapping spectrophotometer (TOMS) (NASA 2003), produce daily global mapping of total air column $O_3$ (including detection capabilities for smoke from biomass burning, desert dust and other aerosols, and sulfur dioxide and ash emitted by volcanoes) and provide a means to document the spatial distribution of airborne substance concentrations. At present, it is not possible to evaluate ecosystem exposure or ground-level concentrations of pollutants using current remote-sensing techniques. New developments are needed to improve estimates of exposure and concentrations at spatially and temporally relevant scales.

Meteorology is one of the major factors contributing to air pollution episodes (Seaman 2003). Over the next 5–20 years, better spatial and temporal resolution of meteorological observations and measurement of additional variables (for example, incident direct and diffuse radiation and water vapor exchange) are required as inputs to meteorological models to advance studies on ecosystem exposure and effects. Reduced uncertainty in these studies will depend on the effective use of new meteorological measurements, improvements in the models, and the use of remote-sensing observations in more advanced data assimilation in the models to adapt them to spatially and temporally relevant scales (Seaman 2003).

## Tracking and Characterizing Ecosystem Effects from Exposure to Air Pollution

### Forest Issues

The direct effects of air pollution on plants has been a primary focus of air pollution research. However, indirect effects on other ecosystem components (for example, soils, surface waters, and estuaries) are less well understood, and the effects on vertebrates are virtually unknown (Innes 2003). Most studies of pollution impacts have concentrated on the responses of plants to single pollutants, although a few studies have looked at potential interactions with one or two additional factors (see Isebrands et al. 2001). Interactions of multiple pollutants and such factors as soil fertility, drought, pests, and pathogens are important (Karnosky et al. 2001). Most risk assessment has similarly been done with individual species or groups of species. However, risk assessments should also address ecosystems with attention to cascading effects (direct and indirect effects, Heck et al. 1998), and at multiple scales (for example, Riitters et al. 1997).

### Forest Soils

Soils promote vegetation growth, control water flow, and filter or retain potentially harmful chemicals that would otherwise enter surface waters and groundwater. These functions are related to specific soil physical, biological, and chemical properties that can be monitored. An NRC review (1993b) identified soil organic matter (SOM) as the single most important indicator of soil quality and productivity (also see Johnson 1993; Bauer and Black 1994; Reeves 1997). SOM strongly influences site productivity because of effects on physical (for example, bulk density and soil water-holding capacity), biological (for example, populations of microbes and soil invertebrates), and chemical properties (for example, cation exchange capacity) of soils.

Acid precipitation probably represents the most serious and widespread threat to sensitive forest soils from air pollution in the United States at this time. Acids from atmospheric oxidation and deposition of sulfur and nitrogen-containing compounds have accumulated in soils in the United States and are causing cascading chemical effects that can ultimately deplete nutrient cations from soil (that is, calcium and magnesium), change community structure and suppress forest growth, degrade water quality, and harm fish  in sensitive regions (Cronan and Schofield 1990; Cronan and Grigal 1995; Bailey et al. 1996; Driscoll and Postek 1996; Likens et al. 1996, 1998; Driscoll et al. 1998; Lawrence et al. 1999; Huntington et al. 2000; Driscoll et al. 2001b; Lilleskov et al. 2002).  These effects could continue for decades and, in some cases, require centuries to recover fully (Stoddard et al. 2000; Driscoll et al. 2001b) (Box 6-6).

---

**BOX 6-6   The Response of Sensitive Ecosystems to Acid Rain Emission Controls**

Despite marked decreases in $SO_2$ emissions and sulfate deposition over the past decade, sensitive areas have not recovered from acid deposition at the rates anticipated at the time of the passage of the 1990 CAA Amendments. In March 2001, the Hubbard Brook Research Foundation released a report concluding that sensitive areas, such as the Adirondack region of New York, are not recovering from the long-term exposure of acid deposition because of the diminished ability of the affected soils to neutralize inputs of strong acids (Driscoll et al. 2001b). The General Accounting Office also reported that affected surface waters in the Adirondacks were not recovering under the current Acid Rain Program (GAO 2000). However, a report by EPA (Stoddard et al. 2003) and research by Driscoll et al. (2003a) suggest some improvements in the acid-base status of surface waters in acid-sensitive regions. Although the most recent trends are promising, many waters remain unresponsive to decreases in emissions, and those showing improvement are recovering at a relatively slow rate. This research suggests that for sensitive ecosystems to recover within the next 20 to 25 years, greater $SO_2$ emission reductions—on the order of 80% beyond the Phase II cap—will be needed. A further concern is the 10.4 million tons of banked $SO_2$ emission allowances that can be withdrawn and thereby lessen Phase II reductions.

More recent scientific work pointing to the importance of nitrates in acid deposition suggests that further reductions in $NO_x$ emissions will also be needed (for example, Castro and Driscoll 2002; Aber et al. 2003; Driscoll et al. 2003b; Fenn et al. 2003 a,b; Galloway et al. 2003). $NO_x$ emission reductions are also desirable in terms of controlling regional ambient $O_3$ concentrations during the warmer months of the year and mitigating eutrophication of coastal waters. In part to address these concerns, recent regulations have been promulgated to bring about further reductions in $NO_x$ emissions. These regulations include the Tier II regulations Ozone Transport Commission $NO_x$ budget program, the EPA $NO_x$ SIP call, and states initiatives in the eastern part of the United States. Several national legislative proposals are also pending in Congress that call for further reductions of $SO_2$ and $NO_x$ emissions in the range of 50% to 75% of current levels.

The nation does not yet have a coordinated program to monitor soil-chemistry perturbations induced by acid precipitation. The merging of the U.S. Department of Agriculture (USDA) Forest Service's Forest Inventory and Analysis (FIA) and the Forest Health Monitoring (FHM) programs, designed to track soil condition and plant status, is a positive step; however, the program is not designed specifically to monitor impacts related to air pollution and acid deposition. Although soil acidity is one of the variables being monitored in the FIA/FHM program, soil acidity can be influenced by a number of factors; atmospheric deposition may or may not have a major role, depending on location and soil type. Extractable soil sulfur measurements, which may be useful for quantifying sulfur deposition, were added to the program in 2001. The "2000 Forest Health Monitoring National Technical Report" will provide information on soil monitoring and summarize some of the soil-chemistry data (including soil pH) that have been analyzed to date (Stolte et al. in press).

### Forest Vegetation

Air pollution has a number of avenues to affect forest vegetation. Gaseous pollutants can enter a plant through the leaf pores (stomata) that remain open during the day to facilitate photosynthesis through the uptake of $CO_2$ from the atmosphere. Pollutants deposited in soils or toxic elements released in the soil from pollutant-induced chemical perturbations can enter the plant through the root system during the uptake of nutrients. Damage can also occur below ground as a result of changes in root activity or microbial processes; however, little is known about these effects. Finally, air pollution may either directly or indirectly stress vegetation and make it more susceptible to drought, insects, or pathogens.

No coordinated monitoring programs exist in the United States that are dedicated to tracking air pollution impacts on forested ecosystems. However, a number of programs monitor forest health through periodic sampling at selected sites. These programs include the FIA/FHM and the National Forest System (NFS) of the USDA Forest Service. A particularly promising aspect of the FIA/FHM program is the use of an $O_3$ bioindicator sampling system to track air pollution impacts. With this system, a variety of plants that respond to high ambient $O_3$ concentrations with distinct visible foliar symptoms are used to detect and quantify $O_3$ stress in the forest environment. Each year, a subset of plots (one plot for every 500,000 acres of forestland) is evaluated for the amount and severity of $O_3$ injury on sensitive plants during the summer months. The foliar injury data are used to monitor changes in relative air quality over time and to examine relationships between $O_3$ stress and tree health. Presumably, similar sampling programs could be introduced to track damage from other environmental stressors, such as wet and dry deposition of sulfates, nitrates, other sources of acidification, and ammonium, as well as to track changing climatic and forest stand conditions.

### Sensitive Surface Waters and Estuarine Systems

Deposition of air pollutants can also perturb the chemistry and ecology of sensitive surface waters. Acid deposition is of concern for acid-sensitive lakes and streams, primarily in the Northeast and the mid-Atlantic highlands (see discussion in Box 6-6). Further, deposition of nitrates and ammonium has been shown to be a major source of nutrients and associated eutrophication of estuarine systems along the eastern seaboard of the United States (Castro and Driscoll 2002). There is also considerable interest in the effects of atmospheric mercury deposition on coastal and other ecosystems.

A number of programs are in place to monitor aspects of this problem. The most comprehensive atmospheric deposition monitoring networks are NADP/NTN, NADP/MDN, CASTNet, and AIRMoN (for example, see Figure 6-6). However, these networks do not track dry gaseous ammonia ($NH_3$) deposition, which could be important in estuaries that drain agricultural watersheds (for example, Delaware Inland Bay, Chesapeake Bay, and Neuse River Estuary). Another shortcoming of these programs is that they do not measure organic nitrogen deposition, which may account for 10–30% of the total nitrogen budget to East Coast estuaries (Peierls and Paerl 1997; Whitall and Paerl 2001).

In addition to the networks noted above, EPA administers the temporally integrated monitoring of ecosystems (TIME) and long-term monitoring (LTM) programs, which monitor lake and stream chemistry and document changes in response to changing emissions and acidic deposition. There also are several national coastal monitoring networks (for example, the EPA national estuarine program and NOAA's National Estuarine Research Reserve System [NERRS]), but none directly assesses the effects of atmospheric deposition of pollutants. The NERRS has 20 coastal sites (and one Great Lakes site) and includes a systemwide monitoring program through which data for pH, salinity, dissolved oxygen, temperature, and turbidity are collected (NERRS 2001). No attempt is made to relate the oxygen status of estuaries directly to nutrient loading from either atmospheric or watershed sources except at very basic levels. Similarly, a study that documents estuarine water quality problems related to runoff was published by NOAA's NERRS (Bricker et al. 1999), but no direct link to atmospheric deposition was made.

The National Estuaries Program (NEP) was established to allow local groups to take responsibility for tracking estuarine function, and perhaps as a result, there is no systemwide effort to document the effects of nitrogen or mercury deposition. Nitrogen loading as a water quality problem has been addressed for several estuaries (for example, Long Island Sound and Tampa Bay) in the program but only to assess inputs from specific major stationary sources of concern.

The best examples of the integration of atmospheric deposition measurements with coastal effects can be found at specific sites where the research community and government agencies are actively engaged in understanding these problems. These efforts are usually multidisciplinary and involve cooperative efforts between university researchers and state or federal scientists to look holistically at a system plagued by problems, such as coastal eutrophication, associated with atmospheric deposition. For example, Neuse River estuary modeling and monitoring program involves researchers from several North Carolina universities as well as cooperation from the North Carolina Department of Environment and Natural Resources and Weyerhauser Corporation. This program has quantified not only the fluxes of atmospherically deposited nitrogen to the system (Whitall and Paerl 2001) but also the ecological effects of nitrogen loading from the atmosphere and from other sources (Peierls and Paerl 1997; Paerl et al. 1998). Similar efforts have been made for the Chesapeake Bay (Russell et al. 1998; CBADS 2001), including the development and application of the Chesapeake Bay eutrophication model (Cerco 2000).

## Agriculture

Because of their direct economic import, the effects of air pollutants on agricultural crops are of particular concern. Probably the most comprehensive assessment of the agricultural losses incurred from exposure to air pollutants was conducted in the 1980s as part of the U.S. National Crop Loss Assessment Network (NCLAN) (Heck et al. 1988; Preston and Tingey 1988). NCLAN goals were (1) to conduct experiments using chambers with tops open to the atmosphere (see Figure 2-12 in Chapter 2) to relate doses of $O_3$ to yields of economically important crops in several major areas in the United States; (2) to estimate actual crop losses over the United States by combining the $O_3$-dose-to-crop-yield information with the data on crop acreage and pollutant levels in each county; (3) to assess dollar losses each year from these pollutant effects; and (4) to create models that relate yields to level of pollutant, water stress, stage of crop development, and temperature, using the results to determine the NAAQS based on injury thresholds.

There are three noteworthy aspects of the NCLAN study: (1) the data from this study are over two decades old and still represent the most comprehensive information on $O_3$ effects on crops and are widely used to assess crop losses in the United States from air pollution; (2) the study found that crop yields are depressed substantially when $O_3$ concentrations reach about

50–70 parts per billion by volume (ppbv), a level not uncommon in rural areas of the United States during the summer; and (3) the study indicated that crops are best protected from $O_3$ damage by an air quality standard that limits the integrated exposure over the 3–4 month period that the crop is growing rather than the relatively short-term 1-hr or 8-hr primary NAAQS used to protect human health (EPA 1996b). In part, because of the NCLAN study, EPA recommended an alternative secondary standard in its 1996 staff paper to review the $O_3$ NAAQS. This standard would regulate $O_3$ in a seasonal, cumulative manner and was designed to be more protective of vegetation. However, it was never implemented (Heck et al. 1998).

The Agricultural Research Service (ARS) in the U.S. Department of Agriculture (USDA) has the Air Quality-Plant Growth and Development Research Unit to carry out the following objectives: determine the separate and combined effects of $O_3$ and elevated $CO_2$ on growth and yield of selected agronomic species; determine whether plant, pest, and pathogen interactions are altered by exposure of plants to these pollutants; and develop techniques for mitigating the problems (USDA-ARS 2002). The effects of $O_3$ and $CO_2$ are studied individually, in combination, and in interaction with other factors associated with changes in global climate, and research is conducted under field, greenhouse, and laboratory conditions. The research unit is also working with crop-growth models for evaluation of air quality impacts on production. The future plans of the unit are to investigate the extent to which rooting media affect soybean response to mixtures of $O_3$ and revised $CO_2$ and to investigate the effects of increased concentrations of $CO_2$ and $O_3$ on a few insect pest populations.

## Integrated Ecosystem Studies

A major problem of impacts research is the difficulty of predicting ecosystem-level responses from short-term studies of young trees grown under controlled conditions. More realistic studies, such as FACE experiments, are needed (Karnosky et al. 2001). To assess the effect of AQM on ecosystems, programs are needed that integrate measurements and analysis across terrestrial and aquatic systems, integrate disciplines, and integrate information across vegetation types and climatic and physiographic regions. To date, a number of research programs have been initiated that attempt to achieve this level of integration. However, for the most part, they are focused on unraveling the links between climate change and ecosystem function and are not considering air pollution per se as a perturbing factor. Some studies of note are described below:

• AmeriFlux is a network of more than 100 sites dedicated to studying the influence of climate variation, vegetation developmental stage, vegetation type, disturbance, and other factors on ecosystem processes controlling the exchange of $CO_2$, water vapor, and energy. Measurements are made of meteorology, $CO_2$, water vapor, and energy exchange, as well as soil and plant processes (respiration, photosynthesis, transpiration, and production) in intact ecosystems. The Free Air $CO_2$ Enrichment (FACE) sites (see discussion in Chapter 2 and Figure 2-13) involve long-term experimental studies on the effects of increased atmospheric $CO_2$ on terrestrial ecosystem processes (for example, photosynthesis, respiration, transpiration, and carbon allocation within plants). Atmospheric $CO_2$ is enriched in natural ecosystems through a pipe distribution system, and ecosystem processes are compared with reference plots. Some of the FACE sites also measure concentrations of $O_3$ or methane or nitrogen deposition.

• National Science Foundation long-term ecological research (LTER) sites (24 sites) have core areas of research on plant production, population distributions representing trophic structure, soil processes, and disturbance patterns and frequency.

A key feature of the first two aforementioned studies is the focus on the effects of changing $CO_2$ concentrations. Comprehensive studies of the effects of air pollutants on terrestrial ecosystems should probably also consider the effects of increased atmospheric $CO_2$. The atmospheric concentrations of $CO_2$ are clearly increasing at a faster rate than has occurred during the evolution of current vegetation (Indermuhle et al. 1999), and it is conceivable that ecosystem response to air pollutant exposures will change as a function of $CO_2$ concentrations. For example, the interaction between increased nitrogen availability from nitrogen deposition, increased atmospheric $CO_2$, and water availability can result in greater effects on carbon uptake and allocation by some species than others (Hungate et al. 1997) and may have implications for changes in plant community composition and biogeochemistry.

## Action Needed for Enhanced Ecosystem Monitoring, Research, and Risk Assessment

Improved and sustained long-term monitoring of ecosystem condition and its relationship to air pollution exposure is essential if the nation's AQM system is to have a credible capability to protect ecosystems and monitor progress. A unified multiagency network to monitor the ecosystem, air quality, and meteorology will likely be required to accomplish that capability (Farrell and Keating 1998). Intensive ecosystem studies to understand the influence of air pollutants on ecosystem processes and community dynamics have been conducted by independent research programs at academic and research institutions but not as part of a larger integrated approach to measuring progress in air quality management. Development of mechanistic ecosystem models that quantify and propagate uncertainty is essential as they represent an integration of direct and indirect effects on ecosystems. Model development should be an integral part of ecosystem studies and modeling. The models should be appropriate for application across regions to estimate response at the scale of air pollutant impacts. Such application requires linking atmospheric models (for example, exposure and meso-scale climate models) with ecosystem process models (for example, carbon, water, nitrogen, and element cycling) and community dynamics models. Finally, selection of measurement end points should be done with the regulatory community to ensure that the measurements will provide a basis for accurate and defensible risk assessment (Laurence and Andersen 2003). For example, end points should go beyond productivity and species composition to include integrity of soil food webs; quantity and quality of water supplied from terrestrial ecosystems; wildlife and recreational values; and transfer and fate of carbon, nutrients, and water within the systems (Laurence and Andersen 2003). In Appendix D, a more detailed discussion of the new and expanded program elements that are needed in ecosystem research and monitoring is presented. The committee notes that similar recommendations can be found in previous reports, such as the EPA "Ecological Research Strategy" (1998e), the White House Office of Science and Technology Policy's Committee on Environment and Natural Resources proposed framework for integrating the nation's environmental monitoring and research networks and programs (NSTC 1997a), the NRC's (2000d) report *Ecological Indicators for the Nation,* and the Heinz Center's (2002) report *The State of the Nation's Ecosystems.*

## ASSESSING THE ECONOMIC BENEFITS OF AIR QUALITY IMPROVEMENTS

### Overview

A final step that can be taken to measure the effectiveness of AQM involves an economic assessment of the costs and benefits of the policies and regulations implemented to manage air quality. Because such assessments require that health and welfare benefits, such as fewer cases of asthma in children and increased visibility in our nation's parks, be assigned a specific monetary value, they are controversial. Nevertheless, economic assessments are intrinsic to the system, because they provide policy-makers with a tangible, quantitative measure of the net gains obtained from AQM (NRC 2002a). Indeed, the 1990 CAA Amendments explicitly direct the EPA administrator to carry out a retrospective assessment followed by biannual prospective assessments of the costs and benefits of implementation of the CAA.

In an economic assessment, costs and benefits are estimated relative to a baseline. For example in EPA's most recent retrospective assessment, the baseline was modeled to replicate conditions in the absence of implementation of the CAA (EPA 1997). Progress in air quality and the resulting human health and welfare benefits were then estimated using the air quality data and methods described in the preceding sections of this chapter. In EPA's first prospective assessment (EPA 1997), conditions in 1990 were chosen as the baseline, and models, instead of air quality data, were used to estimate the air quality benefits from continued implementation of the CAA. A similar approach is taken in the Office of Management and Budget's (OMB's) annual cost-benefit analyses of AQM and other federal regulatory programs (for example, OMB 2003a), in assessments of the Acid Rain Program (Burtraw et al. 1998), and in many regionally specific assessments (for example, Winer et al. 1989; Hall et al. 1989, 1992; SCAQMD, 1996; Krupnick et al. 1996; Lurmann et al. 1999; Burtraw et al. 2001a). An alternative approach uses a shorter-term episode or event to establish a baseline. For example, Ransom and Pope (1995) used air quality and epidemiological data from the Utah Valley during two intervals in the late 1980s when a steel mill was shut down (because of a labor dispute) to estimate the health costs incurred by the 200,000 residents of the Utah Valley because of their exposure to normally increased concentrations of $PM_{10}$.

Most economic assessments of AQM in the United States conclude that the economic benefits of implementation of the CAA exceed the economic costs. For example, EPA estimated that the economic benefits of implementation of the CAA between 1970 and 1990 exceeded the costs by a range of about $5–50 trillion (EPA 1997). The estimates from EPA's prospective analysis for 1990–2010 had smaller but still positive net monetary benefits (EPA 1999a).[6] OMB (2003a) estimated a range in the monetary benefits of regulation for 1992 to 2002 to be approximately $121 to $193 billion and a range in costs to be $23 to $27 billion. In virtually all of these assessments, the major monetary benefit arises from deferred mortality associated with reductions in atmospheric $PM_{10}$ and $PM_{2.5}$. As discussed below, these estimates raise some challenges to the robustness of the estimated net monetary benefits.

---

[6]Some critics have questioned the existing framework of estimating the benefits and costs of environmental improvements. Parry (1995) and Goulder et al. (1999) point to hidden costs of additional regulations in society (see also the special issue of *Journal of Risk and Uncertainty*, Vol 8, No. 1, 1994). Williams (2002) points to the hidden benefits of additional regulations. An important issue for EPA is whether it continues to remain open to these criticisms and new ideas.

## Economic Assessments

Economic assessments (as they are now generally done) attempt to provide a transparent process for relating economic benefits to a series of specific measures to reduce pollutant emissions, population and ecosystem exposure, and adverse health and welfare effects. Consequently, a wealth of information is provided to decision-makers about the probable future gains from better air quality. Aggregate analyses, such as those mandated in the 1990 CAA Amendments and carried out by EPA, also provide a sense of scale and can support assessments of the relative benefits of alternative regulatory policies (for example, focusing resources on reducing ambient concentrations of criteria pollutants versus hot-spot concentrations of HAPs).

The cost-benefit analyses carried out thus far by EPA in response to the directive of the 1990 CAA Amendments have undergone extensive peer review, and EPA's Science Advisory Board (EPA/SAB 1997) concluded that the overall results were sound and useful for broad policy purposes. The qualitative consistency between the retrospective analyses of EPA and OMB also lends credibility to the analysis. However, concerns remain about aspects of the methods used; some are discussed below.

- The estimates of the health benefits from improved air quality were based on the same health-effects literature that was used to develop the NAAQS; thus, the assessment does not represent an independent confirmation that such benefits were attained.
- The dominant contributor to the benefits estimates was the estimate of deferred mortality from reduced population exposure to PM. Thus, the aggregate benefits of the CAA are sensitive to the dose-response relationship adopted for PM and the monetary value assigned to deferred mortality. Different choices for either of those parameters drawn from the mainstream literature can change the net benefits estimated by a factor of nine (Burtraw et al. 2003). The principal study used for benefits estimation to date (Pope et al. 1995) was subjected to intensive reanalysis by an independent team of analysts who in large measure confirmed the results (HEI 2000). However, new analyses of the dose-response relationships conducted in different populations will provide enhanced understanding of these relationships. Several such studies are now under way.
- The measure of effects used in these analyses—the numbers of lives lost to exposure—is controversial. One important source of controversy is the epidemiological evidence that the very young, the old, and the infirm are most susceptible to environmental exposures, and economic valuation of changes in health status is often based on prime-age adults. Economic estimates based on willingness to pay to avoid small risk to health status

indicate that, at least for valuing the benefits of environmental improvements for older populations, the usual economic measures are fairly accurate (Krupnick et al. 2002). Other metrics to monetary valuation hinge on cost-effectiveness analysis, which is designed to compare a set of regulatory actions with the same primary outcome, such as number of years of life lost (for example, see NRC 2002a) or the disability-adjusted life-years (DALYs) lost (Murray and Lopez 1996),[7] but a universally accepted method has yet to emerge.

• Gaps in data and knowledge limited the scope of the analysis. HAPs were excluded from the analysis because of a scarcity of information on exposure and health effects. Benefits related to improvements in ecosystem function were excluded because of uncertainties in the monetary value of such improvements. Similarly, the benefits gained from improvements in lung function in children were omitted because a dollar value could not be reliably attached to this outcome. Finally, because monitoring is concentrated in urban areas, the exposure characterization of rural populations is not robust, and consequently, the benefits to rural populations could be either overvalued or undervalued.

• Any attempt to compare monetized costs with monetized benefits, as in the EPA study, is subject to uncertainty and missing information. Not all costs and benefits can be equally well quantified and monetized. Also, elicitation of estimates can be expensive and analysis often must rely on the transfer of estimates from other settings that are imperfect substitutes. Further, aggregation of measured individual preferences to achieve an estimate of social costs and benefits can involve political decisions that are implicitly controversial.

• The usefulness of economic assessments such as those carried out by EPA could be enhanced if the connection between specific air quality policies and the dollar values of discrete health and welfare benefits were more transparent (Hall 1996; Krupnick and Morgenstern 2002). For example, a source-specific multipollutant approach could be adopted that would assess the costs and monetized benefits associated with implementing emission controls on specific types of sources (for example, utilities and mobile sources). Alternatively, specific individual measures could be evaluated on

---

[7]These kinds of measures have been important in decisions about allocating public health resources but are controversial with regard to measuring the benefits of social programs. Approaches using estimates of life-years lost and disability-adjusted life years (DALY) are controversial, because they associate a nonmonetary value of change in health status that differs for different members of the population. For example, a senior citizen may have fewer life-years lost than a prime-age adult because of an environmental exposure. Similarly, the DALY approach suggests that older citizens or citizens with a preexisting disease, or disabilities, suffer a smaller loss from premature mortality than younger or healthier citizens (Heinzerling 2000).

> ### BOX 6-7  Pollution Abatement Cost and Expenditures (PACE) Survey
>
> The PACE survey was conducted annually by the U.S. Bureau of the Census from 1973 through 1994, when it was suspended by the bureau for budgetary reasons. In 2000, another survey was carried out. However, time and resource constraints have delayed implementation of subsequent PACE survey cycles.
>
> • The PACE survey data provide a distinctive tool for evaluating the costs of compliance with environmental regulations. EPA has used the PACE data in its reports on the cost of clean air, Section 812 clean air retrospective cost analysis, numerous sector-specific studies, regulatory impact analyses, analyses of recycling activities, and national studies of environmental protection activities.
> • The PACE survey provides three levels of data. First, the published PACE survey provides aggregate data on pollution abatement spending, both for new capital expenditures and for operating costs. Second, the PACE survey provides abatement spending data at the industry and state level. EPA has used these data for specific sector studies and regulatory impact analyses. Third, the facility-level data collected for the PACE survey can be linked with other census-collected data for those plants and accessed by researchers working at census research data centers around the nation under strict controls to maintain confidentiality.

a disagregated basis, because large net benefits might be masking ineffective or inefficient programs or regulations. A regional approach could identify control options that produce the largest benefits in specific parts of the country.

• Economic assessment, such as that carried out by EPA, requires a reliable assessment of the costs of implementation. The United States probably leads the world in the availability of information about the cost of environmental compliance through data collected in the Pollution Abatement Cost and Expenditures (PACE) Survey. The survey has provided the basis for a large number of scholarly studies that have contributed to improving the effectiveness of environmental regulation. In recent years, the survey has been intermittent due to interruptions in funding. The PACE survey is discussed further in Box 6-7.

## SUMMARY

### Strengths of Techniques for Tracking Progress in AQM

• Continuous emissions monitoring (CEM) systems on electric utilities regulated under the acid deposition program of the CAA have documented substantial reductions in $SO_2$ emissions.

• National atmospheric deposition monitoring stations have documented a reduction in sulfate deposition in the eastern United States.

• Air quality monitoring networks are a significant national resource and have provided qualitative confirmation of emission-inventory estimates that indicate that pollutant emissions in the United States (and especially in urban areas) have decreased over the past three decades.

• Analyses of short-term episodes with significant changes in pollutant concentrations confirm that health benefits accrue when air quality is improved.

• Biomarkers may provide a useful surrogate for documenting trends in population exposure to pollutants over the longer term.

• EPA's congressionally mandated economic assessments of the costs and benefits of AQM in the United States are peer reviewed and appear to represent the state of the science.

### Limitations of Techniques for Tracking Progress in AQM[8]

• The nation's AQM system has not developed a comprehensive and quantitative program to track emissions and emission trends.

• Although improvements have been made, accessibility to actual data acquired from the monitoring networks is limited.

• The AQM system has not completed a comprehensive program to monitor HAPs so that population exposure and concentration trends can be tracked.

• With the exception of CEM, there is limited ability to quantify stationary sources emissions.

• The air quality monitoring network for criteria pollutants is dominated by urban sites, limiting its ability to address a number of important issues.

• Some of the instruments and methods used in the nation's air quality monitoring networks are inadequate to meet the objectives of the monitoring.

• Methods used by EPA to calculate pollutant trends from data collected from air quality monitoring networks could be improved.

• The AQM system has not developed a method and related program to document independently improvements in health and welfare outcomes achieved from improvements in air quality (see Box 6-8).

• The AQM system has not developed a cohesive program capable of reliably reporting the status of ecosystem effects of air pollution and the response to changing air pollution conditions across regions and the nation.

---

[8]Recommendations are provided in Chapter 7.

---

### BOX 6-8    The State of the Environment Report—
### A Sign of a New Paradigm Emerging at EPA?

The list of limitations in tracking progress at the end of this chapter presents a sobering picture. While significant resources have been expended in the United States to identify air quality problems and to reduce the pollutant emissions believed to be fostering these problems, there appears to have been a far less concerted effort to track and document objectively and comprehensively the real-world benefits of AQM. Fortunately, a new paradigm appears to be emerging at EPA that recognizes the importance of such an effort. As this committee was completing its work, EPA released its *Draft Report on the Environment* (2003n). The purpose of the report was to identify and quantify environmental indicators "to better measure and report on progress toward environmental and human health goals and to ensure the Agency's accountability to the public." The report represents an important addition to an understanding of the state of the environment and of the limitations of data available for assessing progress. Few data are available that can be used to assess the impacts of AQM measures in the United States on specific human and ecological health goals. Even with large data collection efforts, the establishment of trends attributable to AQM will continue to be a formidable task for diseases such as cancer and asthma because of other strong risk factors. For those environmental indicators that have data, a continuous tracking of the indicators over a span of a decade or more will be needed to establish a trend and its relationship to AQM activities. Thus, long-term support for the activity initiated with the production of EPA's *Draft Report on the Environment* is recommended (1) to ensure that EPA is able to produce a series of reports on a biannual or triannual basis, and (2) to provide the scientific and technical basis for environmental indicators that link human and ecological health outcomes with air quality.

---

- Cost-benefit analyses of AQM carried out by EPA and others are limited by a lack of relevant data (for example, on the health effects of HAPs) and a reliance on controversial value judgments.
- Data on costs, such as the PACE survey, are necessary to monitor the costs of CAA compliance and to identify cost-effective policies. However, inconsistent levels of funding in the past have undermined the ability of the PACE survey to provide a long-term data set and to serve as a tool to estimate costs.

# 7

# Transforming the Nation's AQM System to Meet the Challenges of the Coming Decades

## INTRODUCTION

The air quality management (AQM) system has been effective in addressing some of the most serious air quality problems confronting the United States in the latter half of the twentieth century. New technologies and fuels, developed largely in response to Clean Air Act (CAA) requirements, have substantially reduced emissions from mobile, stationary, and other sources. As a result, the U.S. population has experienced large reductions in ambient concentrations of lead (Pb), carbon monoxide (CO), sulfur dioxide ($SO_2$), and, in some regions, ozone ($O_3$) and particulate matter ($PM_{10}$). These reductions in ambient concentrations have come despite substantial economic and population growth in the United States that brought about increases in power generation, vehicle miles traveled, and other activities that are traditionally associated with emissions of air pollutants.

However, significant and perhaps even more difficult challenges are to be met in the coming decades. In this chapter, a number of specific changes to the air quality management (AQM) system are recommended that would improve our ability to meet these challenges effectively. To place these recommendations in an appropriate context, this chapter begins with a brief discussion of some of the air quality challenges that the nation will need to confront in the future and then outlines a set of overarching principles that guided the committee in designing its recommendations (see Figure 7-1).

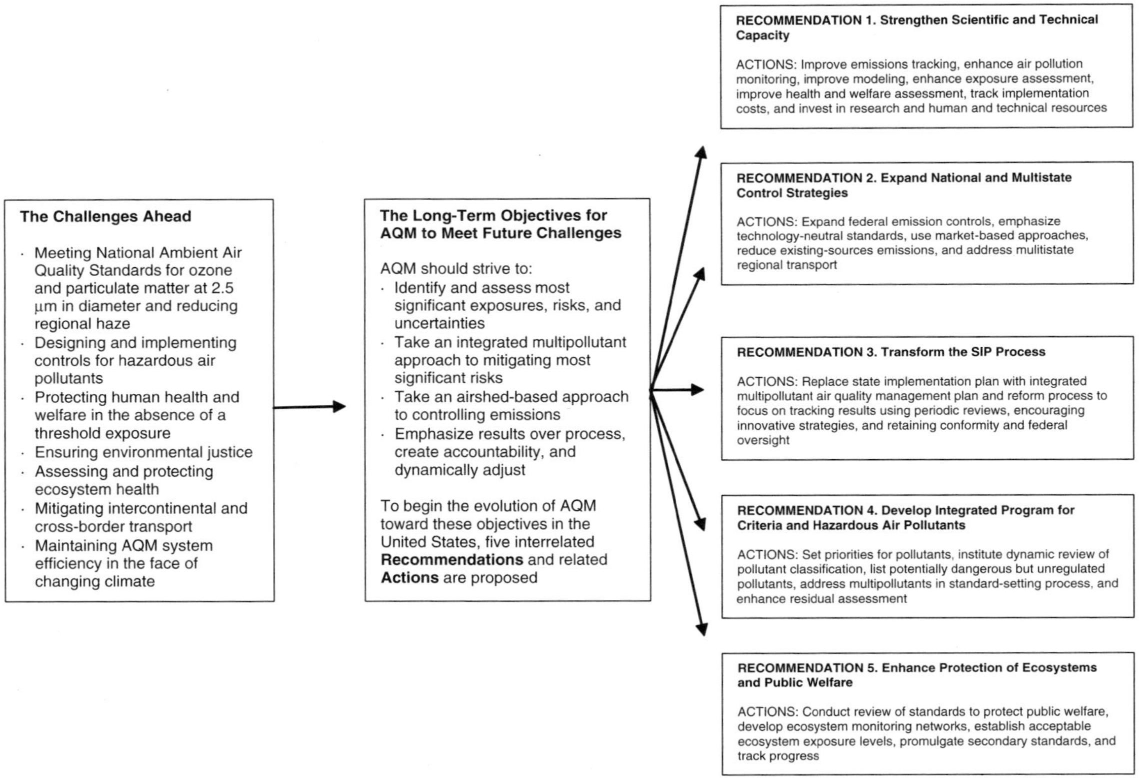

FIGURE 7-1 To meet the major challenges that will face air quality management (AQM) in the coming decade, the committee identified a set of overarching long-term objectives. Because immediate attainment of these objectives is unrealistic, the committee made five interrelated recommendations to be implemented through specific actions.

## THE CHALLENGES AHEAD

## Meeting NAAQS for $O_3$ and $PM_{2.5}$ and Reducing Regional Haze

In 1997, the U.S. Environmental Protection Agency (EPA) promulgated new National Ambient Air Quality Standards (NAAQS) for two criteria pollutants: $O_3$ and PM. These standards were developed because of new scientific data that indicated deleterious health effects from exposure to concentrations of the two pollutants that were below the current NAAQS. Meeting the new standards will require additional reductions in pollutant emissions. Moreover, because $O_3$ and PM are secondary pollutants (produced in the atmosphere from reactions involving primary pollutants), it will be necessary to determine which pollutant emissions to reduce and to devise appropriate monitoring systems to assess progress toward meeting the new standards. All of these tasks will be major challenges for the AQM system in the United States over the next decade.

The standard for $O_3$ has been changed from a maximum 1-hr peak concentration of 120 parts per billion by volume (ppbv) to an 8-hr average concentration of 80 ppbv.[1] It has proved to be extremely difficult to attain the previous (1-hr) $O_3$ NAAQS. In the United States, approximately 56 areas composed of 233 counties have yet to attain it after decades of trying to do so (EPA 2003p). It will probably be even more difficult to meet the new $O_3$ 8-hr standard (NARSTO 2000). Retrospective analysis of air quality data indicates that there will be many more exceedances of the new 8-hr standard than the previous 1-hr standard. More frequent exceedances will occur in areas already in nonattainment of the 1-hr standard, and new exceedances will occur in areas currently in attainment of the 1-hr standard. Many of these new nonattainment areas will be in rural areas that do not have the major sources of the various air pollutants that produce $O_3$. The decline of the 8-hr averaged $O_3$ concentrations (11%) in the United States has been slower than that of the 1-hr averaged $O_3$ concentrations (18%) over the past two decades (EPA 2002a). Therefore, additional pollution control strategies are likely to be needed to meet the new $O_3$ NAAQS— for example, a further departure from the local emission-control approach demanded in the current state implementation plan (SIP) process and the enhanced development of multistate airshed[2] management approaches, such as those embodied by the Ozone Transport Assessment Group (OTAG) and the resulting requirement to submit a revised $NO_x$ SIP.

---

[1]EPA is phasing out the 1-hr 0.12-ppm standards (primary and secondary) and putting in place the 8-hr 0.08-ppm standards. However, the 0.12-ppm standards will not be revoked in a given area until that area has achieved 3 consecutive years of air quality data meeting the 1-hr standard (EPA 2001a).

[2]The geographic extent of a pollutant or its precursor emissions in air is often referred to as an airshed.

Developing effective strategies to attain the new $PM_{2.5}$ standard may also prove to be difficult. Atmospheric PM is a complex mixture of solid and liquid particles suspended in air; $PM_{2.5}$ is a subset of PM in the atmosphere with (aerodynamic) diameters of less than $2.5 \times 10^{-6}$ m (or 2.5 µm). Modern instrumentation capable of characterizing individual particles in the atmosphere confirms that $PM_{2.5}$ within any airshed comprises numerous particles having different sizes, shapes, and chemical components (NARSTO 2003). Some actions are already under way or proposed to reduce these emissions (for example, the existing 2007 heavy-duty on-road diesel requirements and the proposed multipollutant controls on electric utilities). However, for the nation's AQM system to protect human health from $PM_{2.5}$ pollution over the long term, the specific characteristics of $PM_{2.5}$ that negatively affect health need to be identified, the sources of emissions and the atmospheric processes responsible for the ambient concentrations of particles with these characteristics need to be determined, and new control technologies will need to be developed and implemented. These tasks will require a major investment in research (as outlined in NRC 1998b) and close collaboration between the policy-making and scientific and engineering communities in the United States.

Reducing regional haze to improve visibility in scenic areas, such as national parks, is another difficulty that the U.S. AQM system will be confronting for decades to come. As discussed in Chapters 2 and 5, EPA's regional haze regulations call for states to develop strategies that will bring about interim improvements by 2018 but do not project attainment of the stated goal of returning national parks and wilderness areas to their natural visibility conditions until 2065.

$O_3$, $PM_{2.5}$, and regional haze share, to some extent, common precursor emissions and chemical pathways for the generation of these pollutants and are all to greater or lesser extents affected by long-range transport. For those reasons, it is critically important that pollution control strategies targeted for mitigation of $O_3$, $PM_{2.5}$, and regional haze be developed in tandem and on a multistate basis. Such a multipollutant, multistate approach should minimize the possibility that control strategies implemented for one pollutant will inadvertently increase the concentrations of another pollutant[3] and should enhance the ability of policymakers to maximize the cost-effectiveness of their overall air pollution control strategies. The one-pollutant-at-a-time approach that is currently used to develop SIPs may substantially hinder the development of multipollutant control strategies.

---

[3]In some cases, reducing sulfate emissions can increase concentrations of nitrate-containing PM (NARSTO 2003).

## Toxic Air Pollutants

The health risks faced by U.S. citizens from exposure to toxic air pollutants remain an important concern, albeit one that is not well quantified. The National Air Toxics Assessment (NATA) estimates that most Americans face cancer and noncancer risks of public health concern from exposure to hazardous air pollutants (HAPs) (EPA 2002e). Moreover, these estimates do not consider the risks associated with exposures to numerous poorly characterized HAPs, or to the large number of chemicals that are not identified as HAPs but that might pose a health hazard. Although some monitoring data suggest that concentrations of commonly measured HAPs are declining and the implementation of planned maximum achievable control technology (MACT) and other regulations is expected to substantially reduce toxic emissions, significant residual risk is predicted to remain (EPA 2000a). Given the multitude of sources of toxic air pollutants in the nation and the variety and complexity of the risks they pose, protection of human health and ecosystems from exposure to HAPs will continue to challenge the AQM system over the coming decades.

Ideally, control strategies for HAPs would be scaled to the degree, severity, and pervasiveness of the risks posed. The difficulty in developing such strategies has been and will probably continue to be a lack of sufficient information on the sources, atmospheric distribution, and effects of most HAPs. Evidence regarding risks for the majority of HAPs, unlike the criteria pollutants, is often indirect (that is, from animal studies rather than human laboratory or epidemiological studies) and extrapolated from effects reported for HAPs at concentrations much higher than typical ambient concentrations. In addition, current efforts to monitor HAPs fall far short of that needed to characterize HAP exposures adequately. There is a clear need to enhance resources for research, data collection, and analysis efforts on HAPs. However, in the past, priority for these resources has generally been given to criteria pollutants.

A large number of potentially toxic pollutants in the atmosphere are unregulated and, in most cases, poorly characterized in terms of environmental concentrations and subsequent health and ecological effects. An illustration of the enormity of this problem is that while 188 compounds are officially designated as HAPs by EPA, an estimate of approximately 300 compounds with varying tendencies to exist in the atmosphere as gases or particles are introduced into commerce each year by U.S. industries.[4] A major challenge for the nation's AQM system over the coming

---

[4]Information obtained from the Notice of Commencement Database maintained by the inventory section in EPA's Office of Pollution Prevention and Toxics.

decades will be the development of a research and regulatory infrastructure capable of protecting human health and welfare from the increasing number of potentially toxic pollutants in the atmosphere in an effective and timely manner while not unnecessarily impeding economic activity and technological progress.

## Protecting Human Health and Welfare in the Absence of a Threshold Exposure

There is increasing evidence that for some criteria pollutants and subsets of the population, any exposure is harmful—that is, there is no threshold exposure below which harmful effects cease to occur (Daniels et al. 2000). Under these circumstances, there is a tendency to set air quality goals and standards at ever lower concentrations—concentrations so low, in fact, that they approach what might be considered the irreducible background concentration that is unaffected by human pollutant emissions and thus impervious to even the most aggressive air pollution control efforts. To address this challenge, AQM needs to develop a better understanding of the reducible (human-induced) and irreducible components of pollution in the United States (NARSTO 2000). Achievement of this understanding will require a substantial expansion of monitoring networks into rural and remote regions, as well as in cities (see Chapter 6). Enhanced tools will also be required for exposure assessment and health and ecosystem impact analysis to better characterize risks at low levels of exposure. When scientific understanding is improved, it might be necessary to reconsider how to set standards to protect public health from those pollutants with no established thresholds.

## Ensuring Environmental Justice

The CAA Amendments of 1990 make no direct or specific reference to environmental justice. Nevertheless, environmental justice issues clearly can arise in the implementation of the CAA, and for this reason, environmental justice is a goal of the nation's AQM system (see Chapter 2). Addressing environmental justice in the nation's AQM system will require actions on a number of levels. First, environmental justice concerns reinforce the need to monitor and model the distribution of HAPs and other pollutants in microenvironments and to use those results to estimate exposures to the populations in those microenvironments. Second, the concept of environmental justice will need to be incorporated in the earliest stages of air quality planning and management. Implementation planning has a critical role in the CAA, and issues of environmental justice often can be addressed most effectively during development of these plans especially if they begin to address multiple pollutants and hot spots,[5] as proposed in the third and fourth recommendations in Recommendations for an Enhanced AQM System later in this chapter. In one instance, the California Air Resources Board (CARB) led in establishing a framework for incorporating environmental justice into its programs consistent with the directives of state law (CARB 2001). Third, new proposals for pollutant reduction (cap-and-trade programs, for example) should be judged in terms of the environmental justice impact and the efficacy of the reduction and should be designed to include sufficient local emission-control requirements to minimize the possibility that hot spots will result, especially in disadvantaged communities. Fourth, Native American tribes should be given help to develop and implement AQM programs for reasons of environmental justice and tribal self-determination.

The recommendations advanced later in this chapter, specifically those that allow AQM to target the most significant exposures and risks, are designed in part to address these issues.

### Assessing and Protecting Ecosystem Health

The goal of protecting ecosystems is clearly enunciated in the CAA under the proviso to protect public welfare through the promulgation and implementation of secondary standards. The protection and maintenance of ecosystems is critical not only because of a general desire to protect and preserve forests and undeveloped spaces in the United States but also because ecosystems provide invaluable services (for example, water purification, water supply, forest production, and carbon and nitrogen fixation) that are essential to our economy and the public health (Daily 1997; ESA 1997a; Balmford et al. 2002). Indeed, provisions to mitigate ecosystem impairment in wilderness areas and national parks in the 1977 CAA Amendments and to mitigate the effects of acid rain in Title IV of the 1990 CAA Amendments are proactive steps taken by Congress to protect public welfare in the absence of formally established secondary standards.

Despite the mandate in the CAA to protect welfare, protection of ecosystem health has not received adequate attention in the implementation of

---

[5]Hot spots are locales where pollutant concentrations are substantially higher than concentrations indicated by ambient outdoor monitors located in adjacent or surrounding areas. Hot spots can occur in indoor areas (for example, public buildings, schools, homes, and factories), inside vehicles (for example, cars, buses, and airplanes), and outdoor microenvironments (for example, a busy intersection, a tunnel, a depressed roadway canyon, toll plazas, truck terminals, airport aprons, or nearby one or many stationary sources). The pollutant concentrations within hot spots can vary over time depending on various factors including the emission rates, activity levels of contributing sources, and meteorological conditions.

the act (see Chapter 2). Nevertheless, research over the past 30 years suggests that there is a critical need to protect ecosystems from the damaging effects of air pollution. In addition to impairment of visibility, air pollution can have far-reaching effects on ecosystems, including damage to trees and crops; degradation of soil quality (particularly shallow forest soils); acidification of surface waters; and a resulting decrease in the diversity of biota, contamination of fish tissue (for example, by mercury and polychlorinated biphenyls), and eutrophication of coastal waters. Those ecological effects have an impact on humans through decreases in the productivity of forests and crops, increased advisories on consumption of contaminated fish, loss of fisheries in waters located in upland forests and in estuaries, and a deterioration of the quality of recreational activities. A major goal of the nation's AQM system in the coming decades should be to establish an appropriate research and monitoring program that can quantitatively document the links between air pollution and the structure and function of ecosystems and use that information to establish realistic standards and goals for the protection of ecosystems and implement strategies to attain those standards and goals.

### Addressing Multistate, Cross-Border, and Intercontinental Transport

Historically, the primary emphasis of AQM in the United States has been on controlling emissions in and nearby urban and industrial centers where pollutant concentrations are generally the highest; this approach is often referred to as a local pollution control strategy. During the late 1980s and 1990s, it was realized that controlling local emissions alone was insufficient to meet the NAAQS for some air pollutants in some areas. In response, regional planning organizations were created to devise multistate AQM strategies. As discussed in Chapter 3, some of these regional planning organizations were created in response to specific requirements of the 1990 CAA Amendments (for example, the Ozone Transport Commission and the Grand Canyon Visibility Transport Commission), and others were formed on a more ad hoc basis (for example, the Ozone Transport Assessment Group). Whatever the mechanism that led to their formation, these regional planning organizations all shared a common purpose: to fill the gap in the nation's AQM system that has historically fallen between the responsibilities and regulatory authority vested with state governments and those vested with the federal government. To better fill this gap and thereby facilitate the development and implementation of multistate AQM plans in the future, the recommendations section of this chapter proposes that EPA's role in addressing regional problems be enhanced.

As more is learned about the atmosphere, it has become more apparent that air quality in a locale (even an urban locale) can be influenced by even

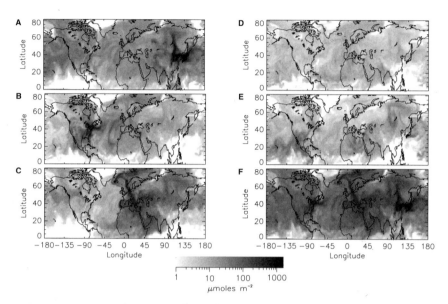

FIGURE 7-2  Contribution to the sulfate column burden for July 15, 1997, at 00UT (vertical integral of the concentration) from different source regions showing intercontinental transport.  Data are in micromoles per square meter.  Source regions are (A) Asia (anthropogenic), (B) North America (anthropogenic), (C) Europe (anthropogenic), (D) biogenic (dimethyl suflide and hydrogen sulfide from natural sources), (E) volcanoes, and (F) total.  Results from the Brookhaven National Laboratory chemical transport and transformation model for sulfate.  SOURCES: Data from Benkovitz et al. 2003; and from C.M. Benkovitz, Brookhaven National Laboratory; S.E. Schwartz, Brookhaven National Laboratory; M.P. Jensen, Columbia University; M.A. Miller, Brookhaven National Laboratory; R.C. Easter, Pacific Northwest National Laboratory; and T.S. Bates, Pacific Marine Environmental Laboratory; unpublished material, 2004.

longer range pollutant transport; namely, transport across national boundaries and even between continents.  Analyses of data sets gathered from space-based and airborne platforms in combination with sophisticated computer models indicate that air pollutants from Central America, Asia, Africa, and Europe reach North America, and, in turn, pollutants from North America reach Europe (Figure 7-2).  The implications for AQM in the United States are 2-fold:

   • International and intercontinental transport of pollutants can significantly degrade air quality in the United States, particularly over the short term.  For example, in April and May 1998, large amounts of smoke were observed by the total $O_3$ mapping spectrometer (TOMS) satellite in plumes emanating from fires in Mexico and extending into Florida, Texas, New Mexico, California, and Wisconsin. Along the Gulf Coast of Texas, a public health alert was issued advising residents to stay indoors to avoid the smoke. Similarly, chemical measurements have documented the import of pollutants transported from the Eurasian continent across the Pacific Ocean into the western part of North America (Jaffe et al. 1999; Wilkening et al. 2000). The rapid industrialization of the Asian continent could conceivably exacerbate this phenomenon.

- Pollutant emissions from North America, Europe, and Asia are probably causing an increase in the so-called background concentrations of pollutants in the northern hemisphere. For example, there is evidence that the background concentration of tropospheric (lower atmospheric) $O_3$ in the northern hemisphere has increased by as much as a factor of 3 in the past 100 years, presumably in response to growing pollutant emissions from throughout the hemisphere, including the United States (Volz and Kley 1988; Staehelin et al. 1994). In addition, the mean summer afternoon concentration in rural areas of the United States (Logan 1988) and Europe (Scheel et al. 1997) have grown by a factor of 4 to 6. As standards for PM and $O_3$ become more stringent and the thresholds for health effects from air pollution are found to be lower or nonexistent, the increasing level of background pollution causes difficulty in separating the effects of local and regional air pollution from global problems.

To address these international aspects of air pollution, the AQM system will need to continue to develop, implement, and utilize sophisticated remote-sensing technology to document and track the phenomena. It will also be necessary for the United States to continue to pursue collaborative projects and enter into agreements and treaties with other nations (especially developing nations) to help minimize the emissions of pollutants that can degrade air quality on continental and intercontinental scales. Examples of past initiatives and treaties undertaken by the United States to mitigate atmospheric problems of international concern include the development and implementation of the Montreal Protocol to address stratospheric $O_3$ depletion, the Convention on the Long-Range Transport of Transboundary Air Pollution (CLTRAP) to mitigate a wide range of air quality problems, and NARSTO to develop a coordinated program of research on the causes of and remedies to ground-level $O_3$ and PM pollution in Canada, Mexico, and the United States.

## Adapting the AQM System to Climate Change

The earth's climate is warming (IPCC 2001). Although uncertainties exist, the general consensus within the scientific community is that this warming trend will continue or even accelerate in the coming decades (IPCC 2001; NRC 2001b). Some forms of air pollution might be exacerbated by these climate changes. For example, precursor emissions and photochemical reactions that result in the production of $O_3$ tend to increase with warmer temperatures (Carter et al. 1979; Tingey 1981; Chock et al. 1982; Halberstadt 1989; Cardelino and Chameides 1990; Bernard et al. 2001). The AQM system must be flexible and vigilant in the coming decades to ensure that pollution mitigation strategies remain effective and sufficient as our climate changes.

At the same time, air pollution and human-induced climate change have one important common characteristic: they are both fostered by the burning of fossil fuels and other anthropogenic activities. Although some emissions contribute to climate warming (for example, $CO_2$, soot, and upper tropospheric $O_3$), others cool the climate (for example, sulfates formed from sulfur oxide emissions) (IPCC 2001; NRC 2001b; Hansen and Sato 2001). Thus, some efforts to mitigate air pollution may also help to mitigate climate warming (for example, reducing $O_3$ precursor emissions), and others may inadvertently exacerbate climate warming (for example, reducing sulfur oxide emissions). If the current trends in climate continue, the air pollution and climate interactions will need to be considered in designing air pollution control strategies. Multipollutant approaches that include mitigation of climate warming as well as air pollution may be desirable, and some states have already considered implementing such programs (STAPPA/ALAPCO 1999).

## PRINCIPLES FOR ENHANCING THE AQM SYSTEM

In the U.S. democratic system, the AQM system is designed and implemented by political decision-makers and is, therefore, greatly influenced by political and economic considerations. However, since its inception, the CAA has recognized that its effectiveness can be substantially enhanced by policies that are informed by and consistent with scientific and technological realities. On the basis of the committee's analysis of the strengths and limitations of the AQM system (in Chapters 2 through 6), as well as the future challenges described above, the committee has identified a set of overarching scientific and system-design principles that should guide the improvement and development of the nation's AQM system.

### One Atmosphere Approach for Assessing and Controlling Air Pollutants

Air pollutants are constituents of the atmosphere and, as such, can be transported and mix freely in atmosphere and transfer to other environmental media. Air pollutants do not have political and statutory boundaries and when humans and ecosystems are exposed to air pollution, they are simultaneously exposed to a complex array of contaminants. Moreover, pollutants that affect humans can affect ecosystems, and, perhaps more important, pollutants that appear to affect ecosystems only can also directly or indirectly affect human health and activities, because society depends on ecosystems for essential environmental services.

In large part because of the federal system of government, efforts to mitigate air pollution in the United States are organized along state boundaries and, within states, along county and metropolitan boundaries. Because of limitations in both resources and knowledge and the structure of the NAAQS process, most mitigation efforts focus on pollutants separately. Although such approaches may be expedient, they are often inadequate to characterize the transport, mixing, and reaction of air pollutants and the exposure of people and ecosystems to these pollutants. In the long run, such approaches may limit the ability of an AQM system to protect human health and welfare most effectively. The air pollution challenges facing the nation over the coming decades are complex; they are likely to require mitigation strategies that are consistent with the "principle of one atmosphere." Such a principle requires understanding the dispersion and interaction of multiple pollutants over multistate or even international airsheds, developing the air pollution control strategies that span multistate airsheds, and understanding and mitigating the impacts on human health and ecosystem condition that arise from simultaneous exposure to multiple pollutants. It will also require better understanding of the range of important emissions from any one set of sources so that facilities and other pollutant emitters have the underlying information needed to develop innovative multipollutant control technologies and pollution prevention practices. Detailed recommendations are provided later in the chapter.

Ultimately, in light of the substantial scientific data documenting the role of air pollution in the effects on water and soil, the AQM system will need to move beyond one atmosphere and address one environment (see Box 7-1). Although complete consideration of the implications of this broader approach was beyond the scope of this committee's charge, the committee notes that a comprehensive assessment of these issues is needed in the future.

## Risk Determined by Actual Exposure

The adverse effects of air pollutants on humans and ecosystems are ultimately determined by the actual exposures and sensitivity of humans and ecosystems to the myriad air pollutants in the atmosphere. However, many of the specific mitigation efforts under way in the United States are guided by standards and goals that target the ambient concentrations or

---

**BOX 7-1 Beyond One Atmosphere to One Environment: Accounting for Cross-Media Pollution**

Pollutants are transported between the atmosphere and other media (water and soil). Although research on the consequences of air pollutants has traditionally focused on the direct effects on human and plant health, materials and atmospheric visibility studies over the past 30 years have demonstrated that air pollutants can accumulate in soil; contaminate groundwaters, surface waters, and estuaries; and, under certain conditions, be re-emitted back to the atmosphere (for example, mercury and nitrogen oxides). The multimedia effects of air pollutants greatly increase the complexity of AQM. Air pollutants may have an impact on human health through contamination of water supplies and food, in addition to the impact through direct exposure and inhalation. A prime example of such phenomena and the serious environmental effects that can be thus engendered is found in an examination of the reactive nitrogen cycle, where nitrogen compounds generated from food and energy production are believed to have profound effects on air and water quality and on ecosystem function and climate (Ambio Special Issue, March 2002). Persistent organic pollutants (POPs), such as polychlorinated biphenyls (PCBs) and polybrominated diphenylethers (PBDEs), also tend to persist in the environment and can accumulate in human tissue. They can appear as air pollutants, but they can also contaminate water, soils, and the food web (Rodan et al. 1999).

  Recognition of multimedia effects has blurred the distinction between air and water pollution, and that may require rethinking the approach for setting air pollution standards and for linking air quality policy with other environmental policies. For example, the contamination of groundwater supplies, surface waters, and estuaries by air pollutants has implications for the Safe Drinking Water Act, the Clean Water Act, and state and local environmental management policies, including fish-consumption advisories. At a minimum, the growing awareness of these multimedia challenges calls for renewed efforts, in the CAA and elsewhere, to provide the incentives to all parties to prevent the emission of pollutants before they are created rather than treating them after they have been produced. Beyond that, a comprehensive, multisector strategy for the mitigation of both air and water pollution may ultimately prove necessary to comprehensively protect human and ecosystem health.

---

emission-control technologies of a few specific pollutants rather than the actual risks borne by people and ecosystems.

  Developing an AQM system that explicitly targets risks would be a challenging task. Pollutant concentrations can vary considerably in time and space, and pollution sources that contribute to exposure may do so to different extents. Of particular concern are the so-called hot spots, where pollutant concentrations are significantly higher than the average ambient concentrations. An additional complication arises when some individuals or groups (such as those with sensitivities due to genetic makeup or preexisting conditions) and certain ecosystems, habitats, or species (such as sensitive crops and estuaries) are more susceptible to the effects of pollutants than the average for that category. Enhanced susceptibility to certain toxicants during certain life stages (for example, in utero, childhood, and old age) is another important concern that has been considered inadequately in the development of regulatory strategies. Some populations, particularly those living or working in and around hot spots, also might be more heavily exposed and thus at greater risk to air pollutants. As discussed in Chapter 2, disadvantaged communities and individuals are often the ones most exposed to pollutants from industrial facilities and transportation. Finally, high exposures can be generated indoors through many sources, including the use and off-gassing of consumer products.

To address these complex and multifaceted exposures and effects, the nation's AQM system would have to be modified substantially. More comprehensive multipollutant monitoring systems would be required for those pollutants posing the most significant risk, and enhanced efforts would be needed to understand the relationship between ambient pollution and the full range of indoor and outdoor exposures of individuals and ecosystems to pollutants. It would also require systematic efforts to assess both human health and ecosystem impacts, especially in susceptible or more highly exposed populations and ecological settings. Changes in statutory requirements and standard-setting procedures may be necessary as well.

### Dynamic AQM in a Constantly Changing Technological Society

The United States is a technological society that is complex and continuously changing. The future trajectory of society is determined by a complex interaction of social, economic, political, and technological forces as well as natural phenomena that occur independently of, and sometimes in response to, human influence (for example, climate change). Moreover, scientific understanding of the causes, consequences, and management options of air pollution and the technology for addressing air pollution are in flux. For those reasons, unforeseen pollutants might someday overshadow air pollutants that are of primary concern today, and air pollution mitigation strategies that are effective today might not apply in the future (for example, as a result of globalization of trade and shifting emissions patterns). The rate of change can cause difficulties: rules and regulations need to provide some level of certainty to regulated parties, but they can also impose substantial inertia on the system, making it difficult to respond to new scientific information and technological developments. If the nation's AQM system is to apply the best knowledge on a continuing basis, the system must be dynamic—a system that can be adjusted and corrected as new information, scientific understanding, and technological advances become available.

## Emphasizing Performance Rather Than the Process

Science advances through iteration. Theory and prediction are used to explain observed phenomena, and observations are used to test theories and predictions. Empirical data and observations of performance ultimately determine the viability of scientific understanding. The AQM system, on the other hand, is inherently regulatory in nature. As a public and political endeavor that imposes requirements and restrictions on parties who would not voluntarily observe them, AQM must attend to a variety of system-design objectives, including fairness, openness, and predictability for regulated parties, stability, legal enforceability, and cost effectiveness along with the overarching goal of improving air quality. In such a system, the promulgation of statutory requirements, rules, and mandated methods and practices can become primary, while the testing of the system's performance in improving air quality becomes secondary. That development is not entirely surprising, because the system is designed to anticipate predictable attempts by regulated entities to circumvent more flexible regulatory requirements.

An overemphasis on procedures can produce an AQM system that is overly complex and rigid and insufficiently focused on measuring performance. Indeed, participants in the existing AQM system at all levels (the public; stakeholders; and local, state, and federal officials) have expressed concern that the current system is sometimes overly driven by rules and procedures, focusing too much time on paperwork and not enough time on tracking the efficacy of the statutes, rules, and methods that were enforced. Insufficient tracking makes it difficult to identify weaknesses and flaws in the technologies and strategies used to control air pollution, and it makes it difficult to help in the development of technologies and strategies that are more cost-effective and thus capable of attracting broader community support.

A stronger performance-oriented approach would potentially be more effective. It would borrow from the scientific approach that emphasizes reconciling expectations and observations, not process. It would also create accountability for achieving results and allow procedures and methods to be adjusted and corrected as data on impacts indicated. Such a system would give regulated entities greater discretion in developing plans for achieving the goals of the AQM system, while holding them accountable for the results. Scientific tools would be used to monitor the impact of control strategies on emissions, air quality, and relevant human health and welfare outcomes and to monitor the method of hypothesis testing through observations to improve the relevant policies and regulations. Automatic rewards for achieving results and automatic sanctions for failing to achieve them might also be included in this system.

## RECOMMENDATIONS FOR AN ENHANCED AQM SYSTEM

Ideally, an *enhanced AQM system* should build upon the current AQM system in the following key ways:

- *Strive to identify the most significant exposures, risks, and uncertainties.* The ideal AQM system would systematically assess the population's exposure to all pollutants and the relative human health, welfare, and ecological impacts of all pollutants. It would set priorities for pollutants according to those exposures and impacts.
- *Strive to take an integrated multipollutant approach to address the most significant exposures and risks.* Foster control strategies that accomplish comprehensive reductions in the most cost-effective manner for all priority pollutants.
- *Strive to take an airshed-based approach.* Address and, where appropriate and feasible, control the full range of emissions arising from local, multistate, national, and international sources.
- *Strive to take a performance-oriented approach.* (1) Track significant results and impacts (for example, the effectiveness of specific control technologies and policies, air quality improvements, and human health and welfare effects); (2) create accountability for the results; and (3) dynamically adjust and correct the system as data on progress are assessed.

The committee sees the above four ways to AQM enhancement as long-range objectives for the nation's AQM system. A rapid transformation of the AQM system to one with those characteristics is unrealistic. Although the scientific community has obtained considerable understanding about air pollution in recent decades, knowledge is not extensive enough to rank pollutants comprehensively on the basis of risk. There is insufficient understanding of the mechanisms by which pollutants affect human beings and the environment and of the incidence and distribution of each pollutant in the atmosphere. Finally, the diversity of health and welfare effects associated with different pollutants further complicates a simple ranking of all pollutants. There also is a severe lack in resources and infrastructure as well as knowledge to comprehensively track the performance of the controls and actions instituted by an enhanced AQM system to best inform public and private decision-makers as they face the challenges ahead.

Nevertheless, the AQM system can begin a steady and continuous development toward an approach that more closely approximates the characteristics enumerated above. In that spirit, five sets of recommendations addressing specific aspects of the nation's AQM system are proposed below. Although each set of recommendations is presented separately, they are interdependent and should be implemented as such. Each recommendation has associated with it a number of specific actions, and each action is designed to help to attain one or more of the long-term objectives for AQM described above (see Figure 7-1).

Although all the recommendations are important, the first set of recommendations to enhance the technical capacity of the AQM system is important to implementing the others. Without substantial progress on the first recommendation, the actions called for in the remaining four recommendations will be more difficult to accomplish.

## Recommendation One

**Strengthen the scientific and technical capacity of the AQM system to assess risk and to track progress.**

### Findings

Over the past 30 years, the nation has developed an extensive system to monitor air quality and a large body of scientific observations concerning the health effects of exposure to air pollution and the impacts of air pollutants on ecosystems. However, because of the continuing challenges to the AQM system, the current system is inadequate to meet the future needs of an enhanced AQM system.

- *Emissions.* The nation's AQM system has not developed a comprehensive program to track emissions and emission trends accurately and, as a result, is unable to verify claimed reductions in pollutant emissions that have accrued as a result of implementation of the CAA (see Chapter 6).
- *Ambient Monitoring.* The nation's air quality monitoring network is dominated by urban sites, limiting its ability to address a number of important issues, such as documenting national air quality trends and assessing the exposure of ecosystems to air pollution (see Chapter 6).
- *Modeling.* Substantial progress has been made in the development of air quality models, but their predictive capabilities and their usefulness to air quality policy-makers are limited by the availability and quality of data needed on meteorological conditions and emissions (see Chapter 3).
- *Assessing Exposure.* Although health and welfare effects are ultimately the product of exposures of populations and ecosystems to mixes of pollutants from specific sources, the nation's AQM system has not invested adequate resources in assessing exposure, relying instead on surrogates (for example, attainment of an ambient NAAQS) to achieve benefits (see Chapter 6).
- *Tracking and Assessing Risks and Benefits to Human Health and Welfare.* The nation's AQM system has not developed a method and

program to independently document improvements in health and welfare outcomes achieved from improvements in air quality (see Chapter 6).

• *Tracking Implementation Costs.* Programs to systematically collect information on the costs of implementation of the CAA have been funded inconsistently and have been limited in their ability to independently validate company estimates of compliance costs (see Chapter 6).

## Proposed Actions

The successful transformation of the AQM system will require renewed assessment of, and investment, in the nation's scientific and technical capacity. Most critical in this regard is the capacity to comprehensively document and monitor pollutant emissions, human and ecosystem exposure, ambient air quality, and human health and welfare outcomes. Because of insufficient resources, transformation of the AQM system should begin with a reappraisal of current resource deployment, identifying opportunities to disinvest in portions of the nation's monitoring and risk assessment system that are less useful and to reinvest those funds in high-priority improvements. Even with the most creative reinvestment of existing resources, however, an enhanced AQM system will require substantial new resources as well. Any investment of new resources would be modest in comparison to the $27 billion of annual compliance costs for the CAA expected by 2010 (EPA 1999a). Even doubling the approximate $200 million in federal funds currently dedicated to air quality monitoring and research at EPA alone would be less than 1% of the costs expended annually to comply with the CAA. Such resources are even smaller when compared with the costs imposed by the deleterious effects of air pollution on human health and welfare.

On the basis of its review of the current ability of the AQM system to assess risk and monitor progress, the committee identified a set of seven priority recommendations:

### Improve Emissions Tracking

EPA should lead a coordinated effort with state, local, and tribal air quality agencies to improve the current system of tracking emissions and their reductions in time and space and estimating overall trends in emission inventories. This undertaking will be challenging because of the large number of emission sources and the changes in their emissions over time. Despite those difficulties, emission inventories should be based on emission measurements whenever possible rather than model calculations. Efforts to achieve that should include the following:

• Investment in the development of new emissions-monitoring techniques (for example, continuous emission monitors for a wider number of sources) to better characterize actual emissions.

• Improvement in current and projected mobile-source emission inventories through expansion of the in-use emission measurement program for on-road vehicles and the various categories of off-road mobile sources.

• Development of better source signatures to facilitate an assessment of the quality of the emission data by source category, using, for example, chemical and isotopic tracers.

• A systematic program of applying the best available emissions measurement and characterization technologies to develop emission factors for every major class of sources.

• Investment in more comprehensive efforts to develop, maintain, and regularly update source inventories.

• Independent efforts, using ambient as well as emissions data, to validate and improve models used in emission inventories.

• Continuous efforts to track the adequacy of emission inventories by reconciling the inventories with ambient measurements.

• Incorporation of more formal uncertainty analysis, based on the validation efforts, in the presentation and use of inventories.

The Emission Inventory Improvement Program, initiated by EPA and the State and Territorial Air Pollution Program Administrators and Association of Local Air Pollution Control Officials, is one example of such efforts that should be vigorously implemented and enhanced to more fully address important shortcomings of the current system.

### Enhance Air Pollution Monitoring

Increasingly complex air quality challenges require a substantially improved multipollutant air quality monitoring system. An effort should be made to ensure that air quality monitoring systems are capable of meeting the increasingly complex challenges and diverse objectives for monitoring in a risk- and performance-oriented AQM system with maximum efficiency. A detailed list of suggested improvements is included in Chapter 6 and Appendix D. The following needs are among the highest priority needs for optimization of the nation's monitoring network:

• A comprehensive review of the various air quality monitoring programs to develop an integrated monitoring strategy that incorporates appropriate measurements of pollutants and their precursors. This review should consider shifting resources. An integrated monitoring strategy should improve the ability of the system to address the objectives that have been often overlooked (attention to emerging air quality problems, especially those relating to HAPs, and exposure of people and sensitive ecosystems; and verification of emission inventories, accountability, and AQM-related atmospheric process studies).

• New monitoring methods to respond to the changes in air quality and monitoring data needs that have occurred over the past 30 years. A more active program of methods development within EPA and deployment through the redesigned network should be initiated. Novel approaches that allow higher spatial resolution (both horizontally and vertically) will be necessary to address the future challenges in AQM.

• Increased number and distribution of air quality monitoring stations in rural, agricultural, and remote forest areas, aided by a statistical design that will improve spatial and temporal estimates of exposure. Colocated long-term measurements of air quality, meteorology, atmospheric deposition, and ecosystem response to air pollutants (for example, along pollution gradients).

• Review of the methods used to determine statistically significant long-term trends in ambient pollutant concentrations to ensure that the results of such analyses are robust.

• Enhanced accessibility of ambient air quality measurement data to the scientific community and the public.

A recent effort by EPA to work with states, tribes, and local air quality agencies to develop the National Core Monitoring Network (NCore), which is assessing the current system and recommending areas for reduced or increased investment, is a valuable first step in enhancing the monitoring network. To achieve substantial enhancement, however, will require sustained agency commitment to implementing such efforts as NCore and applying substantial additional funds to continue key efforts. It will also require innovative programs, including ones that give incentives to the private sector to develop and implement advanced monitoring technologies.

## Improve Modeling

The following key steps would enhance the current models for air quality planning and management:

• Continued and expanded efforts by EPA to develop shared modeling resources by supporting regional modeling centers to train and work with states and multistate organizations on air quality planning. Ideally, these centers would support multiple state-of-the-science models and data analysis.

• Thorough evaluations of models required before they are used for attainment demonstrations of SIPs and other planning purposes. Beyond the evaluation of a model's overall performance in matching observed pollutant concentrations, evaluation of meteorology and emission inputs should be done separately. Thorough evaluation, including improved quanitification of uncertainty, is required, and precursor species (for example, $NO_x$ and volatile organic compounds) should be measured along with secondary pollutants and that chemical constituents and microphysical properties of PM be measured along with total mass.

• Additional short-term campaigns that use government and academic resources from around the country to obtain the detailed measurements needed to develop model inputs and the data needed for model evaluation. Intensive field studies have been conducted in the past decade in Atlanta, Chicago, Denver, Houston, and Los Angeles.

### Enhance Exposure Assessment

A more targeted understanding of the pollutants and the sources that are causing adverse health and welfare effects is needed. To obtain that, a substantial increase in efforts will be required to assess exposure, including the following:

• Development of enhanced techniques for measuring personal and ecosystem exposure.
• Development and application of techniques to measure the portion of exposure (in humans the "intake fraction") that is due to different sources.
• Ultimately, adjustment of regulatory strategies to address the most significant sources of actual exposure in ambient, hot-spot and indoor settings.

### Develop and Implement Processes to Assess Human Health and Welfare Effects

In the final analysis, a performance-oriented AQM system must be able to track progress by improving the understanding of risks and documenting the actual benefits of air pollution control measures on the human health and welfare outcomes for which these measures were adopted. Full accounting for the health and ecosystem impacts and benefits of reducing air pollution will require a systematic approach.

For human health effects, EPA should

• Work with the Centers for Disease Control and Prevention (CDC) to develop a comprehensive suite of health indicators to be measured consistently across the United States and reported on a regular basis. Recent efforts in Congress and at the CDC and EPA have begun to move in that direction (see Chapter 6).

• Develop and implement tools to track the temporal pattern in attributable risk and population-based burden of disease due to short- and long-term exposure to ambient concentrations of criteria pollutants.

• Develop and implement methods and markers for tracking population exposure to, and risk from, the range of other air pollutants, including HAPs. To be successful, all efforts for tracking exposure and risk need to be done over time.

For ecosystem health, EPA should

• Collaborate with other federal agencies (for example, U.S. Department of Agriculture, U.S. Forest Service, U.S. National Oceanic and Atmospheric Administration, and U.S. Geological Survey) on a comprehensive strategy to monitor ecosystem exposure to pollutants and the effects of this exposure on ecosystem structure and function (see detailed recommendations in Chapter 6).

• Analyze data from such a monitoring network or networks to better understand the structural and functional consequences of exposure of ecosystems to air pollution and to identify useful biological and chemical indicators for detecting ecosystem response to pollutants at various levels of biological organization.

For tracking the full range of environmental outcomes, EPA should

• Provide long-term support for the activity initiated with the production of its *Draft Report on the Environment* (2003n) (1) to ensure that EPA is able to produce a series of reports on a biannual or triannual basis, and (2) to provide the scientific and technical basis for environmental indicators that link human and ecological health outcomes with air quality.

### Continue to Track Implementation Costs

As noted earlier, cost effectiveness is a desirable feature of an AQM system and is embedded in many of the specific control requirements of the CAA. In that regard, the nation should maintain and improve its ability to track costs by

• Continuing to support and fund the Pollution Abatement Cost and Expenditures (PACE) Survey on a regular basis to measure the cost of environmental programs for comparison with ex ante estimates and to improve the design of programs.

• Undertaking detailed and periodic retrospective examinations of a subset of past regulatory programs to compare the actual implementation costs with EPA's initial projections of these costs.

### Invest in Research to Facilitate Evolution to a Multipollutant Approach Targeted at the Most Significant Risks

Substantial investments into research and development will be needed if the nation's AQM system is to eventually adopt the scientifically rigorous, risk-focused multipollutant paradigm discussed in this chapter. This will involve enhanced research into the full range of exposures and their potential risks, ultimately resulting in a comprehensive understanding of what sources, pollutant mixtures, and exposures place the public at risk. One component of such an investment would be a review of the risks and adequacy of regulatory protections from pollutant exposures indoors. The review would take a comprehensive look at indoor air pollutant sources, exposures, and risks and regulatory responses to address them, including voluntary programs and labeling requirements on consumer products; evaluate the adequacy of such approaches; and, if warranted, make recommendations for improved approaches.

### Invest in Enhanced Human and Technical Resources to Support the AQM System

The scientific and technical capacity of the nation to attain a fully risk-focused and performance-oriented AQM system rests on the ability of the public and private institutions that constitute the AQM system to train, develop, and retain an adequate and diverse corps of scientists and engineers to conduct research, develop new technologies, and implement policies with the best-available scientific and technical knowledge. The nation today benefits from having many fine institutions that are training future generations for these tasks, and their efforts should be enhanced by

• Providing special programs and incentives to attract and train a new corps of air quality specialists, through federally sponsored training initiatives, including those for nascent and established tribal air quality programs.

• Developing and implementing an environmental extension service to provide both follow-up hands-on training for recent scientific and engineering graduates and a mechanism for more rapid technology and knowledge transfer to local, tribal, and state air quality agencies.

• Enlisting the assistance from professional societies and associations to facilitate the broad-based training of scientists, engineers, and educators needed to fulfill the wide-ranging goals of an enhanced AQM system.

## Recommendation Two

**Expand national and multistate performance-oriented control measures to support local, state, and tribal efforts.**

### Findings

In our federal system of government, the states have been assigned a major role in determining policies needed to achieve air quality objectives and standards. This state role is supported by a philosophic commitment to federalism and a practical recognition that states have substantial administrative capacity, detailed knowledge of local environmental conditions, and understanding of the local political, economic, and social context—all of which equip them to craft policies that can gain local acceptance and be implemented effectively. However, because of the complexity and scale of the nation's economy and the need to maintain open interstate commerce, control of emissions from the many sectors of the economy sometimes cannot be efficiently regulated at the local or state level. In addition, air pollution does not follow political boundaries, and many of the sources affecting a particular area may be outside a particular jurisdiction facing an air quality problem. For that reason, numerous measures that set nationwide emission standards have been promulgated by Congress and EPA. Examples of effective federally mandated emission-control programs include EPA's on-road motor vehicle control program, the phase-out of lead in gasoline, and the acid rain control program. With regard to these national emission standards,

- Emission-control measures promulgated and implemented by EPA have been effective in achieving substantial reductions in emissions of air pollutants on a national level (see Chapters 4, 5, and 6).
- The existence of federally mandated emission-control measures has eased the burden of state and local authorities who prepare the attainment SIPs (see Chapter 3).
- In addition to generating air quality benefits, the national emission-control programs have often provided the drive for technological advancements and, in some instances, have established consistent regulatory requirements for an entire emission-source sector (see Chapters 4 and 5).
- The development of cap and trade has provided the AQM system with a mechanism for achieving substantial emission reductions at reduced costs (see Chapter 5).
- In many instances, the net emission reductions achieved from the promulgation of emission standards that set a limit on the rate of emissions (as opposed to the total amount of emissions) from a vehicle, product, or facility have been substantially offset by concomitant increases in use, demand, or productivity (see Chapters 4 and 5).

• Mobile- and stationary-source emission standards often do not apply to a large fraction of older sources that were "grandfathered" at the time the standards were promulgated. In some cases, the emissions from these grandfathered sources make a major contribution to the total pollutant burden in the nation (see Chapters 4 and 5).

• EPA lacks a sufficient and specific mandate to proactively address multistate airshed aspects of air pollution (see Chapter 3).

## Proposed Actions

### Expand Use of Federal Emission-Control Measures

Additional federal emission-control limits on a variety of nationally distributed products and unregulated as well as underregulated sources will undoubtedly be needed to cost-effectively attain ambient air quality standards for $O_3$ and $PM_{2.5}$ and reduce exposure to various HAPs. To take advantage of economies of scale and the opportunity to address air pollution generated across multistate airsheds, EPA should expand its role in establishing and implementing national emission-control measures on specific sectors of the economy so that states can focus their efforts on local emissions. Among the source categories that should be considered for national emission standards are nonroad mobile sources (for example, aircraft, ships, trains, and construction equipment), more dispersed area sources (for example, wastewater treatment facilities and agricultural practices), and building and consumer products (for example, paints and coatings, cleaners, and other consumer products). In this regard, the recent proposal by EPA for stricter emission standards for nonroad engines is a positive step (68 Fed. Reg. 28328 [2003]). Federal limits and, when appropriate, trading programs to implement those limits, should be established whenever feasible.

States, tribes, local agencies, and stakeholders should be actively involved in identifying and developing those enhanced federal measures. The federal role might include issuing periodic requests to all parties for suggestions for sectors of the economy where increased nationwide or multistate emission controls are needed and having regular consultation throughout the process of developing new measures.

An expanded federal role in the promulgation of control measures should not be carried out in a way that prevents or even discourages innovative initiatives at the state and tribal level. Toward that end, the CAA should continue to allow states to adopt innovative programs that may go beyond the requirements of the federal government, and EPA should expand its efforts to give state and local agencies incentives to undertake, in the context of a revised AQM planning process (see Recommendation Three below), innovative approaches to pollution control.

### Place Emphasis on Technology-Neutral Standards for Emissions Control

Whenever practical, the control measures implemented by EPA at the federal and multistate level should be technology-neutral standards—that is, standards that set clear performance goals without specifying a technological solution—and provide flexibility for sources to achieve those goals in the most cost-effective way possible and provide incentives for developing new technologies. These standards can take at least two forms:

- One form places a cap on the total emissions from a given source or group of sources instead of a limit on the rate of emissions per unit of resource input or product output (for example, a cap on the total $NO_x$ emissions from a power plant instead of a limit on the amount of allowable $NO_x$ emitted per British thermal unit produced by a power plant). This approach is especially effective when the sources or source owners are a relatively defined group (for example, in the hundreds or thousands), and emissions are easily and accurately monitored. This approach can be applied in many ways that would have to be determined as appropriate.

- A second form is a technology-promoting rate-based emissions limit, which sets stringent emission-rate standards that can be met by a mix of still-to-be-determined but foreseeable responses, such as fuel change, combustion enhancement, and add-on controls. An example of this approach is the 2007 highway diesel rule, which sets clear and stringent emission-rate standards that cannot be met by in-use technologies. The rule requires lower sulfur fuel to facilitate new technology and allows manufacturers of vehicles and engines to meet the standards through their own mix of combustion and after-treatment approaches. This type of standard is most often suitable when ownership and use of the source is diffuse (for example, in the hundreds of thousands or millions) and may be more amenable to addressing mobile, consumer, and area sources. Although such a standard is still dependent on the rate of activity and the likely growth in that activity, experience has shown that such standards can be set stringently enough to more than offset growth.

In either case, the emission-cap or emission-rate standard should not be permanent. Instead, mechanisms should be incorporated from the beginning to allow caps and standards to be adjusted. For example, as control technologies evolve and the costs of these technologies fall, further tightening of the standards might be deemed acceptable. Moreover, if new scientific understanding so indicates, tightening of the standards may prove to be necessary.

### Use Market-Based Approaches Whenever Practical and Effective

Congress and EPA should also look for opportunities to pursue the expanded use of market forces and economic incentives to implement federal and multistate emission-control programs. A common feature of current legislative proposals to control multiple pollutants in the electricity sector is their reliance on market-based approaches. This feature indicates a continuing emphasis on achieving environmental goals in a cost-effective manner.

The recommendations in this report are likely, on balance, to impose new costs on the private and public sectors. Therefore, policy-makers must try to achieve these goals using institutions and incentives that will incur the least possible cost (for example, the use of cap-and-trade systems to control $SO_2$ and $NO_x$ emissions). Cap-and-trade programs appear to have saved billions of dollars in capital costs and have realized substantial cost savings compared with conventional regulatory approaches (see Chapter 5). However, cap-and-trade programs have potential pitfalls. Such programs can result in emission trades from one location to another and from one period to another with potentially detrimental consequences, although experience in the acid rain $SO_2$ emissions trading program suggests that it may be possible to design such programs to minimize detrimental consequences. The endorsement of the committee for the expanded use of cap-and-trade programs assumes that future cap-and-trade programs will be designed and implemented to avoid the potential pitfalls. Specific suggestions on how that can be accomplished are presented in Chapter 5.

### Reduce Emissions from Existing Facilities and Vehicles

As noted in Chapters 4 and 5, reducing emissions from older stationary and mobile sources is difficult. Retrofits can be expensive, and substantial economic incentives can keep older (and dirtier) facilities and equipment in operation. Nevertheless, as emission standards on new sources become increasingly tight, emissions from unregulated older sources become increasingly important and begin to hinder further progress. Creative approaches to addressing the problem of grandfathered sources should be developed and implemented. The advent of cap-and-trade programs included existing and new stationary sources under the cap, placing requirements on their owners either to reduce emissions further or to pay for excess reductions that have been made at other facilities. As control costs decline and scientific understanding of emission effects increases, continual reexamination and refinement of the caps can help to ensure that all facilities are participating in efforts to reduce emissions.

For large equipment, such as heavy-duty on-road and nonroad engines, states and EPA have begun to apply a variety of strategies: inspection requirements of in-use engines, financing programs to subsidize replacement programs (especially for public agencies such as school districts and transit agencies), and environmental fines dedicated to support early replacement and implementation of requirements for retrofit (SAE 2003; M.J. Bradley & Associates, Inc. 2002a,b). However, a more comprehensive and systematic effort by EPA and the states to develop heavy-duty vehicle inspection and maintenance programs and to enforce in-use emission requirements, as well as sustained financing from the federal and state level for retrofits and early replacement of older vehicles, will be required if this important source of continuing exposure, especially for inner-city urban populations, is to be brought under control.

### Enhance the Ability and Responsibility of EPA to Address Multistate Regional Transport Problems

Throughout the 1980s and 1990s, there has been a growing recognition that many of the air quality problems facing the nation have a large spatial scale and require a coordinated multistate mitigation effort. When the primary origin of the transported emissions is from sources that are not appropriately or adequately regulated through a nationwide policy, a multistate strategy is the only remaining recourse. This problem poses a special difficulty for areas working to attain the NAAQS that are situated downwind of an area that has significant sources of transported pollution. Here the absence of a regional mandate affects the downwind area's ability to rely on actions in the upwind areas to reduce regional or interstate transported pollution and consequently attain the NAAQS.

The 1990's saw a series of legislative and voluntary efforts in both the eastern and western United States to attempt to address this important issue. However, constitutionally, interstate environmental rules and regulations must be based on federal authority to be effective. Therefore, Congress should provide EPA with the affirmative authority and responsibility to

- Assess multistate air quality issues on an ongoing basis.
- Identify the upwind areas that contribute substantially to SIP nonattainment areas.
- Adopt appropriate regulatory requirements that expeditiously limit emissions from contributing sources.

In addition to focusing on criteria pollutant transport dynamics, the scope of EPA's multistate transport responsibilities should include HAPs and the adoption of mitigation measures to address ecosystem and welfare impacts.

## Recommendation Three

**Transform the SIP process to meet future air quality challenges.**

### Findings

The SIP process has been an important component of the nation's AQM system (see Chapter 3). It allows state and local agencies to account for emission controls adopted at the federal and multistate level and then to choose additional local emission-control measures to attain NAAQS. On balance, this division of responsibility should be appropriate. It can also be the basis of a constructive partnership between the federal and state governments that steadily improves air quality on local, multistate, and national scales. In fact, air quality monitoring data confirm that such improvements have occurred in the past two decades.

Nevertheless, important adjustments to the SIP process are needed if the difficult challenges ahead are to be addressed effectively. As discussed in Chapter 3, the major concerns are described below:

• The SIP process places too much emphasis on the development of a one-time NAAQS "attainment demonstration." Because of the significant scientific, political, and legal uncertainties inherent in such an exercise, it should not be expected to be an accurate predictor of future air quality.

• The SIP process has mandated extended amounts of local, state, and federal agency time and resources in an iterative, often frustrating, proposal and review process that focuses primarily on compliance with intermediate process steps and not on the more germane long-term indicators of performance.

• Each SIP is developed for a single criteria pollutant in isolation from other SIPs developed in the same location for other criteria pollutants, making it difficult for SIPs to pursue multipollutant, source-based strategies.

• The SIP process does not provide mechanisms to integrate control strategies for noncriteria pollutants (HAPs) that can be emitted from many of the sources that emit criteria pollutants.

• The SIP process lacks methods for identifying and acting on air pollution hot spots, where populations are exposed to significantly high concentrations of air pollutants from one or multiple sources.

- In theory, the SIP process contains mechanisms to enable EPA to ensure that all states are making continuous progress toward meeting the NAAQS.[6] In practice, EPA's authority for compliance is limited except in some of the most severe nonattainment areas, and even where it has the authority, it has at times—for lack of resources or other reasons—failed to promulgate the necessary requirements in a timely manner. Although states and other affected persons may file an "action forcing" citizen suit against EPA, that is an expensive and an unwieldy tool for ensuring that EPA attends to its responsibilities in a timely manner.

Addressing these issues will require two basic sets of actions: (1) transform the SIP into a comprehensive air quality management plan (AQMP), and (2) reform the planning and implementation process.

## Proposed Actions

### Transform the SIP into an AQMP

Looking forward, successful AQM in the United States requires significant changes in the scope and the implementation of the SIP process so that it places greater emphasis on performance and results and facilitates development of multipollutant strategies. The committee recommends that this change be accomplished by mandating that each state prepare an AQMP that integrates all relevant air quality measures and activities into a single, internally consistent plan. This change would involve the ultimate elimination of single-pollutant SIPs and their replacement by a single, comprehensive multipollutant AQMP.

Recognizing that the implementation of this recommendation will require a significant change in standard procedure at the federal, state, and local level, we further recommend that implementation, perhaps with incentives, occur in stages over a defined transition period. Currently, there are examples of local air quality agencies that are attempting to create such plans, most notably, that of the South Coast Air Quality Management District in California (SCAQMD 2000).

---

[6]For states that fail to submit adequate plans or that fail adequately to implement approved plans, the statute specifies the consequences (for example, loss of most federal highway funds and two-for-one offset ratios for new sources) that must attend such failures. For states containing areas that fail to achieve the NAAQS by the statutory deadlines, the only consequence (with some modest exceptions for marginal, moderate, and serious $O_3$ nonattainment areas) is a requirement that the state promulgate a new plan capable of meeting the relevant standard by a new 5-year deadline using "technologically achievable" extra measures. For states that fail to make rate-of-progress demonstrations toward attainment, there are no consequences, because EPA has failed for more than a decade to promulgate the necessary implementing regulations.

Today, SIPs are created to address attainment of the NAAQS for individual criteria pollutants. In the future, the scope of planning process through the AQMP should be expanded in three specific ways:

*Integrated Multipollutant Plan:* Given the similarity of sources, precursors, and control strategies, the AQMP should encompass *all* criteria pollutants for which a state has not attained the NAAQS. This approach will be especially important for the many areas that face nonattainment of both $O_3$ and PM NAAQS, because those pollutants share common sources and precursors. Although this change might make development of the plan more complex, the added complexity will likely be offset by the increased efficiency of government and industry in developing and implementing performance-oriented multipollutant control strategies.

*Inclusion of HAPs:* EPA, states, and local agencies should identify key HAPs that have diverse sources or substantial public health impacts or both, which would merit their inclusion in an integrated multipollutant control strategy (for example, benzene; see Recommendation Four). These HAPs should be included and addressed in the AQMP for each state. This initiative will probably take substantial EPA investment to provide the states with the necessary technical basis and resources. Beyond the existing MACT requirements, the level at which a HAP is to be addressed in an AQMP could vary. In some cases, EPA could require that states only develop an emissions inventory and monitoring program for a HAP or take advantage of multipollutant-control opportunities. In other cases, EPA could require addressing a HAP more comprehensively in the AQMP and require an attainment demonstration and a detailed emission-reduction plan. In the latter case, it would also be necessary for EPA to specify a target for attainment; that decision could be based on a local risk assessment and expressed in terms of an ambient concentration or a concentration at specific hot spots.

*Greater Consideration of Hot Spots and Environmental Justice:* Although implicit in current SIPs for individual criteria pollutants, the scope of the AQMP should explicitly identify and propose control strategies for air pollution hot spots to reduce exposures experienced disproportionately by some subset of the population and to provide incentives to do so.

## Reform Planning and Implementation

Substantial reforms will also be necessary to enhance the technical basis and administrative efficiency of the statewide planning process. These reforms are especially important because of the expanded and integrated scope of the AQMP described above. Specific recommendations are discussed below.

*Focus on Tracking and Assessing Performance:* Each SIP is now statutorily required to contain an attainment demonstration in which predictive models and related weight-of-evidence analyses are used to "demonstrate" that the relevant nonattainment area will reach attainment by a certain date as a result of the specific pollution control measures proposed in the SIP. Such an exercise provides useful input to policy-makers and should be retained. However, its use in the current SIP process as a one-time assumed-to-be robust prediction of how air quality in an area will evolve over multiple years or decades is inappropriate because of the uncertainties inherent in the modeling exercise (see Chapter 3) and the false sense of confidence that plans implemented on the basis of the model calculations will achieve the anticipated air quality improvements.

Looking forward, a more useful approach would be to retain the attainment demonstration as a planning tool but to place greater emphasis on follow-up measures to track compliance and progress and on actions to be taken if compliance and progress are not satisfactory. The AQMP process should encourage regulatory agencies to concentrate their resources on tracking and assessing the performance of the strategies that have been implemented rather than on preparing detailed documents to justify the effectiveness of strategies in advance of their implementation. For example, the attainment demonstration and the related improved air quality modeling (see Recommendation One) could be used in the beginning of the planning process to guide policy-makers in the development of a provisional emissions ceiling for the area (that is, a budget for the maximum amount of pollutant emissions an area could contain and still be in attainment of the NAAQS). States could then be required to develop and submit (for approval by EPA) a comprehensive and realistic emission-reduction plan that identified the combination of national, multistate, and local actions to be undertaken within the specified period to bring total emissions in the area in line with the provisional emissions ceiling derived from the attainment demonstration. These emission-reduction plans could then serve as a more practical, performance-oriented metric of state and federal agency execution of their respective responsibilities within a more collaborative and dynamic framework (described below).

*Institute A Dynamic, Collaborative Review:* If attainment demonstrations are imperfect predictions of the future, then it follows that a formal and periodic correcting process of review and reanalysis is needed to identify and implement revisions and adjustments to the plan when progress toward attainment falls below expectations or when conditions change sufficiently to invalidate the underlying assumptions of the plan.[7] Given the

---

[7]The timing of the periodic reviews could be designed to facilitate better conformity between air quality plans and the transportation implementation planning process (see recommendations on conformity below).

large contributions of federal and multistate measures to the success of any SIP, it is essential that this review process be *collaborative and include all relevant federal and state agencies*. Given the large investments of the public and private sectors in AQM, inclusion of these groups in aspects of the review would also be advisable.

Although some aspects of the current CAA "reasonable further progress" reviews seek to be dynamic, the process is often one-sided (with states reporting to EPA) and focuses more on process and administrative steps than on performance. The committee recommends, therefore, that these new reviews be *focused specifically on performance*. This process would entail preparation and exchange of progress reviews by the state and EPA. It would also involve the state and EPA in a collaborative review and agreement on where progress has been made, where more progress is needed, and what specific actions need to be enhanced or replaced. The review should have two components:

- One component would be a review of actual as opposed to modeled emissions in the area to assess compliance with the emission-reduction plan submitted in the AQMP. To the extent that emission trends are not consistent with the emission-reduction plan, the responsible agency (state or federal) would then be required to amend the emission-reduction plan or the implementation of the plan. If EPA determined that a state agency was not fulfilling its responsibilities to meet the emission-reduction plan, then enforcement of effective and timely sanctions and federally imposed air pollution control measures (similar to that envisioned in the current CAA for the federal implementation plan [FIP]) might be appropriate.
- The other component of the review would focus on air quality trends to determine whether the emission reductions implemented in the plan are resulting in the anticipated improvements in air quality. If they are not, more comprehensive changes in the structure of the plan are probably needed. Such a change should include an analysis of why air quality was not improving as predicted, followed by development of a new attainment-demonstration plan and emission-reduction plan. The current requirement in the CAA to increase the seriousness of the categorization of nonattainment for areas that fail to meet attainment by statutorily required dates is a useful approach and should be retained within the proposed AQMP framework.

*Encourage Innovative Strategies:* An additional benefit of a dynamic process will be to enable innovations in portions of the emission-reduction plans in the AQMPs. There would be no predetermined and agreed-upon benefit estimates for the innovations, but they could be tried; evaluated at each review; and continued, amended, or discarded as experience dictated.

For example, innovative strategies might be used to address the role of heat islands in creating pollution, strategies without long track records for estimating benefits but with important opportunities for pollution control in certain locations (see Box 7-2). Expanded efforts by EPA and the state, tribal, and local agencies could facilitate the broad exchange of ideas on innovative control strategies via a web-based inventory of such approaches. The opportunity to experiment with innovative strategies should not, however, excuse the failure to achieve the plan's performance goals.

*Retain and Improve Conformity Requirement:* In recognition of the strong link between transportation infrastructure, mobile-source emissions, and air quality, Congress mandated in the 1990 CAA Amendments and in subsequent transportation legislation that metropolitan transportation planners in NAAQS nonattainment areas endeavor to ensure conformity between regional transportation plans and programs and the applicable SIP (see Chapter 4). Given the stringency of Tier 2 standards, the contribution from automobile emissions to the total of all pollutant emissions will probably decrease in the coming decade, and the need for conformity between air quality and transportation planning will probably subside. However, the long-term effectiveness of the control technologies used to meet Tier 2 standards has yet to be determined. Although overall reductions are likely, there is uncertainty in the degree to which future growth in VMT will offset a portion of the emission reductions anticipated from implementation of the standards. For those reasons, it would be prudent to retain the conformity requirement, thus enabling regions to monitor the probable impact of transportation investments on air quality and make adjustments in their transportation or air quality plans.

Although the conformity requirement should be retained, improvements should be made to ensure that it is technically sound:

• Inconsistencies should be substantially reduced in the data, models, and forecasts used in developing the AQMP and in performing the conformity analysis. Currently, a region's conformity analysis is required to use the most up-to-date planning assumptions available, although the region's SIP may have been written years earlier; thus, the planning assumptions and, hence, the emission budgets in the SIP might be seriously outdated. In the future, regular revisions of AQMPs along with updating its planning data and assumptions should help to reduce the potential for incompatibilities between the SIP and the conformity analysis.

• The planning horizons of the transportation planning process and the air quality regulatory process should be better aligned to ensure that the regulatory process is based on firm technical grounds. Currently, SIPs require emission budgets until the NAAQS attainment year, and the attain-

---

**BOX 7-2  Urban Heat Islands and Other Land-Use Impacts: An Opportunity for Innovation in AQM?**

A variety of experimental and innovative programs could be implemented and assessed in a dynamic, performance-oriented AQM planning process. One such program would attempt to mitigate air pollution in urban areas indirectly by mitigating urban heat islands as described below.

Human activities not only affect atmospheric composition, they also result in widespread land-surface changes though such processes as urban sprawl and deforestation. These land-use changes can strongly affect local and regional climate, which in turn, can influence air quality. One recent study (Pielke et al. 2002) asserts that the redistribution of heat in the atmosphere resulting from widespread land-surface changes may have a greater impact on global climate than the impact due to greenhouse gases. The land-use and climate connection that has been most thoroughly studied, and that is probably most relevant to air pollution concerns, is the urban-heat-island effect.

It has long been observed that on warm summer days, the average temperature in urban centers is often several degrees higher than that in surrounding areas. This heat-island phenomenon is the result of several factors. First, fuel combustion in factories, houses, and cars produces a great deal of waste heat. Also, most urban landscapes are dominated by dark surfaces, such as roads, parking lots, and rooftops, which strongly absorb incoming solar radiation, and re-release that energy to the local environment.

The increased temperatures caused by an urban heat island can substantially exacerbate air pollution within the area (Cardolino and Chameides 1990). The air pollution is caused by a number of factors, including increased demand for cooling energy, which in turn leads to higher power-plant emissions of air pollutants and increased pollutant emissions from temperature-sensitive sources, such as evaporation from motor vehicles and biogenic emissions. In addition, warmer temperatures can directly enhance the formation of secondary pollutants, such as $O_3$. Urban heat islands can also potentially influence local meteorological processes, such as convective storm activity, but those effects are generally not well-understood and could have either positive or negative effects upon air quality.

Urban-heat-island mitigation efforts that reduce ambient temperatures by even a few degrees can potentially have an important role in addressing air quality problems. Programs sponsored by EPA and the Department of Energy have been developed to foster mitigation efforts, such as planting trees and replacing dark asphalt and roof shingles with more reflective surfaces. In other parts of the world (particularly Asia), urban-heat-island mitigation strategies are becoming an integral part of planning for sustainable urban development and air quality management.

---

ment maintenance plans look 10 years forward. Regional transportation plans have a 20-year time horizon. As a result, transportation plans may extend many years beyond the time horizon of the SIP and are governed by the emission budgets set for the period of the SIP. Given the possible consequences of a region failing to demonstrate conformity—for air quality and mobility—and the need for a dynamic regulatory process, these planning horizons should be brought more closely into alignment.

*Enhance Public Agency Performance and Accountability*: At the core of a successful AQM system is creative and committed public agency performance and the ability to hold public agencies accountable for that performance. Although the CAA contains mechanisms for ensuring accountability, those mechanisms tend to focus more on successful implementation of procedural steps rather than on actual emission reductions and air quality improvement, and a lack of resources has hampered the ability of EPA to fully implement them. To remedy this challenge, the committee recommends several steps:

1. Although the goal of our recommendations is to foster a more collaborative and dynamic performance-oriented AQM system, experience shows that not all states have addressed or will equally address their air quality problems, and science has demonstrated increasingly the multistate nature of the problem (thus requiring all states to participate in control). To address this, the CAA should continue to specify deadlines for the attainment of NAAQS and milestones to assess progress along the way, as well as to retain EPA's authority to impose sanctions on states that fail to submit and implement adequate implementation plans. At the same time, as discussed in Recommendation Four, those deadlines and requirements for individual pollutants may need to be adjusted to enable the implementation of fully multipollutant AQMPs.

2. EPA should be provided with adequate resources to write necessary implementing regulations when the implementation of such regulations is a necessary component of AQMPs. As part of the dynamic, collaborative AQMPs, states should not be sanctioned for failing to meet a milestone or deadline if they can demonstrate that the failure was caused by the failure of EPA to promulgate necessary implementing regulations for important multistate or national control programs that are beyond the state's authority.

3. In addition to continued strong requirements and enhanced implementation funding, there is room in a transformed AQMP process for building in incentives. For example, there could be incentives for regions that attain the standards ahead of deadlines or implement particularly creative and effective multipollutant reduction strategies. Such incentives already include easing offset requirements for new development once an area attains a NAAQS, but incentives could also include provisions for reduced oversight or flexibility in implementing some provisions when enhanced or accelerated emission reductions or air quality performance can be shown to have occurred.

4. Existing programs might be combined with economic mechanisms to provide greater enforceability with greater certainty. For example, the CAA

could be amended to provide for additional emission fees when states or regions fail to achieve rates of progress set forth in their regional or state implementation plans.

### Ensuring a Successful Transition to AQMP

In designing a more comprehensive AQM system in stages, several factors should be kept in mind:

• As noted in Recommendation One, a substantial enhancement in the resources, methods, and infrastructure used to track progress in terms of emissions and air quality will be needed to maximize the utility of the proposed review process.

• The development of any plan, and especially a new kind of plan, can take a long time. Special care must be taken to ensure that even as this transition occurs, implementation of recognized and effective control strategies must continue so that the planning does not become a barrier to progress on AQM.

### Recommendation Four

**Develop an integrated program for criteria pollutants and HAPs.**

### Findings

The CAA uses two contrasting classification schemes for air pollutants: criteria pollutants and HAPs. As discussed in Chapter 2, there is a difference in how these two classes are defined: criteria pollutants, and not HAPs, are defined as those whose presence "in the ambient air results from numerous or diverse mobile or stationary sources." Each is managed through a different regulatory framework: criteria pollutants through the setting of NAAQS and through the SIP process and HAPs through the promulgation of MACTs followed by a program to reduce residual risk and through separate efforts to address mobile and area sources.

In the past, this two-pronged approach towards criteria pollutants and HAPs has provided a useful framework for addressing and mitigating some of the nation's most pressing air quality problems. However, as the air quality problems of the coming decades are considered, aspects of this approach appear to be problematic.

First, the system, as currently administered, has resulted in disparate allocation of attention and resources to the different classes:

• Criteria pollutants have received the major share of the management and enforcement priority and resources, as well as the attention directed toward data collection and research. Although pollution emissions have been reduced substantially, the emphasis on criteria pollutants might or might not be justified on a continuing basis in terms of actual human health and ecosystem risk.

• A consequence of the current emphasis on criteria pollutants is that resources to study and characterize HAPs are insufficient. Systematic ambient air monitoring of most HAPs has been nearly absent, further hindering the development of appropriate health assessments and control strategies.

• The list of regulated HAPs has been far too static. There have been no periodic reviews to consider additional compounds for inclusion and to consider the possibility of adding certain HAPs to the criteria pollutant list.[8]

Second, the system, as currently administered, has hindered the development of management strategies that apply a multipollutant approach to addressing the most significant risks:

• The classification has become too rigid and inflexible, creating institutional barriers to change as more is learned about individual pollutants. For example, some pollutants listed as HAPs are ubiquitous in the environment and of substantial health concern (such as benzene) and might be treated more appropriately as criteria pollutants.

• The classification scheme assigns pollutants to regulatory regimes that might or might not be optimal for each pollutant. For example, although mercury is classified as a HAP, it is widely dispersed in the atmosphere, creating diverse exposure, and thus might be more appropriately regulated by a cap-and-trade program (see Chapter 5). On the other hand, many HAPs can create hot spots in specific locations. Mitigation of these types of pollutants might be facilitated by an AQMP regulatory approach with input at the state and local levels.

• The current regulatory framework—with its diverse systems for controlling the two types of pollutants and its failure to require HAPs to be considered at all in local and state air quality planning—makes it difficult to create an integrated multipollutant management approach for criteria pollutants and HAPs, even when they share sources, and local populations and ecosystems are exposed simultaneously to both.

• Even among the criteria pollutants, the practice of setting NAAQS for each pollutant separately, with the attendant differences in the timing of standard setting and attainment deadlines, creates a substantial barrier to integrated multipollutant planning at the state and local levels.

---

[8]On May 30, 2003, EPA proposed to remove the compound methyl ethyl ketone (MEK) from the HAPs list, and on November 12, 2003, it proposed to remove ethylene glycol monobutyl ether from the list.

## Proposed Actions

To address these deficiencies, the nation's AQM system must begin the transition toward a risk-focused multipollutant approach to AQM. Several recommendations to initiate this transition are presented below.

### Develop System to Set HAP Priorities

Many HAPs warrant increased resources for monitoring and research so that the risk HAPs pose to human health and welfare can be more accurately assessed and given the regulatory attention needed to protect human health and welfare. However, the statutory list of HAPs is long and may need to be expanded. It is unrealistic to expect that all HAPs can be monitored on a routine basis or that all HAPs can be placed under an aggressive regulatory framework. To ensure an appropriate allocation of resources and regulatory attention to the most dangerous HAPs, the committee recommends that the current system of setting priorities, embodied in EPA's urban air toxics program (EPA 2000b), be continued and enhanced. One possible approach (using a three-tier system to set priorities) is described in Box 7-3 for illustrative purposes. Other approaches might involve further elaboration of EPA's current list of 33 high-priority HAPs and a focus on ensuring that comprehensive strategies to monitor and address the sources of these pollutants are created and integrated into state and local AQMPs.

### Establish List of Potential Air Toxicants for Regulatory Attention

Beyond the current list of HAPs, little information on a vast array of unregulated emitted substances is an important problem. Examples of such possible toxicants are substitutes for various toxicants, such as bromopropane, used as a substitute for tetrachloroethylene and flame retardant polybrominated diphenyl ethers; atmospheric transformation products, such as formylcinnamaldehyde; peroxyacyl nitrates; other oxides, such as 1,3-butadiene diepoxide and benzoic acid; vehicular emissions, such as 2-methylnapthalene, diesel exhaust mixture, polychlorinated dibenzodioxins, polychlorinated dibenzofurans, isobutylene, and black carbon; and a number of pesticides.

Especially for high-volume emissions and hot spots, some reasonable level of regulatory response appears appropriate to curtail exposure to unregulated chemicals with suspicious but unproved adverse impacts. The committee recommends that suspicious chemicals emitted above a certain threshold concentration be tracked through a listing process and that a system for further addressing such chemicals be explored (see Box 7-4).

---

**BOX 7-3 Example of a Potential Classification Scheme for Hazardous Pollutants**

A number of schemes could be used to aid in setting resource priorities for HAPs on the basis of the relative risks posed to human health and welfare by these pollutants. One example would be a system with three tiers, applied on a national basis or, if a more effective allocation of resources is allowed, on a multi-state airshed basis.

• Tier 1 would most likely contain a few HAPs that, because of their diverse sources, ubiquitous presence in the atmosphere, and exceptionally high risk to human health and welfare, merited treatment similar (although not necessarily identical) to the treatment for criteria pollutants (for example, benzene). These HAPs would probably be drawn from the high-priority list identified in EPA's urban air toxics strategy (EPA 2000b) and be those identified as posing the highest population risk in such assessments as EPA's national air toxics assessment (EPA 2000b). In a small number of cases, these pollutants might be proposed for formal criteria pollutant status. In most cases, however, the Tier 1 HAPs not assigned NAAQS, as is done for criteria pollutants, might reasonably be incorporated into national monitoring programs and required or recommended for inclusion in an AQMP. As is the case for all HAPs, Tier 1 HAPs would be regulated through nationally mandated emission controls.

• Tier 2 HAPs, perhaps initially drawn from the remainder of the list of high-priority urban toxics identified by EPA in its urban air toxics strategy, would receive increased resources for monitoring and research, so that the risk they pose to human health and welfare could be assessed more accurately. In addition to nationally mandated emission controls, incentive programs could be implemented to encourage Tier 2 HAP inclusion in multipollutant AQMPs.

• Tier 3 HAPs, presumably initially drawn from the list of remaining HAPs, would be given the lowest priority for research and monitoring but would still be subject to nationally mandated emission controls.

Beyond these tiers, the committee recommends that a list of potential air toxicants be established and that these toxicants be subject to some minimal level of regulatory review and consideration (see Box 7-4).

---

Possible regulatory approaches include the development of exposure triggers (for example, emission concentrations, volume of use, or high exposures to some urban populations) for suspicious chemicals with sparse test data; some degree of testing and control would be required when the trigger measure was exceeded. Testing might include a minimal battery of tests, such as an expanded version of the current high-production-volume testing program instituted by EPA and the chemical industry. Inclusion on such a list of potential air toxicants might encourage the development of substitutes for those that exhibit initial indications of toxicity. However, a dynamic review of all pollutants, including those not on the current list,

## BOX 7-4   Identifying New Toxicants

A well-funded periodic review of air pollutants and their classification as HAPs and criteria pollutants should include an effort to identify new toxicants that pose a threat to human health and welfare. Some newly identified toxicants should be added to the HAPs list, and others should be listed for low-level regulatory oversight, as discussed below.

### Identifying New HAPs

To conserve resources in reviewing the numerous unregulated air pollutants for potential placement on the HAPs list, an EPA-based program could be established that relied on existing hazard evaluations from other agencies and institutions, as well as new hazard evaluations for chemicals that have not received prior adequate evaluation. Special attention should be given to the implementation of these evaluation processes to ensure that they do not become too protracted or resource intensive and that output of chemical evaluations is sufficient. Candidates could be screened by emission concentrations, and screening-level exposure analyses could be performed. New hazard evaluations should focus on those air pollutants that have not been assessed adequately by other institutions and that have a current or future potential for large exposures, such as chemical substitutes for listed HAPs. The evidentiary threshold for listing a chemical as a HAP is that it can be identified as *reasonably anticipated to cause toxicity*. The number of chemicals undergoing traditional toxicity testing in traditional toxicity studies is diminishing as methods for toxicity screening are evolving. Beyond traditional toxicity tests, EPA should explore adding chemicals to the HAPs list with the use of the full range of analyses. Persistent, bioaccumulative toxins released into the air in relatively small volumes may pose substantial toxic risks (Lunder et al. 2004) and are important candidates for evaluation using non-traditional approaches. With regard to the use of hazard evaluations from other institutions, there are several possible sources for use in identifying HAPs candidates. Examples are chemicals required for reporting in the Toxics Release Inventory Program; chemicals classified in categories 2B, 2A, or 1 by the International Agency for Research on Cancer; chemicals identified as reproductive toxicants of concern by the National Toxicology Program's Center for Evaluation of Reproductive Health Risks and listed on California's Proposition 65 list of chemicals known to cause cancer or reproductive toxicity; chemicals identified by EPA as known or likely to cause cancer (old B2 category and above); chemicals regulated on the basis of adverse health effects by the Occupational Safety and Health Administration; chemicals identified as toxic by the National Institute for Occupational Safety and Health; and chemicals described as emitted into the air with a toxicological profile published by the Agency for Toxic Substances and Disease Registry. A screening analysis for addition of chemicals to the HAPs list has recently been provided (Lunder et al. 2004).

### Identifying Chemicals for Regulatory Oversight

There is a vast array of unregulated emitted substances with sparse or no toxicological data to assess hazard potential adequately, and thus they cannot be placed on the HAPs list. Nevertheless, some attempt should be made to identify those chemicals that have sparse toxicological data but have structural similarities

**BOX 7-4** continued

to known HAPs and thus are likely to have adverse impacts. If chemicals so identified are also emitted in large amounts and have the potential for relatively high hot-spot exposures, they should be listed by EPA for enhanced monitoring and effects research and perhaps some low level of regulatory oversight. Because of the interrelated nature of environmental media (air, soil, water, and biota [see Box 7-1]), new chemicals suspected of being toxic or causing exposure but with few data on environmental fate and effects should be examined for inclusion on the HAPs list based on their production and use qualities and their likelihood of release into the air.

followed by decision-making on controlling exposures to those compounds that pose the most significant risks, is essential to incorporating the as-yet-unlisted chemicals in future AQM strategies.

### Institute a Dynamic Review of Pollutant Classification

EPA, as mandated in the CAA, must undertake a periodic review of the classifications given to pollutants. For example, successful mitigation of some criteria pollutants could logically result in their reclassification as HAPs to address remaining exposure and risk issues, and the proliferation of new technologies and products might require that some HAPs be reclassified as criteria pollutants. As new scientific information becomes available, the tier assigned a given HAP might need to be changed. Especially important is the need to identify and regulate pollutants that pose significant risks to human health and welfare but that are not yet listed as HAPs.

Classifying and setting priorities for air toxicants would be facilitated by the development of benchmark air concentrations. The process for developing such values within EPA is resource intensive and protracted, and benchmark concentrations (for example, a reference air concentration of a pollutant likely to cause a harmful effect in humans) are not available for a number of substances on the HAPs list that have sufficient data for guidance level derivations. A tiered system could be adopted for the development of guidance values. The first tier would be the de novo resource-intensive derivations of guidance values. The second tier would be the adoption of values derived by other EPA programs or federal or state agencies. The third tier would be the development of guidance values by expedited techniques. Different levels of review would apply to each of the tiers.

*Address Multiple Pollutants in the NAAQS Review*
*and Standard-Setting Process*

In current practice, EPA interprets the CAA and its amendments as requiring it to set NAAQS for each criteria pollutant independently from one another. Although the committee does not believe that the science has evolved to a sufficient extent to permit development of multipollutant NAAQS, it would be scientifically prudent to begin to review and develop NAAQS for related pollutants in parallel and simultaneously. Such a practice would facilitate the assessment of the commonality of sources, exposures, and effects among the pollutants, as well as the development of multipollutant AQMPs as recommended in Recommendation Three. Although such a change will require a transition period to be accomplished, it is not unprecedented and should be implemented expeditiously. (Earlier criteria documents address PM and $SO_2$ at the same time, for example.) Thus, we recommend the following:

• The criteria document and staff paper processes should be modified so that a simultaneous review of multiple interrelated pollutants could be developed in these documents. The interrelated NAAQS could then be considered in concert.
• Coordinated recommendations should be made to the EPA administrator with respect to modifications of the existing NAAQS so that new or modified NAAQS could be simultaneously promulgated.
• The implementation plans and attainment deadlines to address these NAAQS should be developed in a coordinated fashion to enable the development of multipollutant AQMPs.

*Enhance Assessment of Residual Risk*

In the current program to reduce emissions of HAPs from stationary sources, EPA is directed to undertake an assessment of residual risk following implementation of MACT and, on the basis of that assessment, decide whether additional controls are necessary. This program is getting under way somewhat slowly, the first completed assessment (on coke oven emissions) is expected in 2004. There are two key ways in which this process can be enhanced:

• The assessment of residual risk is challenging and time consuming. Nevertheless, given the importance of these assessments, EPA should move to accelerate this process to address an increased number of assessments in the years to come. To the extent that EPA is challenged to enhance resources to support risk assessments, residual risk assessments should be enhanced.

•  Although the CAA enables EPA to consider the full range of sources of a particular set of emissions in considering residual risk, in practice, EPA has focused primarily on the emissions from the source categories that were the subject of MACT.  To better address the full range of pollutant exposures in all settings (especially in hot spots and in areas surrounding major stationary sources), EPA should attempt to include other major sources of the same chemicals as much as possible, so that the contribution of the MACT-regulated source is assessed in the context of its contribution to broader exposures.  That assessment might take the form of creating hot-spot scenarios for estimating risk, drawn from actual locations of some of the regulated stationary sources.  Targeted monitoring in areas in which relevant industrial activity is heavily concentrated could be useful in this attempt.

## Recommendation Five

**Enhance protection of ecosystems and other aspects of public welfare.**

### Findings

The CAA was established to protect both human health and welfare, and in one key aspect, the setting of NAAQS, the CAA mandates the establishment of both primary standards to protect public health and secondary standards to protect welfare (including sensitive ecosystems, forests, crops, materials, historical monuments, visibility, and other resources). Indeed, ecosystems provide invaluable services, such as the supply of high-quality water, soils that support the structure and function of ecosystems, forest and crop production, diverse aquatic habitat, and maintenance of fisheries.  A loss or limitation of these services as a result of air pollution can therefore have significant consequences on the economy and quality of life.

However, programs and actions undertaken thus far in response to the CAA have largely focused on the protection of human health, neglecting efforts to protect environmental quality with secondary standards or to take actions to address air pollution impacts on ecosystems and crops.

•  The current practice of using the primary standard to serve as the secondary standard for most criteria pollutants does not appear to be sufficiently protective of sensitive crops and unmanaged ecosystems (see Chapter 2), although in one case EPA did recommend a separate secondary standard that was never implemented (EPA 1996b).
•  Concentration-based standards are inappropriate for some resources at risk, such as soils, groundwater, forests, surface water, and coastal eco-

systems from air pollutants, such as sulfur, nitrogen, or mercury. For such resources, a deposition-based standard would be more appropriate (see Chapter 2).

- EPA should undertake a comprehensive review of the need and use of standards to protect public welfare.
- The nation's AQM system has not been able to build a cohesive program capable of reliably reporting the status and trends in exposure and ecosystem conditions across regions and the nation (see Chapter 6).

### Proposed Actions

Specific activities are recommended that will help EPA to establish measures and actions to more effectively protect public welfare:

- *Develop and implement networks for comprehensive ecosystem monitoring.* Networks for monitoring terrestrial and aquatic ecosystem structure and function are needed to quantify the exposure of natural and managed resources to air pollution and the effects of air pollutants on ecosystems.
- *Establish acceptable exposure levels for natural and managed ecosystems.* On an ongoing basis, EPA should evaluate current research on the effects of air pollutants on ecosystems as a means to establish acceptable exposure levels for both natural and managed resources. In setting these acceptable exposure levels, EPA should consider the relevant geographic dimensions and sensitivity of the various resources to determine if acceptable exposure levels vary regionally. The adequacy of resource-specific acceptable exposure levels should be reviewed and revised, if necessary, at least every 10 years.
- *Promulgate secondary standards.* From the improved understanding gained from the above two actions, secondary standards should be promulgated where appropriate. In some cases, deposition-based secondary standards may be preferable to concentration-based standards. If acceptable exposure levels vary significantly from one region of the nation to another, consideration should be given to the promulgation of regionally distinct secondary standards.[9]
- *Design and implement controls.* Within the context of EPA's recommended enhanced responsibility and authority for addressing multistate air

---

[9]A move to regional secondary standards may require an amendment of the CAA. The courts have held that the primary NAAQS must be met on a nationwide basis. It is possible, however, that a court would find that a standard designed to protect the public welfare did not have to be uniform throughout the country, even though a standard designed to protect public health must be uniform. See Lead Industries Ass'n v. EPA, 647 F.2d 1130 (D.C. Cir. 1980).

pollution problems (see Recommendation 2), the agency should develop regulatory programs and mitigation actions to attain the standards.

• *Track progress toward attainment of secondary standards.* The aforementioned monitoring of ecosystem exposure and function should be used to track progress toward attainment of standards and to determine whether the progress results in the expected improvement in ecosystem function.

## CONCLUSION

In an advanced technological society such as the United States, air is a resource whose quality must be managed through the control of pollutant emissions. However, these controls can be implemented without abandoning technology or dismantling the economy. Experience over the past three decades of air pollution control in the United States has shown that effective AQM can often be accomplished best by encouraging and embracing new technologies as well as by using market forces within a vibrant economy to control emissions. AQM is also more effective when science and engineering have a central role in identifying critical problems, helping to optimize strategies for mitigation, designing systems to implement these strategies, and finally, tracking the success of these systems.

The nation's AQM system has had major successes over the past 30 years, but it must work to complete the task already before it (for example, attainment of the NAAQS for PM and $O_3$) and to face substantial new challenges in the future.

In the committee's view, the AQM system should strive to

• *Target the most significant exposures, risks, and uncertainties.*
• *Take an integrated multipollutant approach.*
• *Be a performance-oriented system.*
• *Take an airshed-based approach.*

In this chapter, the committee has proposed a set of five broad and interrelated recommendations for moving the AQM system in the above direction over the next decade or so. Because the nation's AQM system has been effective in many areas over the past decades, much of the system is good and bears retaining. Thus, the recommendations proposed here are intended to evolve the AQM system incrementally rather than to transform it radically. The recommendations are also not intended to deter the current, on-going AQM activities aimed at improving air quality. Indeed, even as these recommendations are implemented, there can be little doubt that important decisions to safeguard public health and welfare must continue to be made, at times in the face of scientific uncertainty. Moreover, new opportunities and approaches for managing air quality will appear. Even today, looking forward, we can identify several such areas:

• First, the challenge of moving beyond "one atmosphere" to "one environment" (see Box 7-1). The effects of air pollutants in water and soil (as well as air) and the multimedia implications of control strategies for all media have already been demonstrated (for example, air pollutant emissions from wastewater-treatment and site-remediation facilities and impacts on water from fuel additives for enhanced combustion). The remaining challenging task is to develop multimedia approaches and strategies, a task that was beyond the scope of this committee's endeavors but one that will require attention in the years ahead.

• Second, the opportunities presented by rapidly developing and increasingly sophisticated science and technology (Box 7-5). New advances in biotechnology, enhanced analytical and monitoring technologies, and many more such innovations are just beginning to have a part in AQM. These advances offer the prospect of even more targeted and effective air quality strategies in the decades ahead.

• Third, the enhancement of the AQM system by the public and private sectors. Over a longer time horizon, the nation's AQM system would be significantly enhanced by empowering the public and private sectors to undertake pollution prevention activities on their own accord rather than by merely controlling air pollutants after they have been produced.

Although specific recommendations for incorporating these new approaches into the nation's AQM system are not advanced here, the enhancements to the system—with its greater emphasis on performance and its encouragement of innovation—should facilitate their appropriate use in AQM over time.

Implementing this integrated set of recommendations will require the development of a detailed plan and a schedule of steps to be undertaken. Although the committee expects that many of the recommendations can be accomplished within the current CAA, some may require legislative action. A comprehensive analysis will be required to identify recommendations that can be implemented within the existing statutory framework and those that require legislative action—an analysis beyond the charge and expertise of this committee. To ensure timely implementation, the committee urges EPA to convene an implementation task force of experts from the key parties—the states; tribal and local agencies; environmental, industrial, and other stakeholders; and the scientific and technical community—to prepare a detailed implementation plan and an analysis of which, if any, statutory changes may be necessary.

---

### BOX 7-5 Advances in Environmental Instrumentation

Resource constraints have resulted in an undersampling of the environment—temporally, spatially, and with regard to chemical speciation. New developments in biotechnology, engineering, nanotechnology, and information technology provide promise for the development of monitoring networks that will overcome some of these deficits (Steinfeld et al. 2001). Advances in instrumentation that could contribute to the enhancement of the AQM system include the following:

• Advances in solid-state tunable diodes that will enable the translation of powerful optical diagnostic techniques from the research laboratory to the field. Methods that provide capability of real-time measurements and remote sensing are differential absorption LIDAR (DIAL), correlation spectroscopy, and Fourier transform spectroscopy (Steinfeld et al. 2002). These laser diagnostics can be used to monitor fluxes (Shorter et al. 1996), in addition to mapping horizontal and vertical concentration distributions of pollutants (Tittel 1998; NASA 2003).

• Reduction in the size of diagnostic and data acquisition systems that will permit the use of accurate and fast response instruments on mobile platforms (vehicular or airborne) capable of mapping concentration profiles for model testing and quantifying area sources and for identifying hot spots, leaks, and upset conditions. Miniaturization will also facilitate the development of personal exposure monitors of increased sophistication to monitor vital health-related statistics. Miniaturized near-real-time instruments are already available for many gaseous pollutants and are becoming available for particles through the use of field-deployable desorption gas chromatography and mass spectrometry techniques (Jeon et al., 2001) and aerosol time-of-flight mass spectrometers (Noble and Prather 1996; Bhave et al. 2002; Jayne et al. 2000).

• The development of distributed networks of microsensors to monitor HAPs and biotoxins may become feasible as a result of recent developments of "laboratory-on-a-chip" technology initially driven by concerns with homeland security (Frye-Mason et al. 2001; Lindner 2001).

• Perhaps the largest potential impact on the current approaches to monitoring will be the development of methods in biotechnology to rapidly screen for impacts of individual chemicals and mixtures. One example of the developments that will be important for the AQM system is that of DNA microarrays that show the potential of differentiating between exposures to different classes of toxicants and different toxicological outcomes (Bartosiewicz et al 2001).

Problems exist in the transitions of these technologies from research tools to commercial products that meet the needs of robustness, ease of use, cost, and equivalence to federal reference methods, problems that have only been partly alleviated by the Environmental Technology Verification Program. Use of instruments or procedures that do not satisfy a rigorous vetting process should be encouraged when valuable new insight is provided. For example, visual plume opacity readings have proved to be of great value, even though they do not provide the mass or composition of any specific pollutant, and cross-road sensors (Stedman et al. 1997; Jiminez et al. 2000) have proved their value in identifying high-emitting vehicles, even though their use for regulatory purposes is problematic. Some of the new methods might be introduced in the AQM system for specialized purposes, such as identifying hot spots, processing upset conditions for stationary sources, identifying breakdown in the emission control for mobile sources, and mapping spatially and temporally concentration distributions for ambient pollutants.

Implementation of the recommendations will also require additional resources. Although these resources are not insignificant, they should not be overwhelming. Even a doubling of the current EPA commitment to air pollution monitoring and research would be only about 1% of the costs incurred annually to comply with the CAA. Such resources are even smaller when compared with the costs imposed by the deleterious effects of air pollution on human health and welfare (see discussion in Chapter 1).

Fundamental changes will also be needed in aspects of the nation's AQM system to shift the focus to tracking progress. Such a transition will be difficult, but as noted above, it is imperative that actions to further reduce emissions continue even as this transition takes place.

Finally, implementation of these recommendations and meeting the challenges of AQM in the decades to come will require a major commitment from the research and development and scientific communities to provide the human resources and technologies needed to underpin an enhanced AQM system and to achieve clean air in the most expeditious and effective way possible. The committee believes that these communities are ready to respond.

# References

Aber, J.D., C.L. Goodale, S.V. Ollinger, M.L. Smith, A.H. Magill, M.E. Martin, R.A. Hallett, and J.L. Stoddard. 2003. Is nitrogen deposition altering the nitrogen status of northeastern forests? BioScience 53(4):375–389.

AIRData. 2003. AirData: Access to Air Pollution Data. Office of Air and Radiation, U.S. Environmental Protection Agency, Washington, DC. [Online]. Available: http://www.epa.gov/air/data/index.html [accessed March 16, 2003].

AIRMoN. 2002. The Atmospheric Integrated Monitoring Network (AIRMoN). Air Research Laboratory. [Online]. Available: http://www.arl.noaa.gov/research/programs/airmon.html [accessed March 16, 2002].

Alberta Research Council. 2004. Controlled Environment Exposure System, World Class Facilities. Alberta Research Council. [Online]. Available: http://www.arc.ab.ca/corp/facilities.asp [accessed Feb. 02, 2004].

Altshuler, A.A., J.P. Womack, and J.R. Pucher. 1979. The Urban Transportation System: Politics and Policy Innovation. Cambridge, MA: MIT Press.

Ambio. 2002. Ambio 31(2). Royal Swedish Academy of Sciences.

Anderson, J.F., and T.L. Sherwood. 2002. Comparison of EPA and Other Estimates of Mobile Source Rule Costs to Actual Price Changes. SAE 2002-01-1980. Government/Industry Meeting, Session: Complying with Tier 2 Gasoline and Diesel Emissions Standards, May 13–15, 2002, Washington, DC.

Ansmann, A., D. Althausen, U. Wandinger, K. Franke, D. Muller, F. Wagner, and J. Heintzenberg. 2000. Vertical profiling of the Indian aerosol plume with six-wavelength lidar during INDOEX: A first case study. Geophys. Res. Lett. 27(7):963–966.

API (American Petroleum Institute). 1989. Detailed Analysis of Ozone State Implementation Plans in Seven Areas Selected for Retrospective Evaluation of Reasons for State Implementation Plan Failure. API Pub. No. 4502. Washington, DC: American Petroleum Institute.

Apogee Research Inc. 1994. Costs and Effectiveness of Transportation Control Measures (TCMs): A Review and Analysis of the Literature. Washington, DC: National Association of Regional Councils.

Atkinson, R., and A.C. Lloyd. 1984. Evaluation of kinetic and mechanistic data for model-ing of photochemical smog. J. Phys. Chem. Ref. Data 13(2):315–444.

Auto/Oil AQIRP. 1993. Auto/Oil Air Quality Improvement Research Program, Phase I Final Report. Auto/Oil Emission Research, Coordinating Research Council, Alpharetta, GA. May 1993.

Auto/Oil AQIRP. 1997. Auto/Oil Air Quality Improvement Research Program: Program Final Report. Auto/Oil Emission Research, Coordinating Research Council, Alpharetta, GA. Jan. 1997.

Axelrad, D.A., R.A. Morello-Frosch, T.J. Woodruff, and J.C. Caldwell. 1999. Assessment of estimated 1990 air toxics concentrations in urban areas in the United States. Environ. Sci. Pol. 2(4-5):397–411.

Bailey, S.W., J.W. Hornbeck, C.T. Driscoll, and H.E. Gaudette. 1996. Calcium inputs and transport in a base-poor forest ecosystem as interpreted by Sr isotopes. Water Resour. Res. 32(3):707–719.

Balmford, A., A. Bruner, P. Cooper, R. Costanza, S. Farber, R.E. Green, M. Jenkins, P. Jefferies, V. Jessamy, J. Madden, K. Munro, N. Myers, S. Naeem, J. Paavola, M. Rayment, S. Rosendo, J. Roughgarden, K. Trumper, and R.K. Turner. 2002. Economic reasons for conserving wild nature. Science 297(5583):950–953.

Bammi, S., and H.C. Frey. 2001. Quantification of Variability and Uncertainty in Law and Garden Equipment $NO_x$ and Total Hydrocarbon Emission Factors. Abstract No. 580. Proceedings, Annual Meeting of Air & Waste Management Association, June 24–28, 2001, Orlando, FL. [Online]. Available: http://www4.ncsu.edu/~frey/FreyBammiLawn Garden AWMA.pdf [accessed Oct. 3, 2002].

Barth, M., T. Younglove, T. Wenzel, G. Scora, F. Ann, M. Ross, and J. Norbeck. 1997. Analysis of modal emissions from diverse in-use vehicle fleet. Transport. Res. Rec. 1587:73–84.

Bartosiewicz, M., S. Penn, and A. Buckpitt. 2001. Application of gene arrays in environ-mental toxicology: Fingerprints of gene regulation associated with cadmium chloride, benzo(a)pyrene, and trichloroethylene. Environ. Health Perspect. 109(1):71–74.

Bauer, A., and A.L. Black. 1994. Quantification of the effect of soil organic matter content on productivity. Soil Sci. Soc. Am. J. 58(1):185–193.

Benkovitz, C.M., and S.E. Schwartz. 1997. Evaluation of modeled sulfate and $SO_2$ over North America and Europe for four seasonal months in 1986–1987. J. Geophys. Res. 102(21):25305–25338.

Benkovitz, C.M., S.E. Schwartz, and B.G. Kim. 2003. Evaluation of a chemical transport model for sulfate using ACE-2 observations and attribution of sulfate mixing ratios to source regions and formation processes. Geophys. Res. Lett. 30(12):1641–1644. [On-line]. Available: http://www.ecd.bnl.gov/AnnualPubs03.html [accessed June 5, 2003].

Bennett, D., T. McKone, J. Evans, W. Nazaroff, M. Margni, O. Jolliet, and K. Smith. 2002. Defining intake fraction although pollution emissions have been linked to health prob-lems, researchers need to standardize their definitions and models to definitively calcu-late what humans actually take in. Environ. Sci. Technol. 36(9):206A–211A.

Bennett, N. 1996. The National Highway System Designation Act of 1995. Public Roads Online 59(4). [Online]. Available: http://www.tfhrc.gov/pubrds/spring96/p96sp10.htm [accessed Jan. 15, 2003].

Berkowitz, C.M., P.H. Daum, C.W. Spicer, and K.M. Busness. 1996. Synoptic patterns associated with the flux of excess ozone to the western North Atlantic. J. Geophys. Res. Atmos. 101(D22):28923–28933.

Bernard, S.M., J.M. Samet, A. Grambsch, K.L. Ebi, and I. Romieu. 2001. The potential impacts of climate variability and change on air pollution-related health effects in the United States. Environ. Health Perspect. 109(Suppl. 2):199–209.

Bhave, P.K., M.J. Kleeman, J.O. Allen, L.S. Hughes, K.A. Prather, and G.R. Cass. 2002. Evaluation of an air quality model for the size and composition of source-oriented particle classes air quality model predictions that track the size, composition, and concentration of representative particles within different source-oriented classes are compared with ambient single particle measurements. Environ. Sci. Technol. 36(10): 2154–2163.

Bishop, G.A., S.S. Pokharel, and D.H. Stedman. 2000. On-Road Remote Sensing of Automobile Emissions in the Los Angeles Area: Year 1. Department of Chemistry and Biochemistry, University of Denver, Denver, CO. Prepared for Coordinating Research Council, Inc., Alpharetta, GA.

Blau, S., and C. Agren. 2001. The Large Combustion Plants Directive, Comments to the Commission's Proposed Revision. The European Environmental Bureau, the European Federation for Transport and Environment, the Swedish NGO Secretariat on Acid Rain. [Online]. Available: http://www.acidrain.org/lcp.pdf [accessed Apr. 1, 2003].

Boynton, W.R., J.H. Garber, R. Summers, and W.M. Kemp. 1995. Inputs, transformations and transport of nitrogen and phosphorus in Chesapeake Bay and selected tributaries. Estuaries 18(1B):285–314.

Bricker, S.B., C.G. Clement, D.E. Pirhalla, S.P. Orlando, and D.R.G. Farrow. 1999. National Estuarine Eutrophication Assessment: Effects of Nutrient Enrichment in the Nation's Estuaries. NOAA, National Ocean Service, Special Projects Office and the National Centers for Coastal Ocean Science, Silver Spring, MD. 71 pp.

Brook, J.R., E. Vega, and J.G. Watson. 2003. Receptor methods. Chapter 7 in Particulate Matter Science for Policy Makers, A NARSTO Assessment, Part 2. EPRI 1007735. Palo Alto, CA: EPRI.

Brown, P. 1995. Race, class, and environmental health: A review and systematization of the literature. Environ. Res. 69(1):15–30.

Brown, G., T.A. Cameron, M.L. Cropper, A.M. Freeman III, D. Fullerton, J.V. Hall, P. Lioy, R. Schmalensee, and T. Tietenberg. 2001. Misrepresenting ACCACA. Regulation 24 (3):4–5. [Online]. Available: http://www.cato.org/pubs/regulation/regv24n3/forthe record.pdf [accessed Jan. 12, 2004].

Burnett, R.T., S. Cakmak, S. Bartlett, D. Stieb, B. Jessiman, M. Raizenne, P. Blagden, J.R. Brook, P.R. Samson, and T. Dann. In press. Measuring progress in the management of ambient air quality: The case for population health. J. Toxicol. Environ. Health.

Burtraw, D., and D.A. Evans. 2003. The Evolution of $NO_x$ Control Policy for Coal-Fired Power Plants in the United States. Discussion Paper 03-23. Washington DC: Resources for the Future.

Burtraw, D., and E. Mansur. 1999. Environmental effects of $SO_2$ trading and banking. Environ. Sci. Technol. 33(20):3489–3494.

Burtraw, D., A.J. Krupnick, E. Mansur, D. Austin, and D. Farrell. 1998. Costs and benefits of reducing air pollutants related to acid rain. Contemp. Econ. Policy 16(4):379–400.

Burtraw, D., K. Palmer, R. Bharvirkar, and A. Paul. 2001a. Cost-effective reduction of $NO_x$ emissions from electricity generation. J. Air Waste Manage. 51(10):1476–1489.

Burtraw, D., K. Palmer, R. Bharvirkar, and A. Paul. 2001b. Effect of Allowance Allocation on the Cost of Carbon Emission Trading. Discussion Paper 01-30. Washington, DC: Resources for the Future. [Online]. Available: http://www.rff.org/disc_papers/abstracts/0130.htm [accessed Sept. 24, 2002].

Burtraw, D., R. Bharvirkar, and M. McGuinness. 2003. Net benefits of emissions reductions of nitrogen oxides. Land Econ. 79(3):382–401.

Cackette, T. 1998. The Cost of Emission Controls Motor Vehicles and Fuels: Two Case Studies, July 1998, Massachusetts Institute of Technology, Cambridge.

Campbell, S.J., G. Smith, P. Temple, J. Pronos, R. Rochefort, and C. Andersen. 2000. Monitoring for Ozone Injury in West Coast (Oregon, Washington, California) Forests in 1998. Gen. Tech. Rep. PNW-GTR-495. U.S. Department. of Agriculture, Forest Service, Pacific Northwest Research Station, Portland, OR.

Canfield, R.L., C.R. Henderson Jr., D.A. Cory-Slechta, C. Cox, T.A. Jusko, and B.P. Lanphear. 2003. Intellectual impairment in children with blood lead concentrations below 10 microg per deciliter. N. Engl. J. Med. 348(16):1517–1526.

CARB (California Air Resources Board). 1994. The California State Implementation Plan for Ozone. Adopted Nov. 15, 1994. California Environmental Protection Agency Air Resources Board. [Online]. Available: http://www.arb.ca.gov/sip/sip.htm.

CARB (California Air Resources Board). 2001. Policy and Actions for Environmental Justice, Approved on Dec. 13, 2001. California Environmental Protection Agency Air Resources Board. [Online]. Available: http://www.arb.ca.gov/ch/ejpolicies_121301.pdf [accessed Apr. 11, 2002].

CARB (California Air Resources Board). 2003a. California Reformulated Gasoline Regulations. Updated Jan. 1, 2003. California Environmental Protection Agency Air Resources Board. [Online]. Available: http://www.arb.ca.gov/cbg/carfg3regs.pdf [accessed May 28, 2003]

CARB (California Air Resources Board). 2003b. Title V Permit: 2003-04-17 Acme Fill Corp.-Landfill. California Environmental Protection Agency Air Resources Board [Online]. Available: http://www.arb.ca.gov/ch/ejpolicies_121301.pdf [accessed May 15, 2003].

Cardelino, C.A., and W.L. Chameides. 1990. Natural hydrocarbons, urbanization, and urban ozone. J. Geophys. Res. 95(9):13971–13979.

Cardelino, C.A., and W.L. Chameides. 2000. The application of data from photochemical assessment monitoring stations to the observation-based model. Atmos. Environ. 34(12): 2325–2332.

Carlson, C., D. Burtraw, M. Cropper, and K.L. Palmer. 2000. Sulfur dioxide control by electric utilities: What are the gains from trade? J. Politic. Econ. 108(6):1292–1326.

Carter, W.P.L., A.M. Winer, K.R. Darnall, and J.N. Pitts, Jr. 1979. Smog chamber studies of temperature effects in photochemical smog. Environ. Sci. Technol. 13 (9):1094–1100.

CASTNet. 2002. Clean Air Status and Trends Network. Clean Air Markets, U.S. Environmental Protection Agency. [Online]. Available: http://www.epa.gov/castnet/ [accessed March 16, 2002].

Castro, M.S., and C.T. Driscoll. 2002. Atmospheric nitrogen deposition to estuaries in the mid-Atlantic and northeastern United States. Environ. Sci. Technol. 36(15):3242–3244.

CBADS (Chesapeake Bay Atmospheric Deposition Study). 2001. Chesapeake Bay Atmospheric Deposition Study. The Power Plant Research Program, Maryland Department of Natural Resources. [Online]. Available: http://esm.versar.com/pprp/features/aciddep/cbads.htm [accessed Oct. 21, 2002].

CCRP (California Comparative Risk Project). 1994. Toward the 21st Century: Planning for the Protection of California's Environment, Final Report. Office of Environmental Health Hazard Assessment, Pesticide and Environmental Toxicology Section, California Environmental Protection Agency, Sacramento, CA.

CDC (Centers for Disease Control and Prevention). 2001. National Report on Human Exposure to Environmental Chemicals. Centers for Disease Control and Prevention, National Center for Environmental Health, Atlanta, GA. March 2001. [Online]. Available: http://www.cdc.gov/nceh/dls/REPORT/index.htm [accessed June 11, 2003].

CDC (Centers for Disease Control and Prevention). 2002. Biomonitoring: Elemental Analysis, Aug. 8, 2002. [Online]. Centesr for Disease Control and Prevention, Atlanta, GA. Available: http://www.aphl.org/docs/Metals.pdf [accessed March 10, 2004].

CDC (Centers for Disease Control and Prevention). 2003a. CDC's Environmental Public Health Tracking Program—Background. National Center for Environmental Health, Centers for Disease Control and Prevention, Atlanta, GA. [Online]. Available: http://www.cdc.gov/nceh/tracking/background.htm [accessed Nov. 19, 2003].

CDC (Centers for Disease Control and Prevention). 2003b. Second National Report on Human Exposure to Environmental Chemicals. NCEH 02-716. National Center for Environmental Health, Centers for Disease Control and Prevention, Atlanta, GA. [Online]. Available: http://www.cdc.gov/exposurereport/ [accessed June 11, 2003].

Cerco, C.F. 2000. Chesapeake Bay eutrophication model. Pp. 363–404 in Estuarine Science: A Synthetic Approach to Research and Practice, J.E. Hobbie, ed. Washington, DC: Island Press.

Chameides, W.L., R.W. Lindsay, J. Richardson, and C.S. Kiang. 1988. The role of biogenic hydrocarbons in urban photochemical smog: Atlanta, Georgia, U.S.A. as a case study. Science 241(4872):1473–1475.

Chameides, W.L., R.D. Saylor, and E.B. Cowling. 1997. Ozone pollution in the rural United States and the new NAAQS. Science 276(5314):916.

Chang, J., R. Brost, I. Isaksen, S. Madronich, P. Middleton, W. Stockwell, and C. Walcek. 1987. Three-dimensional Eulerian acid deposition model: Physical concepts and formulation. J. Geophys. Res. 92:14681–14700.

Chang, M.E., and C. Cardelino. 2002. Air Quality Forecasting in Atlanta, GA. Presentation at Regional Ozone Forecasting Workshop, Feb. 6, 2002, Chattanooga, TN.

Chia, N.C., and S.Y. Phang. 2001. Motor vehicle taxes as an environmental management instrument: The case of Singapore. Environ. Econ. Policy Stud. 4(4):67–94.

Chock, D.P., S. Kumar, and R.W. Herrmann. 1982. An analysis of trends in oxidant air quality in the South Coast Air Basin of California, U.S.A. Atmos. Environ. 16(11):2615–2624.

Chow, J.C., J.D. Bachmann, S.S. Wierman, C.V. Mathai, W.C. Malm, W.H. White, P.K. Mueller, N. Kumar, and J.G. Watson. 2002a. 2002 Critical review discussion—Visibility: Science and regulation. J. Air Waste Manage. Assoc. 52(9):973–999.

Chow, J.C., J.P. Engelbrecht, J.G. Watson, W.E. Wilson, N.H. Frank, and T. Zhu. 2002b. Designing monitoring networks to represent outdoor human exposure. Chemosphere 49(9):961–978.

Christ, M.J., C.T. Driscoll, and G.E. Likens. 1999. Watershed- and plot-scale tests of the mobile anion concept. Biogeochemistry 47(3):335–353.

Clancy, L, P. Goodman, H. Sinclair, and D.W. Dockery. 2002. Effect of air-pollution control on death rates in Dublin, Ireland: An intervention study. Lancet 360(9314): 1210–1214.

Clayton, C.A., E.D. Pellizzari, and J.J. Quackenboss. 2002. National Human Exposure Assessment Survey: Analysis of exposure pathways and routes for arsenic and lead in EPA Region 5. J. Expo. Anal. Environ. Epidemiol. 12(1):29–43.

CPDEP (Commonwealth of Pennsylvania Department of Environmental Protection). 1998. Final State Implementation Plan (SIP) Revision for the Attainment and Maintenance of the National Ambient Air Quality Standard for Ozone; Meeting the Requirements of the Alternative Ozone Attainment Demonstration Policy: Phase I Ozone SP Submittal, Aug. 1, 1998. [Online]. Available: http://www.dep.state.pa.us/dep/deputate/airwaste/aq/plans/plans/rop/Phase_I_Final.pdf [accessed Jan. 21, 2003].

Crandall, R.W., H.K. Gruenspecht, T.E. Keeler, and L.B. Lave. 1986. Regulating the Automobile. Washington, DC: Brookings Institution.

Cronan, C.S., and D.F. Grigal. 1995. Use of calcium/aluminum ratios as indicators of stress in forest ecosystems. J. Environ. Qual. 24(2):209–226.

Cronan, C.S., and C.L. Schofield. 1990. Relationships between aqueous aluminum and acidic deposition in forested watersheds of North America and northern Europe. Environ. Sci. Technol. 24(7):1100–1105.

Croote, T. 1999. Cost-Benefit Analysis of EPA Regulations: An Overview. CRC Report for Congress CRC RL 30326, Sept. 16, 1999. [Online]. Available: http://cnie.org/NLE/CRSreports/Risk/rsk-42.cfm [accessed Oct. 28, 2002].

Daily, G.C., ed. 1997. Natures Services: Societal Dependence on Natural Ecosystems. Washington, DC: Island Press.

Daniels, M.J., F. Dominici, J.M. Samet, and S.L. Zeger. 2000. Estimating particulate matter-mortality dose-response curves and threshold levels: An analysis of daily time-series for the 20 largest U.S. cities. Am. J. Epidemiol. 152(5):397–406.

Davis, S.C., and S.W. Diegel. 2002. Transportation Energy Data Book, 22nd Ed. ORNL-6967. Oak Ridge, TN: Oak Ridge National Laboratory.

Deakin, E.A. 1978. Air quality considerations in transportation planning. Pp. 65–69 in Air Quality Analysis in Transportation Planning. Transportation Research Record 670. Washington, DC: National Academy of Sciences.

D'Elia, C.F., L.W. Harding, M. Meffler, and G.B. Mackiernan. 1992. The role and control of nutrients in Chesapeake Bay. Water Sci. Technol. 26(12):2635–2644.

DeLucia, E.H., J.G. Hamilton, S.L. Naidu, R.B. Thomas, J.A. Andrews, A. Finzi, M. Lavine, R. Matamala, J.E. Mohan, G.R. Hendrey, and W.H. Schlesinger. 1999. Net primary production of a forest ecosystem with experimental $CO_2$ enrichment. Science 284(5417): 1177–1179.

Demerjian, K.L. 2000. A review of national monitoring networks in North America. Atmos. Environ. 34(12/14):1861–1884.

Dimitriades, B. 1977. An alternative to the appendix-J method for calculating oxidant- and $NO_2$-related control requirements. Pp. 871–879 in Proceedings of International Conference on Photochemical Oxidant Pollution and Its Control, Vol. 2. EPA 600/3-77-001b. Environmental Science Research Laboratory, Office of Research and Development, U.S. Environmental Protection Agency, Research Triangle Park, NC.

DOC (U.S. Department of Commerce). 1999. 1997 Economic Census: Vehicle Inventory and Use Survey. EC97TV-US. U.S. Census Bureau, U.S. Department of Commerce. [Online]. Available: http://www.census.gov/prod/ec97/97tv-us.pdf [accessed Jan. 15, 2003].

Dodge, M.C. 1977. Combined use of modeling techniques and smog chamber data to derive ozone–precursor relationships. Pp. 881–889 in Proceedings, International Conference on Photochemical Oxidant Pollution and Its Control, Vol. II, B. Dimitriades, ed. EPA/600/3-77-001a. PB 264232. Environmental Sciences Research Laboratory, Research, U.S. Environmental Protection Agency, Triangle Park, NC.

DOE (U.S. Department of Energy). 2001. Predicting Effects of Elevated Carbon Dioxide Concentrations. Office of Biological and Environmental Research, U.S. Department of Energy. [Online]. Available: http://www.er.doe.gov/feature_articles_2001/June/Decades/98.html [accessed Nov. 7, 2002].

DOE (U.S. Department of Energy). 2003. Annual Energy Outlook 2003 with Projections to 2025. Appendix A Reference Case Forecast (2000–2025) Tables [Online]. Available: http://www.eia.doe.gov/oiaf/aeo/aeoref_tab.html [accessed Jan. 13, 2003].

Drake, B.G., and P.W. Leadley. 1991. Canopy photosynthesis of crops and native plant communities exposed to long-term elevated $CO_2$: Commissioned review. Plant Cell Environ. 14(8):853–860.

Driscoll, C.T., and K.M. Postek. 1996. The chemistry of aluminum in surface waters. Pp. 363–418 in The Environmental Chemistry of Aluminum, 2nd Ed., G. Sposito, ed. Boca Raton, FL: Lewis.

Driscoll, C.T., C.P. Cirmo, T.J. Fahey, V.L. Blette, P.A. Bukaveckas, D.J. Burns, C.P. Gubala, D.J. Leopold, R.M. Newton, D.J. Raynal, C.L. Schofield, J.B. Yavitt, and D.B. Porcella. 1996. The Experimental Watershed Liming Study (EWLS): Comparison of lake and watershed neutralization strategies. Biogeochemistry 32(3):143–174.

Driscoll, C.T., K.M. Postek, D. Mateti, K. Sequeira, J.D. Aber, W.J. Kretser, M.J. Mitchell, and D.J. Raynal. 1998. The response of lake water in the Adirondack region of New York to changes in acidic deposition. Environ. Sci. Pol. 1(3):185–198.

Driscoll, C.T., G.B. Lawrence, A.J. Bulger, T.J. Butler, C.S. Cronan, C. Eagar, K. Fallon Lambert, G.E. Likens, J.L. Stoddard, and K.C. Weathers. 2001a. Acidic deposition in the northeastern United States: Sources and inputs, ecosystem effects, and management strategies. Bioscience 51(3):180–198.

Driscoll, C.T., G.B. Lawrence, A.J. Bulger, T.J. Butler, C.S. Cronan, C. Eagar, K.F. Lambert, G.E. Likens, J.L. Stoddard, and K.C. Weathers. 2001b. Acid Rain Revisited: Advances in Scientific Understanding Since the Passage of the 1970 and 1990 Clean Air Act Amendments. Hanover, NH: Science Links Publications, Hubbard Brook Research Foundation. 20 pp.

Driscoll, C.T., K.M. Driscoll, K.M. Roy, and M.J. Mitchell. 2003a. Chemical response of lakes in the Adirondack Region of New York to declines in acidic deposition. Environ. Sci. Technol. 37(10):2036–2042.

Driscoll, C.T., D. Whitall, J. Aber, E. Boyer, M. Castro, C. Cronan, C.L. Goodale, P. Groffman, C. Hopkinson, K. Lambert, G. Lawrence, and S. Ollinger. 2003b. Nitrogen pollution in the northeastern United States: Sources, effects, and management options. BioScience 53(4):357–374.

Dye, T.S., C.P. MacDonald, and C.B. Anderson. 2000. Air Quality Forecasting for the SPARE the Air Program in Sacramento, California: Summary of Four Years of Ozone Forecasting. Paper 16.4A. Presentation at the 11th Joint Conference on the Applications of Air Pollution Meteorology with the Air and Waste Management Association, Long Beach, CA. [Online]. Available: http://ams.confex.com/ams/annual2000/11airpollut/index.html [accessed June 13, 2003].

EHTPT (Environmental Health Tracking Project Team). 2000. America's Environmental Health Gap: Why the Country Needs a Nationwide Health Tracking Network, Technical Report. Johns Hopkins School of Hygiene and Public Health, Sept. 2000. [Online]. Available: http://pewenvirohealth.jhsph.edu/html/reports/pewtrackingtechnical.pdf [Oct. 17, 2002].

EIA (Energy Information Administration). 2000. Annual Energy Review, 1999. Energy Information Administration, Office of Energy Markets and End Use, U.S. Department of Energy, Washington, DC.

Ellerman, A.D., P.L. Joskow, R. Schmalensee, J.P. Montero, and E.M. Bailey. 2000. Markets for Clean Air: The U.S. Acid Rain Program. Cambridge, UK: Cambridge University Press.

Environmental Defense. 2002. Scorecard. Pollution Locator. Caveats on Scorecard's Hazardous Air Pollution Data. [Online]. Available: http://www.scorecard.org/env-releases/def/hap_caveats.html [Sept. 23, 2002].

EPA (U.S. Environmental Protection Agency). 1971. Air Quality Criteria for Nitrogen Oxides. AP-84. U.S. Environmental Protection Agency, Washington, DC. Jan.

EPA (U.S. Environmental Protection Agency). 1973. EPA Requires Phase-Out of Lead in All Grades of Gasoline. EPA Press Release, Nov. 28, 1973. [Online]. Available: http://www.epa.gov/history/topics/lead/03.htm [accessed May 30, 2003].

EPA (U.S. Environmental Protection Agency). 1985. Cost and Benefits of Reducing Lead in Gasoline: Final Regulatory Impact Analysis. EPA-230-05-85-006. Office of Policy Analysis and Review, U.S. Environmental Protection Agency, Washington, DC.

EPA (U.S. Environmental Protection Agency). 1987. Unfinished Business: A Comparative Assessment of Environmental Problems. Office of Policy, Planning and Evaluation, U.S. Environmental Protection Agency, Washington, DC.

EPA (U.S. Environmental Protection Agency). 1991. Nonroad Engine and Vehicle Emission Study-Report. EPA 460/3-91-02. Office of Air and Radiation, U.S. Environmental Protection Agency, Research, Washington, DC. [Online]. Available: http://www.epa.gov/otaq/regs/nonroad/nrstudy.pdf [accessed Jan.15, 2003].

EPA (U.S. Environmental Protection Agency). 1992. Principles of Environmental Enforcement. EPA 300/F-93-001. Office of Enforcement, U.S. Environmental Protection Agency, Washington, DC. July 15, 1992.

EPA (U.S. Environmental Protection Agency). 1995. Compilation of Air Pollutant Emission Factors, AP-42, 5th Ed., Vol. 1: Stationary Point and Area Sources. U.S. Environmental Protection Agency, Research Triangle Park, NC. Jan. 1995. [Online]. Available: http://www.epa.gov/ttn/chief/ap42/index.html [accessed Nov. 17, 2003].

EPA (U.S. Environmental Protection Agency). 1995a. Users Guide for the Industrial Source Complex (ISC3) Dispersion Models, Vol. 2. Description of Model Algorithms. EPA-454/B-95-003b. Office of Air Quality Planning and Standards, U.S. Environmental Protection Agency, Research Triangle Park, NC. [Online]. Available: http://www.epa.gov/scram001/tt22.htm [accessed March 20, 2003].

EPA (U.S. Environmental Protection Agency). 1995b. Users Guide for the Calpuff Dispersion Model. EPA-454/B-95-006. Office of Air Quality Planning and Standards, U.S. Environmental Protection Agency, Research Triangle Park, NC.

EPA (U.S. Environmental Protection Agency). 1996a. Review of National Ambient Air Quality Standards for Ozone. Assessment of Scientific and Technical Information OAQPS Staff Paper. EPA-452/R-96-007. Office of Air Quality Planning and Standards, U.S. Environmental Protection Agency, Research Triangle Park, NC.

EPA (U.S. Environmental Protection Agency). 1996b. Air Quality Criteria for Ozone and Related Photochemical Oxidants. EPA/600/P-93/004a-cF. National Center for Environmental Assessment, Office of Research and Development, U.S. Environmental Protection Agency, Washington, DC. July. [Online]. Available: http://cfpub.epa.gov/ncea/cfm/ozone.cfm [accessed Feb. 5, 2003].

EPA (U.S. Environmental Protection Agency). 1997. Benefits and Costs of the Clean Air Act, 1970 to 1990. Final Report to U.S. Congress. EPA 410/R-97-002. Office of Air Quality Planning and Standards, U.S. Environmental Protection Agency, Research Triangle Park, NC [Online]. Available: http://www.epa.gov/oar/sect812/contsetc.pdf [accessed July 10, 2003].

EPA (U.S. Environmental Protection Agency). 1998a. Chemical Hazard Data Availability Study. What Do We Really Know About the Safety of High Production Volume Chemicals? Office of Pollution Prevention and Toxics, U.S. Environmental Protection Agency, Washington, DC. [Online]. Available: http://www.epa.gov/chemrtk/hazchem.pdf [accessed Nov. 17, 2003].

EPA (U.S. Environmental Protection Agency). 1998b. Handbook for Air Toxics Emission Inventory Development, Vol. 1. Stationary Sources. EPA 454/B-98-002. Office of Air Quality Planning and Standards, Research Triangle Park, NC. [Online]. Available: http://www.epa.gov/ttn/chief/eidocs/airtoxic.pdf [accessed Oct. 1, 2002].

EPA (U.S. Environmental Protection Agency). 1998c. Evaluation of Emissions from the Open Burning of Household Waste in Barrels. EPA/600-SR-97/134. National Risk Management Research Laboratory, Cincinnati, OH. March 1998. [Online]. Available: http://www.epa.gov/ttn/atw/burn/trashburn1.pdf [accessed Jan. 15, 2002].

EPA (U.S. Environmental Protection Agency). 1998d. Commonly Asked Questions about ORVR (Onboard Refueling Vapor Recovery). Office of Transportation and Air Quality, U.S. Environmental Protection Agency, Washington, DC. [Online]. Available: http://www.epa.gov/otaq/regs/ld-hwy/onboard/orvrq-a.txt [accessed Jan. 15, 2003].

EPA (U.S. Environmental Protection Agency). 1998e. Ecological Research Strategy. EPA/600/R-98/086. Office of Research and Development, U.S. Environmental Protection Agency, Washington, DC. June 1998.

EPA (U.S. Environmental Protection Agency). 1999a. Final Report to Congress on Benefits and Costs of the Clean Air Act, 1990 to 2010. EPA 410-R-99-001. Office of Air and Radiation, U.S. Environmental Protection Agency, Washington, DC. Nov. 1999.

EPA (U.S. Environmental Protection Agency). 1999b. Residual Risk Report to Congress. EPA-453/R-99-001. Office of Air Quality Planning and Standards, U.S. Environmental Protection Agency, Research Triangle Park, NC. March 1999. [Online]. Available: http://www.epa.gov/ttnatw01/risk/risk_rep.pdf [accessed Jan. 31, 2003].

EPA (U.S. Environmental Protection Agency). 1999c. Progress Report on the EPA Acid Rain Program. EPA 430-R-99-011. Office of Air and Radiation, U.S. Environmental Protection Agency, Washington, DC. Nov. 1999. [Online]. Available: http://www.epa.gov/airmarkets/progress/arpreport/acidrainprogress.pdf [accessed Oct. 16, 2002].

EPA (U.S. Environmental Protection Agency). 1999d. Ozone and Your Health. EPA 452/F-99-003. Office of Air and Radiation, U.S. Environmental Protection Agency, Washington, DC. [Online]. Available: http://www.epa.gov/airnow/brochure.html [accessed Apr. 1, 2003].

EPA (U.S. Environmental Protection Agency). 1999e. Emissions Inventory Guidance for Implementation of Ozone and Particulate Matter National Ambient Air Quality Standards (NAAQS) and Regional Haze Regulations. EPA-454-99-006. Office of Air Quality Planning and Standards, U.S. Environmental Protection Agency, Research Triangle Park, NC. [Online]. Available: http://www.epa.gov/ttn/chief/eidocs/eidocfnl.pdf [accessed Oct. 17, 2003].

EPA (U.S. Environmental Protection Agency). 1999f. Achieving Clean Air and Clean Water: The Report of the Blue Ribbon Panel on Oxygenates in Gasoline. EPA420-R-99-021. Blue Ribbon Panel for Reviewing Use of MTBE, Office of Transportation and Air Quality, U.S. Environmental Protection Agency, Washington, DC. [Online]. Available: http://www.epa.gov/otaq/consumer/fuels/oxypanel/r99021.pdf [accessed Jan. 21, 2003].

EPA (U.S. Environmental Protection Agency). 1999g. Economic Analysis of Alternate Methods of Allocating $NO_x$ Emission Allowances, Draft. Prepared by ICF Consulting, for the Acid Rain Division, Office of Air and Radiation, U.S. Environmental Protection Agency, Washington, DC. [Online]. Available: http://www.epa.gov/airmarkets/fednox/alloc-rprt.pdf [accessed Oct. 14, 2002].

EPA (U.S. Environmental Protection Agency). 2000a. Taking Toxics Out of the Air. Progress in Setting "Maximum Achievable Control Technology" Standards Under the Clean Air Act. EPA/452/K-00-002. Office of Air Quality Planning and Standards, U.S. Environmental Protection Agency, Research Triangle Park, NC. [Online]. Available: http://www.epa.gov/oar/oaqps/takingtoxics/index_small.html [accessed Sept. 23, 2002].

EPA (U.S. Environmental Protection Agency). 2000b. National Air Toxics Program: The Integrated Urban Strategy. Report to Congress. EPA-453/R-99-007. Office of Air Quality Planning and Standards, U.S. Environmental Protection Agency, Research Triangle Park, NC. July 2000. [Online]. Available: http://www.epa.gov/ttn/atw/urban/natprpt.pdf [accessed Apr. 11, 2003].

EPA (U.S. Environmental Protection Agency). 2000c. National Air Pollutant Emission Trends: 1900–1998. Trend Charts. EPA-454/R-00-002. Office of Air Quality Planning and Standards, U.S. Environmental Protection Agency, Research Triangle Park, NC. [Online]. Available: http://www.epa.gov/ttn/chief/trends/trends98/trendcharts.pdf [accessed Feb. 11, 2003].

EPA (U.S. Environmental Protection Agency). 2000d. Draft Technical Support Document: Control of Emissions of Hazardous Air Pollutants from Motor Vehicles and Motor Vehicle Fuels. EPA420-D-00-003. Assessment and Standards Division, Office of Transportation and Air Quality, U.S. Environmental Protection Agency, Washington, DC. [Online]. Available: http://www.epa.gov/otaq/regs/toxics/d00003.pdf [accessed May 6, 2003].

EPA (U.S. Environmental Protection Agency). 2000e. National Air Quality and Emissions Trends Report, 1998. EPA 454/R-00-003. Office of Air Quality Planning and Standards, Emissions Monitoring and Analysis Division, Air Quality Trends Analysis Group, U.S. Environmental Protection Agency, Research Triangle Park, NC.

EPA (U.S. Environmental Protection Agency). 2001a. National Air Quality and Emissions Trends Report, 1999. EPA 454/R-01-004. Office of Air Quality Planning and Standards, U.S. Environmental Protection Agency, Research Triangle Park, NC. [Online]. Available: http://www.epa.gov/oar/aqtrnd99/ [accessed Oct. 14, 2002].

EPA (U.S. Environmental Protection Agency). 2001b. Introduction to Area Source Emission Inventory Development, Vol. 3, Ch. 1, Revised Final. Emission Inventory Improvement Program, Emission Factor and Inventory Group, U.S. Environmental Protection Agency, Research Triangle Park, NC. [Online]. Available: http://www.epa.gov/ttn/chief/eiip/techreport/volume03/ [accessed Jan. 13, 2003].

EPA (U.S. Environmental Protection Agency). 2001c. Draft Guidance for Demonstrating Attainment of Air Quality Goals for $PM_{2.5}$ and Regional Haze. U.S. Environmental Protection Agency. [Online]. Available: http://vistas-sesarm.org/tech/draftpm.pdf [accessed Oct. 17, 2003].

EPA (U.S. Environmental Protection Agency). 2001d. Emission Scorecard 2000. Clean Air Markets, Office of Air and Radiation, U.S. Environmental Protection Agency, Washington, DC. June 2001. [Online]. Available: http://www.epa.gov/airmarkt/ [accessed Oct. 14, 2002].

EPA (U.S. Environmental Protection Agency). 2002a. Summary of Air Quality and Emissions Trends. Six Principal Pollutants. Air Trends, Office of Air and Radiation, U.S. Environmental Protection Agency, Washington, DC. [Online]. Available: http://www.epa.gov/airtrends/sixpoll.html [accessed March 14, 2003].

EPA (U.S. Environmental Protection Agency). 2002b. Air Quality Criteria for Particulate Matter, Third External Review Draft, Vol. 1 and 2. EPA/600/P-99/002aC. National Center for Environmental Assessment, Office of Research and Development, U.S. Environmental Protection Agency, Washington, DC. [Online]. Available: http://cfpub.epa.gov/ncea/cfm/partmatt.cfm [accessed March 25, 2003].

EPA (U.S. Environmental Protection Agency). 2002c. Maps of Air Quality, Deposition, Emissions, and Population Changes, February 2002. The Ambient Monitoring Technology Information Center (AMTIC). Technology Transfer Network, Office of Air and Radiation, U.S. Environmental Protection Agency, Washington, DC. [Online]. Available: http://www.epa.gov/ttn/amtic/files/ambient/monitorstrat/detailmp.pdf [accessed Feb. 10, 2003].

EPA (U.S. Environmental Protection Agency). 2002d. What Is an Ambient Air Quality Standard? Office of Air Quality Planning and Standards, U.S. Environmental Protection Agency, Research Triangle Park, NC. [Online]. Available: http://www.epa.gov/air/oaqps/ozpmbro/ambient.htm [accessed Feb. 10, 2003].

EPA (U.S. Environmental Protection Agency). 2002e. The National-Scale Air Toxics Assessment. Technology Transfer Network, Air Toxics Website. Office of Air and Radiation, U.S. Environmental Protection Agency, Washington, DC. [Online]. Available: http://www.epa.gov/ttn/atw/nata [accessed Feb. 04, 2003].

EPA (U.S. Environmental Protection Agency). 2002f. Air, EPA and State Progress in Issuing Title V Permits. Report No. 2002-P-00008. Evaluation Report. Office of Inspector General, U.S. Environmental Protection Agency, Washington, DC. March 29, 2002. [Online]. Available: http://www.epa.gov/air/oaqps/permits/maps/mapslink.html [accessed May 15, 2003].

EPA (U.S. Environmental Protection Agency). 2002g. AP-42: Compilation of Air Pollutant Emission Factors. [Online]. Available: http://www.epa.gov/otaq/ap42.htm [accessed Oct. 1, 2002].

EPA (U.S. Environmental Protection Agency). 2002h. Draft Design and Implementation Plan for EPA's Multi-Scale Motor Vehicle and Equipment Emission System (MOVES). EPA 420-P-02-006. Office of Transportation and Air Quality, U.S. Environmental Protection Agency, Washington, DC. [Online]. Available: http://www.epa.gov/otaq/models/ngm/p02006.pdf [accessed Oct. 17, 2003].

EPA (U.S. Environmental Protection Agency). 2002i. Final Regulatory Support Document: Control of Emissions from Unregulated Nonroad Engines. EPA420-R-02-022 Assessment and Standards Division, Office of Transportation and Air Quality, U.S. Environmental Protection Agency, Washington, DC. [Online]. Available: http://www.epa.gov/otaq/regs/nonroad/2002/r02022.pdf [accessed Feb. 3, 2004].

EPA (U.S. Environmental Protection Agency). 2002j. List of Federal Reformulated Gasoline Program Areas. Office of Transportation and Air Quality, U.S. Environmental Protection Agency, Washington, DC. [Online]. Available: http://www.epa.gov/otaq/regs/fuels/rfg/rfgarea.wpd [accessed May 30, 2003].

EPA (U.S. Environmental Protection Agency). 2002k. Bulletins & Memos. New Sources Review Announcement, Nov. 22, 2002. Technology Transfer Network, Office of Air and Radiation, U.S. Environmental Protection Agency, Washington, DC. [Online]. Available: http://www.epa.gov/ttn/nsr/bull_mem.html [accessed Jan. 15, 2003].

EPA (U.S. Environmental Protection Agency). 2002l. Acid Rain Program 2001 Progress Report. EPA 430-R-02-009. Clean Air Markets Program, Office of Air and Radiation, U.S. Environmental Protection Agency, Washington, DC. Nov. 2002. [Online]. Available: http://www.epa.gov/airmarkets/cmprpt/arp01/2001report.pdf [accessed March 31, 2003].

EPA (U.S. Environmental Protection Agency). 2002m. An Evaluation of the South Coast Air Quality Management District's Regional Clean Air Incentives-Lessons in Environmental Markets and Innovation, Nov. 2002. EPA's Evaluation of the RECLAIM Program in the South Coast Air Quality Management District. Air Programs, Region 9, U.S. Environmental Protection Agency, San Francisco. [Online]. Available: http://www.epa.gov/region09/air/reclaim/ [accessed Nov. 6, 2003].

EPA (U.S. Environmental Protection Agency). 2002n. Evaluation of Air Emissions Credits Results in Reduced Credits and Penalty Assessment in New Jersey. Pp. 2–3 in Quarterly Report, Apr. 1–June 30, 2002, Fiscal Year 2002. Office of Inspector General, U.S. Environmental Protection Agency, Washington, DC. [Online]. Available: http://www.epa.gov/oigearth [accessed Apr. 4, 2003].

EPA (U.S. Environmental Protection Agency). 2002o. PM Supersite Projects (Phase I and II) Information. Technology Transfer Network, Ambient Monitoring Technology Information Center. Office of Air and Radiation, U.S. Environmental Protection Agency, Washington, DC. [Online]. Available: http://www.epa.gov/ttn/amtic/ssprojec.html [accessed March 16, 2003].

EPA (U.S. Environmental Protection Agency). 2002p. National Ambient Monitoring Strategy, Summary Document. Sept. 1, 2002, Final Draft for Comment. National Air Monitoring Strategy Information, Draft Monitoring Strategy, Ambient Monitoring Technology Information Center, Office of Air and Radiation, U.S. Environmental Protection Agency, Washington, DC. [Online]. Available: http://www.epa.gov/ttn/amtic/stratdoc. html [accessed Apr. 8, 2003].

EPA (U.S. Environmental Protection Agency). 2002q. All Designated Methods. Technology Transfer Network, Ambient Monitoring Technology Information Center, Office of Air and Radiation, U.S. Environmental Protection Agency, Washington, DC. [Online]. Available: http://www.epa.gov/ttn/amtic/all.html [accessed Oct. 17, 2002].

EPA (U.S. Environmental Protection Agency). 2002r. Limitations in the 1996 National-Scale Air Toxic Assessment. The National Scale-Air Toxics Assessment, Technology Transfer Network, Office of Air and Radiation, U.S. Environmental Protection Agency, Washington, DC. [Online]. Available: http://www.epa.gov/ttn/atw/nata/natsalim2.html [accessed Apr. 8, 2003].

EPA (U.S. Environmental Protection Agency). 2002s. Interim Report on Personal and Residential Particulate Matter Gaseous Co-pollutant Air Concentrations for Potentially Sensitive Individuals and Their Association with Ambient Sources. National Exposure Research Laboratory, Office of Research and Development, U.S. Environmental Protection Agency, Research Triangle Park, NC [Online]. Available: http://www.epa.gov/ nerl/research/2002/g1-1.html. [accessed June 11, 2003].

EPA (U.S. Environmental Protection Agency). 2003. Latest Findings on National Air Quality, 2002 Status and Trends. EPA 454/K-03-001. Emissions Monitoring and Analysis Division, Office of Air Quality Planning and Standards, U.S. Environmental Protection Agency, Research Triangle Park, NC.

EPA (U.S. Environmental Protection Agency). 2003a. Average Annual Emissions, All Criteria Pollutants Years Including 1980, 1985, 1989–2001, February 2003. Emission Factor and Inventory Group, TTNWeb, Office of Air and Radiation, U.S. Environmental Protection Agency, Washington, DC. [Online]. Available: http://www.epa.gov/ttn/ chief/trends/trends01/trends2001.pdf [accessed May 14, 2003].

EPA (U.S. Environmental Protection Agency). 2003b. Fact Sheet: Final Amendments to the "General Provisions" of National Emission Standards for Hazardous Air Pollutant Emissions. Final Amendments to Clean Air Act's "Section 112(j) Rule." Office of Air and Radiation, U.S. Environmental Protection Agency, Washington, DC. May 14, 2003. [Online]. Available: http://www.epa.gov/airlinks/factsheet5803.pdf [accessed May 14, 2003].

EPA (U.S. Environmental Protection Agency). 2003c. Environmental Justice. Compliance and Enforcement, U.S. Environmental Protection Agency, Washington, DC. [Online]. Available: http://www.epa.gov/compliance/environmentaljustice/index.html [accessed Apr. 1, 2003].

EPA (U.S. Environmental Protection Agency). 2003d. FY 2002 Annual Report, Annual Performance Reports. Office of Chief Financial Officer, Office of Planning, Analysis, and Accountability, U.S. Environmental Protection Agency, Washington, DC. [Online]. Available: http://www.epa.gov/ocfo/finstatement/2002ar/2002ar.htm [accessed May 28, 2003].

EPA (U.S. Environmental Protection Agency). 2003e. Categorical Grants Program. Pp. A-4 in FY 2004, Summary of EPA's Budget. Office of Chief Financial Officer, Office of Planning, Analysis, and Accountability, U.S. Environmental Protection Agency, Washington, DC. [Online]. Available: http://www.epa.gov/ocfo/budget/2004/2004bib.pdf [accessed May 29, 2003].

EPA (U.S. Environmental Protection Agency). 2003f. Emissions Inventories. Technology Transfer Network Clearinghouse for Inventories and Emission Factors, U.S. Environmental Protection Agency, Washington, DC. [Online]. Available: http://www.epa.gov/ttn/chief/eiinformation.html [accessed May 16, 2003].

EPA (U.S. Environmental Protection Agency). 2003g. Hourly Emissions Data. Clean Air Markets—Data and Maps. Office of Air and Radiation, U.S. Environmental Protection Agency, Washington, DC. [Online]. Available: http://cfpub.epa.gov/gdm/ [accessed Jan. 17, 2003].

EPA (U.S. Environmental Protection Agency). 2003h. Green Book Nonattainment Areas for Criteria Pollutants. Office of Air Quality Planning and Standards, U.S. Environmental Protection Agency, Research Triangle Park, NC. [Online]. Available: http://www.epa.gov/oar/oaqps/greenbk/ [accessed May 29, 2003].

EPA (U.S. Environmental Protection Agency). 2003i. Reducing Nonroad Diesel Emissions. Low Emission Program Summary. Fact Sheet. EPA 402-F-03-008. Office of Transportation and Air Quality, U.S. Environmental Protection Agency, Washington, DC. [Online]. Available: http://www.epa.gov/nonroad/f03008.htm [accessed Apr. 22, 2003].

EPA (U.S. Environmental Protection Agency). 2003j. Title V Permit Issuance Status: United States. Office of Air Quality Planning and Standards, U.S. Environmental Protection Agency, Research Triangle Park, NC. [Online]. Available: http://www.epa.gov/oar/oaqps/permits/maps/index.html [accessed May 15, 2003].

EPA (U.S. Environmental Protection Agency). 2003k. Performance Specifications and Other Monitoring Information. Technology Transfer Network, Emission Measurement Center, Office of Air and Radiation, U.S. Environmental Protection Agency, Washington, DC. [Online]. Available: http://www.epa.gov/ttn/emc/monitor.html [accessed Jan. 22, 2003].

EPA (U.S. Environmental Protection Agency). 2003l. FTIR Technology. Technology Transfer Network, Emission Measurement Center, Office of Air and Radiation, U.S. Environmental Protection Agency, Washington, DC. [Online]. Available: http://www.epa.gov/ttn/emc/ftir.html [accessed Jan. 22, 2003].

EPA (U.S. Environmental Protection Agency). 2003m. 2001 Toxics Release Inventory, Executive Summary. EPA 260-S-03-001. Office of Environmental Information, U.S. Environmental Protection Agency, Washington, DC. July 2003. [Online]. Available: http://www.epa.gov/tri/tridata/tri01 [accessed Nov. 20, 2003]

EPA (U.S. Environmental Protection Agency). 2003n. Draft Report on the Environment. Office of Research and Development and the Office of Environmental Information, U.S. Environmental Protection Agency, Washington, DC. [Online]. Available: http://www.epa.gov/indicators/roe/html/roeTOC.htm [accessed Nov. 12, 2003].

EPA (U.S. Environmental Protection Agency). 2003o. Trends in Wet Sulfate Deposition Following Implementation of Phase I of the Acid Rain Program: 1989–1991 vs. 1997–1999 (CASTNet and NADP/NTN Data) Clean Air Markets-Progress and Results. Office of Air and Radiation, U.S. Environmental Protection Agency, Washington, DC. [Online]. Available: http://www.epa.gov/airmarkets/cmap/mapgallery/mg_wetsulfate phase1.html#desc [accessed Apr. 27, 2004].

EPA (U.S. Environmental Protection Agency). 2003p. Ozone Nonattainment Area Summary, Feb. 06, 2003. Green Book Nonattainment Areas for Criteria Pollutants. Office of Air Quality Planning and Standards, U.S. Environmental Protection Agency, Research Triangle Park, NC. [Online]. Available: http://www.epa.gov/oar/oaqps/greenbk/onsum. html [accessed Apr. 11, 2003].

EPA (U.S. Environmental Protection Agency). 2003r. Average Annual Emissions, All Criteria Pollutants Years Including 1970–2001. National Emission Inventory (NEI) Air Pollutant Emission Trends. Technology Transfer Network, Clearinghouse for Inventories and Emission Factors, Office of Air and Radiation, U.S. Environmental Protection Agency, Washington, DC. [Online]. Available: http://www.epa.gov/ttn/chief/trends/ [accessed Jan. 21, 2004].

EPA/SAB (U.S. Environmental Protection Agency Science Advisory Board). 1995. RE: CASAC Closure on the Primary Standard Portion of the Stuff Paper for Ozone. EPA-SAB-CASAC-LTR-96-002. Letter to Carol M. Browner, Administrator, U.S. Environmental Protection Agency, Washington, DC, from George T. Wolff, Chair Clean Air Scientific Advisory Committee, U.S. Environmental Protection Agency, Washington, DC, Nov. 30, 1995. [Online]. Available: http://www.epa.gov/sab/pdf/casac02.pdf [accessed May 9, 2003].

EPA/SAB (U.S. Environmental Protection Agency Science Advisory Board). 1997. Draft Retrospective Study Report to Congress Entitled "The Benefits and Costs of the Clean Air Act, 1970 to 1990" Letter Report. EPA-SAB-COUNCIL-LTR-97-008. Science Advisory Board, U.S. Environmental Protection Agency, Washington, DC. [Online]. Available: http://www.epa.gov/sab/pdf/coul9708.pdf [accessed March 24, 2003].

EPA/SAB (U.S. Environmental Protection Agency Science Advisory Board). 2001. NATA-Evaluating the National-Scale Air Toxics Assessment for 1996—An SAB Advisory, An Advisory by the EPA Science Advisory Board. EPA-SAB-EC-ADV-02-001. Science Advisory Board, U.S. Environmental Protection Agency, Washington DC. Dec. [Online]. Available: http://www.epa.gov/science1/fiscal02.htm [accessed Sept. 23, 2002].

Eriksson, P., E. Jakobsson, and A. Fredriksson. 2001. Brominated flame retardants: A novel class of developmental neurotoxicants in our environment? Environ. Health Perspect. 109(9):903–908.

ESA (Ecological Society of America). 1997a. Ecosystem Services: Benefits Supplied to Human Societies by Natural Ecosystems. Issues in Ecology No. 2. Ecological Society of America, Washington, DC. [Online]. Available: http://www.esa.org/sbi/issue2.pdf [accessed Nov. 14, 2002].

ESA (Ecological Society of America). 1997b. Sustainable Biosphere Initiative Project. Office Workshop Report. Atmospheric Nitrogen Deposition to Coastal Watersheds, 1997. [Online]. Available: http://esa.sdsc.edu/sbindep1.htm [accessed Oct. 21, 2002].

Farrell, A., and T.J. Keating. 1998. Multi-Jurisdictional Air Pollution Assessment: A Comparison of the Eastern United States and Western Europe. Belfer Center for Science & International Affairs (BCSIA) Discussion Paper E-98-12. Environmental and Natural Resources Program, Kennedy School of Government, Harvard University, Cambridge, MA. [Online]. Available: http://environment.harvard.edu/gea/pubs/e%2D98%2D12. html [accessed Oct. 14, 2002].

Fenn, M.E., J.S. Baron, E.B. Allen, H.M. Rueth, K.R. Nydick, L. Geiser, W.D. Bowman, J.O. Sickman, T. Meixner, D.W. Johnson, and P. Neitlich. 2003a. Ecological effects of nitrogen deposition in the western United States. BioScience 53(4):404–420.

Fenn, M.E., R. Haeuber, G.S. Tonnesen, J.S. Baron, S. Grossman-Clarke, D. Hope, D.A. Jaffe, S. Copeland, L. Geiser, H.M. Rueth, and J.O. Sickman. 2003b. Nitrogen emissions, deposition, and monitoring in the western United States. BioScience 53(4):391–403.

Fiore, A.M., D.J. Jacob, J.A. Logan, and J.H. Yin. 1998. Long-term trends in ground level ozone over the contiguous United States, 1980–1995. J. Geophys. Res. 103(D1):1471–1480.

Fisher, D.C., and M. Oppenheimer. 1991. Atmospheric nitrogen deposition and the Chesapeake Bay estuary. Ambio 20(3/4):102–108.

Fishman, J., A.E. Wozniak, and J.K. Creilson. 2003. Global distribution of tropospheric ozone from satellite measurements using the empirically corrected tropospheric ozone residual technique: Identification of the regional aspects of air pollution. Atmos. Chem. Phys. Discuss. 3:1453–1476.

Fitz, D.R., A.M. Winer, S. Colome, E. Behrentz, L.D. Sabin, S.J. Lee, K. Wong, K. Kozawa, D. Pankratz, K. Bumiller, D. Gemmill, and M. Smith. 2003. Characterizing the Range of Children's Pollutant Exposure During School Bus Commutes. Final Report. Prepared for the California Air Resources Board by the College of Engineering, Center for Environmental Research and Technology, University of California, Riverside, CA, and Department of Environmental Health Sciences, School of Public Health, University of California, Los Angeles, CA. Oct. 10, 2003. [Online]. Available: ftp://ftp.arb.ca.gov/carbis/research/schoolbus/report.pdf [accessed March 3, 2004].

Fresno Bee (California). 2002. Smog Check fails to get gross polluters. Last Gasp, A Fresno Bee Special Report on Valley Air Quality, Dec. 15, 2002. Fresno Bee, California [Online]. Available: http://www.fresnobee.com/special/valley_air/part4/story1/ [accessed Apr. 22, 2003].

Frey, H.C., R.R. Bharvirkar, and J. Zheng. 1999. Quantitative Analysis of Variability and Uncertainty in Emission Estimation. Paper No. 99-267. Proceedings of the 92$^{nd}$ Annual Meeting, June 20–24, 1999, St. Louis, MO, Air and Waste Management Association, Pittsburgh, PA. (CD-Room) [Online]. Available: http://www4.ncsu.edu:8030/~frey/FreyBharvirkarZhengAWMA1999.pdf [accessed Oct. 15, 2002].

Friedman, M.S., K.E. Powell, L. Hutwagner, L.M. Graham, and W.G. Teague. 2001. Impact of changes in transportation and commuting behaviors during the 1996 Summer Olympic Games in Atlanta on air quality and childhood asthma. JAMA *285*(7):897–905.

Frye-Mason, G., R. Kottenstette, C. Mowry, C. Morgan, R. Manginell, P. Lewis, C. Matzke, G. Dulleck, L. Anderson, and D. Adkins. 2001. Expanding the capabilities and applications of gas phase miniature chemical analysis systems (μChemLabTM). Pp. 658–660 in Micro Total Analysis Systems 2001: Proceedings of the μTAS 2001 Symposium, Monterey, CA, Oct. 21–25, 2001, J.M. Ramsey, and A. van den Berg, eds. Dordrecht: Kluwer.

Galloway, J.N., J.D. Aber, J.W. Erisman, S.P. Seitzinger, R.H. Howarth, E.B. Cowling, and B.J. Cosby. 2003. The nitrogen cascade. BioScience *53*(4):341–356.

GAO (U.S. General Accounting Office). 1997. Air Pollution: Limitations of EPA's Motor Vehicle Emissions Model and Plans to Address Them: Report to the Chairman, Subcommittee on Oversight and Investigations, Committee on Commerce, House of Representatives. GAO/RCED-97-210. U.S. General Accounting Office, Washington, DC.

GAO (General Accounting Office). 2000. Acid Rain: Emissions Trends and Effects in the Eastern United States. GAO/RCED-00-47. U.S. General Accounting Office, Washington, DC.

GAO (General Accounting Office). 2001a. Air Pollution: EPA Should Improve Oversight of Emissions Reporting by Large Facilities. GAO-01-46. U.S. General Accounting Office, Washington, DC.

GAO (General Accounting Office). 2001b. Environmental Protection: Wider Use of Advanced Technologies Can Improve Emissions Monitoring. GAO-01-313. U.S. General Accounting Office, Washington, DC.

Gbondo-Tugbawa, S.S., and C.T. Driscoll. 2002. Evaluation of the effects of future controls on sulfur dioxide and nitrogen oxide emissions on the acid-base status of a northern forest ecosystem. Atmos. Environ. 36(10):1631–1643.

Gbondo-Tugbawa, S.S., and C.T. Driscoll. 2003. Factors controlling long-term changes in soil pools of exchangeable basic cations and stream acid neutralizing capacity in a northern hardwood forest ecosystem. Biogeochemistry 63(2):161–185.

Gbondo-Tugbawa, S.S., C.T. Driscoll, J.D. Aber, and G.E. Likens. 2001. Evaluation of an integrated biogeochemical model (PnET-BGC) at a northern hardwood forest ecosystem. Water Resour. Res. 37(4):1057–1070.

Getches, D.H., C.F. Wilkinson, and R.A. Williams Jr. 1998. Cases and Materials on Federal Indian Law, 4th Ed. St. Paul, MN: West Group.

Goldman, B.A. 1994. Not Just Prosperity: Achieving Sustainability with Environmental Justice. Prepared for National Wildlife Federation. Feb. 1994.

Goulder, L.H., I.W.H. Parry, and D. Burtraw. 1997. Revenue-raising vs. other approaches to environmental protection: The critical significance of pre-existing tax distortions. RAND J. Econ. 28(4):708–731.

Goulder, L.H., I.W.H. Parry, R.C. Williams III, and D. Burtraw. 1999. The cost-effectiveness of alternative instruments for environmental protection in a second-best setting. J. Public Econ. 72(3):329–360.

GPMN. 2002. Gaseous Pollutant Monitoring Network (GPMN). Air Quality and Deposition Monitoring, National Park Service. [Online]. Available: http://www.aqd.nps.gov/ard/gas.

Greenbaum, D.S., J.D. Bachmann, D. Krewski, J.M. Samet, R. White, and R.E. Wyzga. 2001. Particulate air pollution standards and morbidity and mortality: Case study. Am. J. Epidemiol. 154(12 Suppl.):S78-S90.

Grennfelt, P., and J. Nilsson, eds. 1988. Critical Loads for Sulphur and Nitrogen, Workshop, March 19–24, 1988, Skokloster, Sweden. Report 1988:15. ISBN 87-7303-248-4. The Nordic Council of Ministers, Copenhagen, Denmark.

Gwynn, R.C., and G.D. Thurston. 2001. The burden of air pollution: Impacts among racial minorities. Environ. Health Perspect. 109(Suppl. 4):501–506.

Hahn, R.W., and G.L. Hester. 1989a. Marketable permits: Lessons for theory and practice. Ecol. Law Q. 16(2):361–406.

Hahn, R.W., and G.L. Hester. 1989b. Where did all the markets go? An analysis of EPA's emissions trading program. Yale J. Regul. 6:109–153.

Hahn, R.W., and R.N. Stavins. 1991. Incentive-based environmental regulation: A new era from an old idea? Ecol. Law Q. 18(1):1–42.

Hahn, R.W., and R.N. Stavins. 1992. Economic incentives for environmental protection: Integrating theory and practice. Am. Econ. Rev. 82(2):464–468.

Halberstadt, M.L. 1989. Effective control of highway vehicle hydrocarbon emissions. JAPCA 39(6):862.

Hale, R.C., M.J. La Guardia, E.P. Harvey, M.O. Gaylor, T.M. Mainor, and W.H. Duff. 2001a. Flame retardants. Persistent pollutants in land-applied sludges. Nature 412(6843):140–141.

Hale, R.C., M.J. La Guardia, E.P. Harvey, T.M. Mainor, W.H. Duff, and M.O. Gaylor. 2001b. Polybrominated diphenyl ether flame retardants in Virginia freshwater fishes (USA). Environ. Sci. Technol. 35(23):4585–4591.

Hall, J.V. 1996. Estimating environmental health benefits: Implications for social decision making. Int. J. Soc. Econ. 23(4/5):282–295.

Hall, J.V., A.M. Winer, M.T. Kleinman, F.W. Lurmann, V. Brajer, S.D. Colome, R. Rowe, L. Chestnut, D. Foliart, L. Coyner, and A. Horwatt. 1989. Economic Assessment of the Health Benefits from Improvements in Air Quality in the South Coast Air Basin. Final Report to the South Coast Air Quality Management District, Diamond Bar, CA. Fullerton, CA: California State University Fullerton Foundation.

Hall, J.V., A.M. Winer, M.T. Kleinman, F.W. Lurmann, V. Brajer, and S.D. Colome. 1992. Valuing the health benefits of clean air. Science 255(5046):812–817.

Hall, R.J., G.E. Likens, S.B. Fiance, and G.R. Hendrey. 1980. Experimental acidification of a stream in the Hubbard Brook Experimental Forest, New Hampshire. Ecology 61(4):976–989.

Hallmark, S.L., R. Guensler, and I. Fomunung. 2001. Characterizing on-road variables that affect passenger vehicle modal operation. Transport. Res. D7(2):81–89.

Hamonou, E., P. Chazette, D. Balis, F. Dulac, X. Schneider, E. Galani, G. Ancellet, and A. Papayannis. 1999. Characterization of the vertical structure of Saharan dust export to the Mediterranean basin. J. Geophys. Res. 104(D18):22257–22271.

Hansen, J.E., and M. Sato. 2001. Trends of measured climate forcing agents. Proc. Natl. Acad. Sci. USA 98(26):14778–14783.

Harrington, W., A. Howitt, A.J. Krupnick, J. Makler, P. Nelson, and S.J. Siwek. 2003. Exhausting Options: Assessing SIP-Conformity Interactions. RFF Report, Jan. 2003. Washington, DC: Resources for the Future. [Online]. Available: http://rff.org/reports/PDF_files/ExhaustingOptions.pdf [accessed March 18, 2003].

Hausker, K. 1992. The politics and economics of auction design in the market for sulfur dioxide pollution. J. Policy Anal. Manage. 11(4):553–572.

Heck, W.W., O.C. Taylor, and D.T. Tingley, eds. 1988. Assessment of Crop Loss from Air Pollution. New York: Elsevier Applied Science.

Heck, W.W., C.S. Furiness, E.B. Cowling, and C.K. Sims. 1998. Effects of ozone on crop, forest, and natural ecosystems: Assessment of research needs. EM (Oct.):11–22.

Hedin, L.O., G.E. Likens, K.M. Postek, and C.T. Driscoll. 1990. A field experiment to test whether organic acids buffer acid deposition. Nature 345(6278):798–800.

HEI (Health Effects Institute). 1995. Diesel Exhaust: Critical Analysis of Emissions, Exposure, and Health Effects. Cambridge, MA: Health Effects Institute.

HEI (Health Effects Institute). 1999. Diesel Emissions and Lung Cancer: Epidemiology and Quantitative Risk Assessment. A Special Report of the Institute's Diesel Epidemiology Expert Panel. June 1999. [Online]. Available: http://www.healtheffects.org/Pubs/DieselEpi-C.pdf [accessed Nov. 12, 2003].

HEI (Health Effects Institute). 2000. Science in Action: 1998–1999 Annual Report. Cambridge, MA: Health Effects Institute. Feb. 2000. [Online]. Available: http://www.healtheffects.org/pubs.htm [accessed Apr. 25, 2003].

HEI (Health Effects Institute). 2002. Measuring the Health Impact of Actions That Improve Air Quality. RFA 02-1. Request for Applications, Winter 2002 Research Agenda. Cambridge, MA: Health Effects Institute [Online]. Available: http://www.healtheffects.org/RFA/RFA-archive.htm [accessed March 12, 2003].

Heinsohn, R.J., and R.L. Kabel. 1998. Sources and Control of Air Pollution, 1st Ed. Upper Saddle River, NJ: Prentice Hall.

Heinz (H. John Heinz III Center for Science, Economics, and the Environment). 2002. The State of the Nation's Ecosystems: Measuring the Lands, Waters, and Living Resources of the United States. New York: Cambridge University Press. [Online]. Available: http://www.heinzctr.org/ecosystems/report.html [accessed March 13, 2003].

Heinzerling, L. 2000. The rights of statistical people. Harvard Environ. Law Rev. 24(1):189–207.

Hermanson, M.H., C.A. Scholten, and K. Compher. 2003. Variable air temperature response of gas-phase atmospheric polychlorinated biphenyls near a former manufacturing facility. Environ. Sci. Technol. 37(18):4038–4042.

Hollander, M., and D.A. Wolfe. 1973. Nonparametric Statistical Methods. New York: Wiley.

Holmes, K.J., and R.J. Cicerone. 2002. The road ahead for vehicle emission inspection and maintenance programs. EM (July):15–23.

Hooper, K., and T.A. McDonald. 2000. The PBDEs: An emerging environmental challenge and another reason for breast-milk monitoring programs. Environ. Health Perspect. 108(5):387–392.

Horowitz, J. 1978. Integrated planning and management of transportation and air quality. Pp. 60–62 in Air Quality Analysis in Transportation Planning. Transportation Research Record 670. Washington, DC: National Academy of Sciences.

Howitt, A.M., and A. Altshuler. 1999. The politics of controlling auto air pollution. Pp. 223–255 in Essays in Transportation Economics and Policy, J.A. Gómez-Ibáñez, W.B. Tye, and C. Winston, eds. Washington, DC: Brookings Institution Press.

Howitt, A.M., and E.M. Moore. 1999a. Implementing the Transportation Conformity Regulations. TR News 202(May-June):15–23, 41.

Howitt, A.M., and E.M. Moore. 1999b. Linking Transportation and Air Quality Planning Implementation of the Transportation Conformity Regulations in 15 Nonattainment Areas. EPA420-R-99-011. Prepared for U.S. Environmental Protection Agency and the Federal Highway Administration, by Taubman Center for State and Local Government, John F. Kennedy School of Government, Harvard University, Cambridge, MA.

Hungate, B.A., E.A. Holland, R.B. Jackson, F.S. Chapin III, H.A. Mooney, and C.B. Field. 1997. The fate of carbon in grasslands under carbon dioxide enrichment. Nature 388(6642):576–579.

Hunter, A.J.R., J.R. Morency, C.L. Senior, S.J. Davis, and M.E. Fraser. 2000. Continuous emissions monitoring using spark-induced breakdown spectroscopy. J. Air Waste Manage. Assoc. 50(1):111–117.

Huntington, T.G., R.P. Hooper, C.E. Johnson, B.T. Aulenbach, R. Cappellato, and A.E. Blum. 2000. Calcium depletion in a southeastern United States forest ecosystem. Soil Sci. Soc. Am. J. 64(5):1845–1858.

ICF Kaiser Inc. 1999. Compendium of Sensing Technologies to Detect and Measure VOCs and HAPs in the Air, Final Report. ICF Kaiser Inc., Washington, DC.

Ikonomou, M.G., S. Rayne, and R.F. Addison. 2002. Exponential increases of the brominated flame retardants, polybrominated diphenyl ethers, in the Canadian Arctic from 1981 to 2000. Environ Sci Technol. 36(9):1886–1892.

IMPROVE. 2002. Interagency Monitoring of Protected Visual Environments (IMPROVE). U.S. Environmental Protection Agency and National Park Service [Online]. Available: http://vista.cira.colostate.edu/improve/ [accessed March 16, 2003].

IMRC (California Inspection and Maintenance Review Committee). 1995a. An Analysis of the USEPA's 50-Percent Discount for Decentralized I/M Programs. Prepared by J. Schwartz, California Inspection and Maintenance Review Committee, Sacramento, CA. Feb. 24, 1995.

IMRC (California Inspection and Maintenance Review Committee). 1995b. Reply to EPA Summary Response to the California Committee Report on I/M Effectiveness. Prepared by J. Schwartz, California Inspection and Maintenance Review Committee, Sacramento, CA. March 20, 1995.

Indermuhle, A., T.F. Stocker, F. Joos, H. Fischer, H.J. Smith, M. Wahlen, B. Deck, D. Mastroianni, J. Tschumi, T. Blunier, R. Meyer, and B. Stauffer. 1999. Holocene carbon-cycle dynamics based on $CO_2$ trapped in ice at Taylor Dome, Antarctica. Nature 398(6723):121–126.

Innes, J. 2003. Introduction to the ecological issues section. Environ. Int. 29(2):137–140.

IPCC (Intergovernmental Panel on Climate Change). 2001. Climate Change 2001: The Scientific Basis, J.T. Houghton, Y. Ding, D.J. Griggs, M. Noguer, P.J. van der Linden, X. Dai, K. Maskell, and C.A. Johnson, eds. Cambridge: Cambridge University Press.

Isebrands, J.G., E.P. McDonald, E. Kruger, G. Hendrey, K. Pregitzer, K. Percy, J. Sober, and D.F. Karnosky. 2001. Growth responses of Populus tremuloides clones to interacting elevated carbon dioxide and tropospheric ozone. Environ. Pollut. 115(3):359–371.

ITEP (Institute for Tribal Environmental Professionals). 2001. An Assessment of Tribal Air Quality Data and Programs in the Western United States. Prepared for Tribal Data Development Working Group by the Institute for Tribal Environmental Professionals, Northern Arizona University. [Online]. Available: http://www4.nau.edu/itep/assets/docs/wga_data_report.pdf [accessed March 25, 2003].

Jacoby, H.D., J.D., Steinbruner, et al. 1973. Clearing the Air: Federal Policy on Automotive Emissions Control. Cambridge, MA: Ballinger.

Jaffe, D., T. Anderson, and I. Uno. 1999. Transport of Asian air pollution to North America. Geophys. Res. Lett. 26(6):711–714.

Jayne, J.T., D.C. Leard, X. Zhang, P. Davidovits, K.A. Smith, C.E. Kolb, and D.R. Worsnop. 2000. Development of an aerosol mass spectrometer for size and composition analysis of submicron particles. Aerosol Sci. Technol. 33(1-2):49–70.

Jeffries, D.S., ed. 1997. 1997 Canadian Acid Rain Assessment, Vol. 3. The Effects on Canada's Lakes, Rivers, and Wetlands. Ottawa: Environment Canada.

Jenkins, A., R.C. Helliwell, P.J. Swingewood, C. Seftron, M. Renshaw, and R.C. Ferrier. 1998. Will reduced sulphure emissions under the Second Sulphure Protocol lead to recovery of acid sensitive sites in UK? Environ. Pollut. 99(3):309–318.

Jeon, S.J., H.L.C. Meuzelaar, S.A.N. Sheya, J.S. Lighty, W.M. Jarman, C. Kasteler, A.F. Sarofim, and B.R.T. Simoneit. 2001. Exploratory studies of $PM_{10}$ receptor and source profiling by GC/MS and principal component analysis of temporally and spatially resolved ambient samples. J. Air Waste Manage. 51(5):766–784.

Jimenez, J.L., G.J. McRae, D.D. Nelson, M.S. Zahniser, and C.E. Kolb. 2000. Remote sensing of NO and $NO_2$ emissions from heavy-duty diesel trucks using tunable diode lasers. Environ. Sci. Technol. 34(12):2380–2387.

Johnson, D.W. 1993. Carbon in forest soils—research needs. N.Z. J. For. Sci. 23(3):354–366.

Johnson, D.W., and S.E. Lindberg, eds. 1992. Atmospheric Deposition and Forest Nutrient Cycling: A Synthesis of the Integrated Forest Study. New York: Springer-Verlag.

Johnson, J.H. 1988. Automotive emission. Pp. 39–76 in Air Pollution, the Automobile and Public Health, A.Y. Watson, R.R. Bates, and D. Kennedy, eds. Cambridge, MA: Health Effects Institute.

Karnosky, D.F., G. Gielen, R. Ceulemans, W.H. Schlesinger, R.J. Norby, E. Oksanen, R. Matyssek, and G.R. Hendrey. 2001. FACE systems for studying the impacts of greenhouse gases on forest ecosystems. Pp. 297–324 in Impact of Carbon Dioxide and Other Greenhouse Gases on Forest Ecosystems, Report No. 3 of the IUFRO Task Force on Environmental Change, D.F. Karnosky, G.E. Scarascia-Mugnozza, R. Ceulemans, and J.L. Innes, eds. Wallingford, Oxfordshire, UK: CABI Publishing.

Kean, A.J., E. Grosjean, D. Grosjean, and R.A. Harley. 2001. On-road measurement of carbonyls in California light-duty vehicle emissions. Environ. Sci. Technol. 35(21):4198–4204.

Kerr, S., and R. Newell. 2001. Policy-Induced Technology Adoption: Evidence from the U.S. Lead Phasedown. Discussion Paper 01-14. Resources for the Future, Washington, DC. [Online]. Available: http://www.rff.org/disc_papers/2001.htm [accessed June 13, 2003].

Kleinman, L.I., and P.H. Daum. 1991. Vertical distribution of aerosol particles, water vapor and insoluble trace gases in convectively mixed air. J. Geophys. Res. 96(1):991–1005.

Krewski, D., R.T. Burnett, M.S. Goldberg, K. Hoover, J. Siemiatycki, M. Jerrett, M. Abrahamowicz, and W. White. 2000a. Reanalysis of the Harvard Six Cities Study and the American Cancer Society Study of Particulate Air Pollution and Mortality: A Special Report of the Institute's Particle Epidemiology Reanalysis Project. Cambridge, MA: Health Effects Institute.

Krewski, D., R.T. Burnett, M.S. Goldberg, K. Hoover, J. Siemiatycki, M. Jerrett, M. Abrahamowicz, and W. White. 2000b. Reanalysis of the Harvard Six Cities Study and the American Cancer Society Study of Particulate Air Pollution and Mortality. Investigators' Report, Part 2: Sensitivity Analysis. Cambridge, MA: Health Effects Institute. [Online]. Available: http://www.healtheffects.org/Pubs/Rean-part2.pdf [accessed Apr. 2, 2003].

Krupa, S.V., A.E.G. Tonneijck, and W.J. Manning. 1998. Ozone. Pp. 2-1–2-28 in Recognition of Air Pollution Injury to Vegetation: A Pictorial Atlas, 2nd Ed, R.B. Flagler, A.H. Chappelka, W.J. Manning, P. McCool, and S.R. Shafer, eds. Pittsburgh, PA: Air & Waste Management Association.

Krupnick, A., and R. Morgenstern. 2002. The future of benefit-cost analyses of the Clean Air Act. Annu. Rev. Public Health 23:427–448.

Krupnick, A., D. Austin, D. Burtraw, T. Green, and E. Mansur. 1996. Scenario benefits module. In Tracking and Analysis Framework (TAF) Model Documentation and User's Guide, an Interaction Model for Integrated Assessment of Title IV of the Clean Air Act Amendments. ANL/DIS/TM-36. Decision and Information Sciences Division, Argonne National Laboratory. Dec. 1996. [Online]. Available: http://www.lumina.com/taf/index.html [accessed July 24, 2003].

Krupnick, A., A. Alberini, M. Cropper, N. Simon, B. O'Brien, R. Goeree, and M. Heintzelman. 2002. Age, health and the willingness to pay for mortality risk reductions: A contingent valuation survey of Ontario residents. J. Risk Uncertainty 24:161–186.

Kyle, A.D., C.C. Wright, J.C. Caldwell, P.A. Buffler, and T.J. Woodruff. 2001. Evaluating the health significance of hazardous air pollutants using monitoring data. Public Health Rep. 116(1):32–44.

Laden, F., L.M. Neas, D.W. Dockery, and J. Schwartz. 2000. Association of fine particulate matter from different sources with daily mortality in six U.S. cities. Environ. Health Perspect. 108(10):941–947.

Laris, M. 2002. Smog dog to sniff out emissions violators. Washington Post. Pp. C01. Jan. 13, 2002.

Laurence, J.A., and C.P. Andersen. 2003. Ozone and natural systems: Understanding exposure, response, and risk. Environ. Int. 29(2):155–160.

Lawrence, G.B., M.B. David, and A.W. Thompson. 1999. Soil calcium status and the response of stream chemistry to changing acidic deposition rates. Ecol. Appl. 9(3):1059–1072.

Likens, G.E., and F.H. Bormann. 1995. Biogeochemistry of a Forested Ecosystem, 2nd Ed. New York: Springer-Verlag.

Likens, G.E., C.T. Driscoll, and D.C. Buso. 1996. Long-term effects of acid rain: Response and recovery of a forest ecosystem. Science 272(5259):244–246.

Likens, G.E., C.T. Driscoll, D.C. Buso, T.G. Siccama, C.E. Johnson, G.M. Lovett, T.J. Fahey, W.A. Reiners, D.F. Ryan, C.W. Martin, and S.W. Bailey. 1998. The biogeochemistry of calcium at Hubbard Brook. Biogeochemistry 41(2):89–173.

Lilleskov, E.A., T.J. Fahey, T.R. Horton, and G.M. Lovett. 2002. Belowground ectomycorrhizal fungal community change over a nitrogen deposition gradient in Alaska. Ecology 83(1):104–115.

Lin, C.Y., D.J. Jacob, and A.M. Fiore. 2001. Trends in exceedences of the ozone air quality standard in the continental United States, 1980–1998. Atmos. Environ. 35(19):3217–3228.

Lindner, D. 2001. The μChemLabTM project: Micro total analysis system R&D at Sandia National Laboratories. The Royal Society of Chemistry- Lab on a Chip 1:15N-19N. [Online]. Available: http://www.rsc.org/is/journals/current/loc/lc001001.htm [accessed Feb. 4, 2003].

Linthurst, R.A., A.L. Mulkey, M.W. Slimak, G.D. Veith, and B.M. Levinson. 2000. Ecological research in the Office of Research and Development at the U.S. Environmental Protection Agency: An overview of new directions. Environ. Toxicol. Chem. 19(4 Part 2):1222–1229.

Liroff, R.A. 1986. Reforming Air Pollution Regulations: The Toil and Trouble of EPA's Bubble. Washington, DC: Conservation Foundation.

Liu, G., C. Chang, and C. Adams. 1999. China: Vehicle Emissions Control Technology. Tradeport Industry Sector Analysis. [Online]. Available: http://www.tradeport.org/ts/countries/china/isa/isar0028.html [accessed Jan. 15, 2003].

Logan, J.A. 1988. The ozone problem in rural areas of the United States. Pp. 327–344 in Tropospheric Ozone: Regional and Global Scale Interactions, I.S.A. Isaksen, ed. NATO ASI Series C, Vol. 227. Dordrecht, Holland: D. Reidel Publishing.

Lomborg, B. 2001. The Skeptical Environmentalist: Measuring the Real State of the World. New York: Cambridge University Press.

Lunder, S., T.J. Woodruff, and D.A. Axelrad. 2004. An analysis of candidates for addition to the Clean Air Act list of hazardous air pollutants. J. Air Waste Manage. Assoc. 54(2):157–171.

Lurman, F.W., J.W. Hall, M. Kleinman, L.R. Chinkin, V. Brajer, D. Meacher, F. Mummery, R.L. Arndt, T.L. Haste-Funk, S.B. Hurwitt, and N. Kumar. 1999. Assessment of the Health Benefits of Improving Air Quality in Houston, Texas. STI-998460-1875-DFR. Prepared for City of Houston, Office of the Mayor. Nov.

Lutter, R., and R. Belzer. 2000. EPA pats itself on the back. Regulation 23(3):23–28. [Online]. Available: http://www.cato.org/pubs/regulation/regv23n3/reg23n3.html [accessed Nov. 12, 2003].

Mauzerall, D.L., and X. Wang. 2001. Protecting agricultural crops from the effects of tropospheric ozone exposure: Reconciling science and standard setting in the United States, Europe and Asia. Annu. Rev. Energy Environ. 26:237–268.

Mazdai, A., N.G. Dodder, M.P. Abernathy, R.A. Hites, and R.M. Bigsby. 2003. Polybrominated diphenyl ethers in maternal and fetal blood samples. Environ. Health Perspect. 111(9):1249–1252.

McClenny, W.A., E.J. Williams, R.C. Cohen, and J. Stutz. 2002. Preparing to measure the effects of the $NO_x$ SIP Call-methods for ambient air monitoring of NO, $NO_2$, $NO_y$ and individual $NO_z$ species. J. Air Waste Manage. Assoc. 52(5):542–562.

McDonald, T.A. 2002. A perspective on the potential health risks of PBDEs. Chemosphere 46(5):745–755.

McHenry, J.N., N. Seaman, C. Coats, D. Stauffer, A. Lario-Gibbs, J. Vukovich, E. Hayes, and N. Wheeler. 2000. The NCSC-PSU Numerical Air Quality Prediction Project: Initial Evaluation, Status, and Prospects. Paper 16.7. Presentation at the 11th Joint Conference on the Applications of Air Pollution Meteorology with the Air and Waste Management Association, Long Beach, CA [Online]. Available: http://ams.confex.com/ams/annual2000/11airpollut/index.html [accessed June 13, 2003].

Melnick, R.S. 1983. Regulation and the Courts: The Case of the Clean Air Act. Washington, DC: Brookings Institution.

Mendelsohn, M.L., L.C. Mohr, and J.P. Peeters, eds. 1998. Biomarkers: Medical and Workplace Applications. Washington, DC: Joseph Henry Press.

Milanchus, M.L., S.T. Rao, and I.G. Zurbenko. 1998. Evaluating the effectiveness of ozone management in the presence of meteorological variability. J. Air Waste Manage. Assoc. 48(3):201–215.

Mitra, S., C. Feng, N. Zhu, and G. McAllister. 2001. Application and field validation of a continuous nonmethane organic carbon analyzer. J. Air Waste Manage. Assoc. 51(6): 861–868.

M.J. Bradley & Associates, Inc. 2002a. Heavy-duty vehicle: New technologies for the 21st century. Environ. Energy Insights 5(4). Apr. 2002.

M.J. Bradley & Associates, Inc. 2002b. Heavy-Duty Vehicle Technology Conference, May 2002, New York. Proceedings on CD. M.J. Bradley & Associates, Inc, Concord, MA.

MODMON. 2001. The Neuse River Estuary MODeling and MONitoring Project. North Carolina Department of Environment and Natural Resources, and University of North Carolina, Water Resources Research Institute [Online]. Available: http://www.marine.unc.edu/neuse/modmon/ [accessed March 18, 2003].

Murray, C.J.L., and A.D. Lopez, eds. 1996. The Global Burden of Disease: A Comprehensive Assessment of Mortality and Disability from Diseases, Injuries, and Risk Factors in 1990 and Projected to 2020. Cambridge: Harvard School of Public Health on behalf of the World Health Organization and the World Bank.

NADP/MDN (National Atmospheric Deposition Program/Mercury Deposition Network). 2002. Mercury Deposition Network: A NADP Network. National Atmospheric Deposition Program. [Online]. Available: http://nadp.sws.uiuc.edu/mdn/ [accessed March 16, 2003].

NADP/NTN (National Atmospheric Deposition Program). 2002. National Atmospheric Deposition Program/National Trends Network. A Cooperative Research Support Program of the State Agricultural Experiment Stations (NRSP-3), Federal and State Agencies and Non-Governmental Research Organizations. [Online]. Available: http://nadp.sws.uiuc.edu/ [accessed March 16, 2003].

NAE/NRC (National Academy of Engineering and National Research Council). 2003. The Carbon Dioxide Dilemma: Promising Technologies and Policies. Washington, DC: The National Academies Press.

NAM (National Association of Manufacturers). 2002. New Source Review, Sept. 2002. Air Quality Issues. Resources and Environmental Policy, National Association of Manufacturers, Washington, DC. [Online]. Available: http://www.nam.org/tertiary_print.asp?TrackID=&CategoryID=790&DocumentID=25247 [accessed March 11, 2003].

NAMS/SLAM. 2002. National Air Monitoring (NAMS) Network. State and Local Monitoring (SLAMs) Network. The Ambient Air Monitoring Program, Air Quality Planning and Standards, U.S. Environmental Protection Agency. [Online]. Available: http://www.epa.gov/oar/oaqps/qa/monprog.html#SLAMS [accessed March 16, 2003].

NAPA (U.S. National Academy of Public Administration). 2000. Environment.gov: Transforming Environmental Protection for the 20th Century: A Report. U.S. National Academy of Public Administration, Washington, DC.

NAPAP (U.S. National Acid Precipitation Assessment Program). 1991a. Acidic Deposition: State of Science and Technology, Summary Report of the U.S. National Acid Precipitation Assessment Program, P.A. Irving, ed. U.S. National Acid Precipitation Assessment Program, Washington, DC.

NAPAP (U.S. National Acid Precipitation Assessment Program). 1991b. Acidic Deposition: State of Science and Technology, Vol. 1. Emissions, Atmospheric Processes, and Deposition, P.A. Irving, ed. U.S. National Acid Precipitation Assessment Program, Washington, DC.

NAPAP (U.S. National Acid Precipitation Assessment Program). 1998. NAPAP Biennial Report to Congress. National Science and Technology Council, Committee on Environment and Natural Resources, Washington, DC.

NARSTO. 2000. An Assessment of Tropospheric Ozone Pollution: A North American Perspective. Palo Alto, CA: EPRI.

NARSTO. 2003. Particulate Matter Science for Policy Makers, A NARSTO Assessment. EPRI 1007735. Palo Alto, CA: EPRI.

NASA (National Aeronautics and Space Administration). 2003. TOMS. Total Ozone Mapping Spectrometer. Code 916: Atmospheric Chemistry and Dynamics Branch, National Aeronautics and Space Administration. [Online]. Available: http://toms.gsfc.nasa.gov/ [accessed Apr. 10, 2003].

National Commission on Air Quality. 1981. To Breathe Clean Air. Washington, DC: U.S. Government Printing Office.

NCRARM (National Commission on Risk Assessment and Risk Management). 1997. Report on the Accomplishments of the Commission on Risk Assessment and Risk Management. [Online]. Available: http://www.riskworld.com/nreports/1997/riskrpt/miscinfo/nr7mi002.htm#summary [accessed Jan. 31, 2003].

NCS (National Children's Study). 2002. Health, Growth, Environment. Spring 2002. Study Assembly Meeting of the National Children's Study, April 8, 2002. The National Children's Study, an Interagency Project, U.S. Department of Health and Human Services, NICHD, CDC, NIEHS, and EPA. [Online]. Available: http://nationalchildrensstudy. gov/about/042002.cfm [accessed Oct. 17, 2002].

NERRS (National Estuarine Research Reserve System). 2001. The National Estuarine Research Reserve System. [Online]. Available: http://www.ocrm.nos.noaa.gov/nerr/ [accessed Oct. 22, 2002].

NOAA (National Oceanic and Atmospheric Administration). 2001. Air Quality Forecasting: A Review of Federal Programs and Research Needs. Boulder, CO: NOAA Aeronomy Laboratory.

NOAA (National Oceanic and Atmospheric Administration). 2003. CMDL Ozonesondes. Climate Monitoring and Diagnostics Laboratory, Ozone & Water Vapor, National Oceanic and Atmospheric Administration, U.S. Department of Commerce. [Online]. Available: http://www.cmdl.noaa.gov/ozwv/ozsondes/ [accessed Apr. 9, 2003].

Noble, C.A., and K.A. Prather. 1996. Real-time measurement of correlated size and composition profiles of individual atmospheric aerosol particles. Environ. Sci. Technol. 30(9): 2667–2680.

Norton, S.A., J.S. Kahl, I.J. Fernandez, L.E. Rustad, J.P. Schofield, and T.A. Haines. 1994. Response of the West Bear Brook watershed, Maine, USA, to the addition of $(NH_4)_2SO_4$: 3-year results. For. Ecol. Manage. 68(1):61–73.

NPS (National Park Service). 2002a. Good Air Quality. Big Bend National Park, Rio Grande, TX. [Online]. National Park Service, Washington, DC. Available: http://www.nps.gov/bibe/goodaq.htm [accessed Feb. 10, 2003].

NPS (National Park Service). 2002b. Bad Air Quality. Big Bend National Park, Rio Grande, TX. [Online]. National Park Service, Washington, DC. Available: http://www.nps.gov/bibe/goodaq.htm [accessed Feb. 10, 2003].

NPS (National Park Service). 2002c. Air Quality & Visibility in Big Bend National Park. National Park Service, Washington, DC. [Online]. Available: http://www.nps.gov/bibe/aqvis.htm [accessed Feb. 10, 2003].

NRC (National Research Council). 1991. Rethinking the Ozone Problem in Urban and Regional Air Pollution. Washington, DC: National Academy Press.

NRC (National Research Council). 1993a. Protecting Visibility in National Parks and Wilderness Areas. Washington, DC: National Academy Press.

NRC (National Research Council). 1993b. Soil and Water Quality: An Agenda for Agriculture. Washington, DC: National Academy Press.

NRC (National Research Council). 1994. Science and Judgment in Risk Assessment. Washington, DC: National Academy Press.

NRC (National Research Council). 1996. Understanding Risk. Informing Decisions in a Democratic Society. Washington, DC: National Academy Press.

NRC (National Research Council). 1998a. The Atmospheric Sciences: Entering the Twenty-First Century. Washington, DC: National Academy Press.

NRC (National Research Council). 1998b. Research Priorities for Airborne Particulate Matter. 1. Immediate Priorities and a Long-Range Research Portfolio. Washington, DC: National Academy Press.

NRC (National Research Council). 1999a. Research Priorities for Airborne Particulate Matter. 2. Evaluating Research Progress and Updating the Portfolio. Washington, DC: National Academy Press.

NRC (National Research Council). 1999b. Ozone-Forming Potential of Reformulated Gasoline. Washington, DC: National Academy Press.

NRC (National Research Council). 2000a. Toxicological Effects of Methylmercury. Washington, DC: National Academy Press.

NRC (National Research Council). 2000b. Modeling Mobile-Source Emissions. Washington, DC: National Academy Press.

NRC (National Research Council). 2000c. Waste Incineration and Public Health. Washington, DC: National Academy Press.

NRC (National Research Council). 2000d. Ecological Indicators for the Nation. Washington, DC: National Academy Press.

NRC (National Research Council). 2001a. Global Air Quality: An Imperative for Long-Term Observational Strategies. Washington, DC: National Academy Press.

NRC (National Research Council). 2001b. Climate Change Science: An Analysis of Some Key Questions. Washington, DC: National Academy Press.

NRC (National Research Council). 2001c. Evaluating Vehicle Emissions Inspection and Maintenance Programs. Washington, DC: National Academy Press.

NRC (National Research Council). 2001d. Research Priorities for Airborne Particulate Matter. 3. Early Research Progress. Washington, DC: National Academy Press.

NRC (National Research Council). 2002a. Estimating the Public Health Benefits of Proposed Air Pollution Regulations. Washington, DC: The National Academies Press.

NRC (National Research Council). 2002b. The Ongoing Challenge of Managing Carbon Monoxide Pollution in Fairbanks, Alaska, Interim report. Washington, DC: National Academy Press.

NRC (National Research Council). 2002c. Effectiveness and Impact of Corporate Average Fuel Economy (CAFE) Standards. Washington, DC: National Academy Press.

NRC (National Research Council). 2002d. Making the Nation Safer: The Role of Science and Technology in Countering Terrorism. Washington, DC: The National Academies Press.

NRC (National Research Council). 2003a. Novel Approaches to Carbon Management: Separation, Capture, Sequestration, and Conversion to Useful Products—Workshop Report. Washington, DC: The National Academies Press.

NRC (National Research Council). 2003b. Managing Carbon Monoxide in Meteorological and Topographical Problem Areas. Washington, DC: The National Academies Press.

NRC (National Research Council). 2003c. Air Emissions from Animal Feeding Operations: Current Knowledge, Future Needs, Final Report. Washington, DC: The National Academies Press.

NSTC (U.S. National Science and Technology Council). 1997. Interagency Assessment of Oxygenated Fuels. Washington, DC: National Science and Technology Council, Committee on Environment and Natural Resources.

NSTC (U.S. National Science and Technology Council). 1997a. Integrating the Nation's Environmental Monitoring and Research Networks and Programs: A Proposed Framework. Washington, DC: National Science and Technology Council, Committee on Environment and Natural Resources. March 1997.

NTEC (National Tribal Environmental Council). 2002. Comments Submitted on Behalf of the National Tribal Air Association re the Proposed Settlement of Litigation over 8-hour Ozone Designations. Working Documents of the National Tribal Air Association, Dec. 20, 2002. [Online]. Available: http://www.ntec.org/NTAC/ozsettle.html [accessed March 24, 2003].

Nussbaum, B.D. 1992. Phasing down lead in gasoline in the U.S.: Mandates, incentives, trading, and banking. Pp. 25–40 in Climate Change: Designing a Tradeable Permit System. Paris: Organisation for Economic Co-operation and Development.

Oates, W.E., P.R. Portney, and M. McGartland. 1989. The net benefits of incentive-based regulation: A case study of environmental standard setting. Am. Econ. Rev. 79(5):1233–1242.

Olsen, A.R., J. Sedransk, D. Edwards, C.A. Gotway, W. Liggett, S. Rathbun, K.H. Reckhow, and L.J. Young. 1999. Statistical issues for monitoring ecological and natural resources in the United States. Environ. Monit. Assess. 54(1):1–45.

OMB (Office of Management and Budget). 1997. Report to Congress on the Cost and Benefits of Federal Regulations, Sept. 30, 1997. Office of Management and Budget, Office of Information and Regulatory Affairs, the Executive Office of the President. [Online]. Available: http://www.whitehouse.gov/omb/inforeg/rcongress.html [accessed Oct. 31, 2002].

OMB (Office of Management and Budget). 2003a. Informing Regulatory Decisions: 2003 Report to Congress on the Costs and Benefits of Federal Regulations and Unfunded Mandates on State, Local, and Tribal Entities. Office of Information and Regulatory Affairs, Office of Management and Budget, Washington, DC. [Online]. Available: http://www.whitehouse.gov/omb/inforeg/2003_cost-ben_final_rpt.pdf [accessed Nov. 12, 2003].

OMB (Office of Management and Budget). 2003b. Fiscal Year 2004 Budget. [Online]. Office of Management and Budget, Washington, DC. Available: http://w3.access.gpo.gov/usbudget/fy2004/budget.html [accessed Apr. 1, 2003].

Oreskes, N., K. Shrader-Frechette, and K. Belitz. 1994. Verification, validation, and confirmation of numerical models in the earth sciences. Science 263(5147):641–646.

Ostro, B.D., M.J. Lipsett, and R. Das. 1998. Particulate matter and asthma: A quantitative assessment of the current evidence. Appl. Occup. Environ. Hyg. 13(6):453–460.

OTA (Office of Technology Assessment). 1989. Catching Our Breath: Next Steps for Reducing Urban Ozone. U.S. Congress, Office of Technology Assessment, Washington, DC.

Pacala, S.W, E. Bulte, J.A. List, and S.A. Levin. 2003. False alarm over environmental false alarms. Science 301(5637):1187–1188.

Paerl, H.W. 1997. Coastal eutrophication and harmful algal blooms: Importance of atmospheric deposition and groundwater as "new" nitrogen and other nutrient sources. Limnol. Oceanogr. 42(5 Part 2):1154–1165.

Paerl, H.W., J.L. Pinckney, J.M. Fear, and B.L. Peierls. 1998. Ecosystem responses to internal and water-shed organic matter loading: Consequences for hypoxia in the eutrophying Neuse River Estuary, North Carolina, USA. Mar. Ecol. Progr. Ser. 166(17): 17–25.

PAMS. 2002. Enhanced Ozone Monitoring (PAMS). Air Quality Planning and Standards, U.S. Environmental Protection Agency, Washington, DC. [Online]. Available: http://www.epa.gov/air/oaqps/pams/ [accessed Oct. 16, 2002].

Parrish, D.D., M. Trainer, V. Young, P.D. Goldan, W.C. Kuster, B.T. Jobson, F.C. Fehsenfeld, W.A. Lonneman, R.D. Zika, C.T. Farmer, D.D. Reimer, and M.O. Rodgers. 1998. Internal consistency test for evaluation of measurements of anthropogenic hydrocarbons in the troposphere. J. Geophys. Res. 103(D17):22339–22359. [Online]. Available: http://www-personal.engin.umich.edu/~sillman/web-publications/Parrish 98.pdf [accessed Nov. 12, 2003].

Parry, I.W.H. 1995. Pollution taxes and revenue recycling. J. Environ. Econ. Manage. 29(3):S64–S77.

Peierls, B.L., and H.W. Paerl. 1997. Bioavailability of atmospheric organic nitrogen deposition to coastal phytoplankton. Limnol. Oceanogr. 42(8):1819–1823.

Pellizzari, E.D., R.L. Perritt, and C.A. Clayton. 1999. National human exposure assessment survey (NHEXAS): Exploratory survey of exposure among population subgroups in EPA Region V. J. Expo. Anal. Environ. Epidemiol. 9(1):49–55.

Perlin, S.A., D. Wong, and K. Sexton. 2001. Residential proximity to industrial sources of air pollution: Interrelationships among race, poverty and age. J. Air Waste Manage. Assoc. 51(3):406–421.

Pielke, R.A., and M. Uliasz. 1998. Use of meteorological models as input to regional and mesoscale air quality models—limitations and strengths. Atmos. Environ. 32(8):1455–1468.

Pielke Sr., R.A., G. Marland, R.A. Betts, T.N. Chase, J.L. Eastman, J.O. Niles, D.S. Niyogi, and S.W. Running. 2002. The influence of land-use change and landscape dynamics on the climate system: Relevance to climate-change policy beyond the radiative effect of greenhouse gases. Phil. Trans. R. Soc. Lond. A 360:1705–1719. [Online]. Available: http://blue.atmos.colostate.edu/ [accessed Feb. 5, 2003].

Pierson, W.R., W.W. Brachaczek, T.J. Korniski, T.J. Truex, and J.W. Butler. 1980. Artifact formation of sulfate, nitrate, and hydrogen ion on backup filters: Allegheny Mountain experiment. J. Air Poll. Control Assoc. 30:30–34.

Pierson, W.R., A.W. Gertler, and R.L. Bradow. 1990. Comparison of the SCAQS Tunnel Study with other on-road vehicle emission data. J. Air Waste Manage. Assoc. 40(11): 1495–1504.

Pizer, W.A. 2002. Combining price and quantity controls to mitigate global climate change. J. Public Econ. 85(3):409–434.

Placet, M., C.O. Mann, R.O. Gilbert, and M.J. Niefer. 2000. Emissions of ozone precursors from stationary sources: A critical review. Atmos. Environ. 34(12): 2183–2204.

Pope, C.A., M.J. Thun, M.M. Namboodiri, D.W. Dockery, J.S. Evans, F.E. Speizer, and C.W. Heath Jr. 1995. Particulate air pollution as a predictor of mortality in a prospective study of U.S. adults. Am. J. Respir. Crit. Care Med. 151(3 Pt 1):669–674.

Preston, E.M., and D.T. Tingey. 1988. The NCAN program for crop loss assessment. Pp. 45–62 in Assessment of Crop Loss from Air Pollutants: Proceedings of an International Conference, W.W. Heck, O.C. Taylor, and D.T. Tingey, eds. New York: Elsevier Applied Science.

Ransom, M.B., and A.C. Pope III. 1995. External health costs of a steel mill. Contemp. Econ. Policy 13(2):86–97.

Reddy, P.J. 2000. Analysis of the probability of the carbon monoxide exceedance in Denver during the first week of February for the years 2002 through 2013 and two possible levels of oxygenates in automotive fuels based on historical carbon monoxide for 1975 through 1994. Pp. 575–580 in Technical Support Document: Carbon Monoxide

Redesignation Request and Maintenance Plan for the Denver Metropolitan Area. Colorado Department of Public Health and Environment, Air Pollution Control Division, Denver, CO. Jan. 4, 2000. [Online]. Available: http://apcd.state.co.us/documents/techdocs.html. [accessed March 26, 2003].

Reeves, D.W. 1997. The role of soil organic matter in maintaining soil quality in continuous cropping systems. Soil Till. Res. 43(1):131–167.

Regens, J.L., and R.W. Rycroft. 1988. The Acid Rain Controversy. Pittsburgh, PA: University of Pittsburgh Press.

Reynolds, S.D., P.M. Roth, and J.H. Seinfeld. 1973. Mathematical modeling of photochemical air pollution: 1. Formulation of the model. Atmos. Environ. 7:1033–1061.

Riitters, K.H., R.V. O'Neill, and K.B. Jones. 1997. Assessing habitat suitability at multiple scales: A landscape-level approach. Biol. Conserv. 81(1/2):191–202.

Rodan, B.D., D.W. Pennington, N. Eckley, R.S. Boethling. 1999. Screening for persistent organic pollutants to provide a scientific basis for POPs criteria in international negotiations. Environ. Sci. Technol. 33(20):3482–3488. [Online]. Available: http://pubs.acs.org/cgi-bin/gap.cgi/esthag/1999/33/i20/html/es980060t.html [accessed Jan. 13, 2004].

Rodes, C.E. 2001. Indoor Particles—Their Sources, Fate, and Contributions in Personal Exposure. Presentation at NRC Workshop on Research Priorities for Airborne Particulate Matter, Section 2, Nov. 9, 2001. [Online]. Available: http://dels.nas.edu/best/pm/exp_wkshp.html [accessed March 21, 2003].

Roth, P.M. 1999. A qualitative approach to evaluating the anticipated reliability of a photochemical air quality simulation model for a selected application. J. Air Waste Manage. Assoc. 49(9):1050–1059.

Roth, P.M., S.D. Ziman, and J.D. Fine. 1993. Tropospheric ozone. Pp. 39–90 in Keeping Pace with Science and Engineering: Case Studies in Environmental Regulation, M.F. Uman, ed. Washington, DC: National Academy Press.

Russell, A. 1997. Regional photochemical air quality modeling: Model formulations, history, and state of the science. Annu. Rev. Energy Environ. 22:537–588.

Russell, A., and R. Dennis. 2000. NARSTO critical review of photochemical models and modeling. Atmos. Environ. 34(12):2283–2324.

Russell, K.M, J.N. Galloway, S.A. Macko, J.L. Moody, and J.R. Scudlark. 1998. Sources of nitrogen in wet deposition to the Chesapeake Bay region. Atmos. Environ. 32(14/15):2453–2465.

SAE (Society of Automotive Engineers). 2003. Diesel Exhaust Emission Control. SP 1754. Warrendale, PA: Society of Automotive Engineers.

Samet, J.M., Z.L. Zeger, F. Domenici, F. Curriero, I. Coursac, D.W. Dockery, J. Schwartz, and A. Zanobetti. 2000. The National Morbidity, Mortality, and Air Pollution Study, Part 2. Morbidity and Mortality from Air Pollution in the United States, Research Report 94. Cambridge, MA: Health Effects Institute. June 2000 [Online]. Available: http://www.healtheffects.org/Pubs/Samet2.pdf. [accessed Jan. 30, 2003].

SCAQMD (South Coast Air Quality Management District). 1996. Final 1997 Air Quality Management Plan, South Coast Air Quality Management District, Diamond Bar, CA.

SCAQMD (South Coast Air Quality Management District). 1999. Rule 1186. $PM_{10}$ Emissions from Paved and Unpaved Roads, and Livestock Operations. South Coast Air Quality Management District, Diamond Bar, CA. [Online]. Available: http://www.aqmd.gov/rules/html/r1186.html [accessed Oct. 17, 2003].

SCAQMD (South Coast Air Quality Management District). 2000. Multiple Air Toxics Exposure Study in the South Coast Air Basin. MATES-II. South Coast Air Quality Management District, Diamond Bar, CA [Online]. Available: http://www.aqmd.gov/matesiidf/matestoc.htm [accessed Sept. 23, 2002].

SCAQMD (South Coast Air Quality Management District). 2001. Regulation XX, Regional Clean Air Incentives Market (RECLAIM) Rule 2001 (Amended May 11, 2001). South Coast Air Quality Management District, Diamond Bar, CA. [Online]. Available: http://www.aqmd.gov/rules/html/tofc20.html [accessed March 11, 2003].

SCAQMD (South Coast Air Quality Management District). 2003. Preview of the Proposed 2003 Air Quality Management Plan for the South Coast Air Basin. South Coast Air Quality Management District, Diamond Bar, CA. [Online]. Available: http://www. aqmd.gov/aqmp/docs/Preview2003AQMP.pdf [accessed Apr. 4, 2003].

Schecter, A., M. Pavuk, O. Päpke, J.J. Ryan, L. Birnbaum, and R. Rosen. 2003. Polybrominated diphenyl ethers (PBDEs) in U.S. mothers' milk. Environ. Health Perspect. 111(14): 1723–1729.

Scheel, H.E., H. Areskoug, H. Geiss, B. Gomiscek, K. Granby, L. Haszpra, L. Klasinc, D. Kley, T. Laurila, A. Lindskog, M. Roemer, R. Schmitt, P. Simmonds, S. Solberg, and G. Toupance. 1997. On the spatial distribution and seasonal variation of lower-troposphere ozone over Europe. J. Atmos. Chem. 28(1/3):11–28.

Schindler, D.W., K.H. Mills, D.F. Malley, S. Findlay, J.A. Shearer, I.J. Davis, M.A. Turner, G.A. Lindsey, and D.R. Cruikshank. 1985. Long-term ecosystem stress: Effects of years of experimental acidification on a small lake. Science 228(4706):1395–1401.

Schwartz, J., S. Sample, and R. McIlvaine. 1994. Continuous emissions monitors—issues and predictions. J. Air Waste Manage. Assoc. 44(1):16–20.

Seagrave, J.C., J.D. McDonald, A.P. Gigliotti, K.J. Nikula, S.K. Seilkop, M. Gurevich, and J.L. Mauderly. 2002. Mutagenicity and in vivo toxicity of combined particulate and semivolatile organic fractions of gasoline and diesel engine emissions. Toxicol. Sci. 70(2):212–226.

Seaman, N.L. 2003. Future directions of meteorology related to air-quality research. Environ. Int. 29(2):245–252.

Seigneur, C. 2001. Current status of air quality models for particulate matter. J. Air Waste Manage. Assoc. 51(11):1508–1521.

Seigneur, C., B. Pun, P. Pai, J. Louis, P.A. Solomon, C. Emery, R. Morris, M. Zahniser, D. Worsnop, P. Koutrakis, W.H. White, and I. Tombach. 2000. Guidance for the performance evaluation of three-dimensional air quality modeling systems for particulate matter and visibility. J. Air Waste Manage. Assoc. 50(4):588–599.

Seinfeld, J.H., and S.N. Pandis. 1998. Atmospheric Chemistry and Physics: From Air Pollution to Climate Change. New York: Wiley.

Selevan, S.G., D.C. Rice, K.A. Hogan, S.Y. Euling, A. Pfahles-Hutchens, and J. Bethel. 2003. Blood lead concentration and delayed puberty in girls. N. Engl. J. Med. 348(16):1527–1536.

Seltzer, M.D. 2000. Performance testing of a multimetals continuous emissions monitor. J. Air Waste Manage. Assoc. 50(6):1010–1016.

Sexton, K., H. Gong Jr., J.C. Bailar, J.G. Ford, D.R. Gold, and M.J. Utell. 1993. Air pollution health risks: Do class and race matters? Toxicol. Ind. Health 9(5):843–878.

Shorter, J.H., J.B. McManus, C.E. Kolb, E.J. Allwine, B.K. Lamb, B.W. Mosher, R.C. Harriss, U. Partchatka, H. Fischer, G.W. Harris, P.J. Crutzen, and H.J. Karbach. 1996. Methane emission measurements in urban areas in eastern Germany. J. Atmos. Chem. 24(2):121–140.

Simó, R., and C. Pedrós-Alió. 1999. Role of vertical mixing in controlling the oceanic production of dimethyl sulfide. Nature 402(6760):396–399.

Sisler, J.F., and W.C. Malm. 2000. Trends of $PM_{2.5}$ and Reconstructed Visibility from the IMPROVE Network for the Years 1988–1998. Paper No. 442. Presentation at the 93rd Annual Conference Air and Waste Management Association's, June 18–22, 2000, Salt Lake City, UT.

Solomon, B.D., and R. Lee. 2000. Emissions trading systems and environmental justice. Environment 42(8):32–45.

Solomon, G.M., and P.M. Weiss. 2002. Chemical contaminants in breast milk: Time trends and regional variability. Environ. Health Perspect. 110(6):A339–A347.

Sperling, D. 1991. The Future of the Motor Vehicles in OECD Countries. Proceedings, World Motor Conference, Sept. 11–12, 1991, Frankfurt, Germany.

Staehelin, J., J. Thudium, R. Buehler, A. Volz-Thomas, and W. Graber. 1994. Trends in surface ozone concentrations at Arosa (Switzerland). Atmos. Environ. 28(1):75–78.

STAPPA/ALAPCO (State and Territorial Air Pollution Program Administrators/Association of Local Air Pollution Control Officers). 1999. Reducing Greenhouse Gases & Air Pollution, a Menu of Harmonized Options. Executive Summary and Case Studies. State and Territorial Air Pollution Program Administrators/Association of Local Air Pollution Control Officers, Washington, DC. [Online]. Available: http://www.4cleanair.org/comments/execsum.pdf [accessed June 11, 2003].

STAPPA/ALAPCO/EPA (State and Territorial Air Pollution Program Administrators/ Association of Local Air Pollution Control Officers/ US EPA Air Toxics Steering Committee). 2001. Air Toxics Monitoring Newsletter, Oct. 11, 2001. [Online]. Available: http://www.4cleanair.org/ATMN5.pdf [accessed Sept. 24, 2002].

Stavins, R.N. 2002. Lesson from the American experiment with market-based environmental policies. Pp. 173–200 in Market-Based Governance: Supply Side, Demand Side, Upside, and Downside, J.D. Donahue, and J.S. Nye, Jr., eds. Washington, DC: The Brookings Institution. [Online]. Available: http://www.ksg.harvard.edu/cbg/research/stavins.htm [accessed July 23, 2003].

Stedman, D.H., G.A. Bishop, P. Aldrete, and R.S. Slott. 1997. On-road evaluation of an automobile emission test program. Environ. Sci. Technol. 31(3):927–931.

Steinfeld, J., C.E. Kolb, and G. Thomas. 2001. Instrumentation for Environmental Science, Report of a Workshop and Symposium, Massachusetts Institute of Technology, Cambridge, MA.

Steinfeld, J.I, P. Sheehy, and S. Witonsky. 2002. Sampling and Analysis of Components of the Terrestial Atmosphere. Presentation at the 224th ACS National Meeting, Aug. 18–22, 2002, Boston, MA.

Stockwell, W.R., R.R. Artz, J.F. Meagher, R.A. Petersen, K.L. Schere, G.A. Grell, S.E. Peckham, A.F. Stein, R.V. Pierce, J.M. O'Sullivan, and P.Y. Whung. 2002. The scientific basis of NOAA's air quality forecasting program. EM (Dec.):20–27.

Stoddard, J.L., D.S. Jeffries, A. Lukewille, M. Forsius, J. Mannio, and A. Wilander. 2000. Environmental chemistry: Is acidification still an ecological threat? Nature 407(6806): 856–857.

Stoddard, J.L., J.S. Kahl, F.A. Deviney, D.R. DeWalle, C.T. Driscoll, A.T. Herlihy, J.H. Kellogg, P.S. Murdoch, J.R. Webb, and K.E. Webster. 2003. Response of Surface Water Chemistry to the Clean Air Act Amendments of 1990. EPA/620/R-03/001. Office of Research and Development, U.S. Environmental Protection Agency, Research Triangle Park, NC. [Online]. Available: http://www.epa.gov/airmarkets/articles/index.html [accessed March 12, 2003].

Stolte, K.W., J.W. Coulston, M.A. Ambrose, K.R. Ritters, B.L. Conkling, and W.D. Smith. In press. 2000 FHM National Technical Report. General Technical Report. U.S. Department of Agriculture, Forest Service, Southern Research Station.

Suhrbier, J.H., and E.A. Deakin. 1988. Environmental considerations in a 2020 transportation plan: Constraints or opportunities. Pp. 232–259 in A Look Ahead: Year 2020: Proceedings of the Conference on Long-Range Trends and Requirements for the Nation's Highway and Public Transit Systems. Special Report 220. Transportation Research Board, National Research Council, Washington, DC.

Swift, B. 2000. Allowance trading and potential hot spots: Good news from the acid rain program. Environmental Reporter 31(19):954-959. [Online]. Available: http://www.epa.gov/AIRMARKET/articles/so2trading-hotspots_charts.pdf [accessed Oct. 22, 2002].

Swift, B. 2001. Command without control? Why cap-and-trade should replace rate standards for regional pollutants. Environ. Law Rep. News Anal. 31(3):10330-10341.

Taylor Jr., G.E., D.W. Johnson, and C.P. Anderson. 1994. Air pollution and forest ecosystems: A regional to global perspective. Ecol. Appl. 4(4):662-689.

Tietenberg, T.H. 1990. Economic instruments for environmental regulation. Oxford Rev. Econ. Pol. 6(1):17-33.

Tietenberg, T. 2002. The tradable permits approach to protecting the commons: What have we learned? Pp. 197-232 in The Drama of the Commons, E. Ostrom, T. Dietz, N. Dolsak, P.C. Stern, S. Stonich, and E.U. Weber, eds. Washington, DC: National Academy Press.

Tingey, D.T. 1981. The effect of environmental factors in the emission of biogenic hydrocarbons for live oak and slash pine. Pp. 53–80 in Atmospheric Biogenic Hydrocarbons, Vol.1, J.J. Bufalini, and R.R. Arnts, eds. Ann Arbor, MI: Ann Arbor Science.

Tittel, F.K., ed. 1998. Special Issue on "Environmental Trace Gas Detection Using Laser Spectroscopy". Appl. Phys. B- Lasers O 67:273–532.

TNRCC (Texas Natural Resource Conservation Commission). 2000. Revisions to the State Implementation Plan (SIP) for the Control of Ozone Air Pollution: Post-1999 Rate-of-Progress and Attainment Demonstration SIP for the Houston/ Galveston Ozone Nonattainment Area, Inspection/Maintenance SIP for the Houston/Galveston Ozone Nonattainment Area. Texas Natural Resource Conservation Commission, Austin, TX, Dec. 6, 2000. [Online]. Available: http://www.tnrcc.state.tx.us/oprd/rule_lib/adhgsip.pdf [accessed Mar. 20, 2003].

TNRCC (Texas Natural Resource Conservation Commission). 2001. Revisions to the State Implementation Plan (SIP) for the Control of Ozone Air Pollution: Post-1999 Rate-of-Progress and Attainment Demonstrations Follow-up SIP for the Houston/Galveston Ozone Nonattainment Area. Texas Natural Resource Conservation Commission, Austin, TX, Sept. 26, 2001. [Online]. Available: http://www.tnrcc.state.tx.us/oprd/sips/01007sip_ado.pdf [accessed Jan. 21, 2003].

TNRCC (Texas Natural Resource Conservation Commission). 2003. Accelerated Science Evaluation of Ozone Formation in the Houston-Galveston Area: Emission Inventories. Texas Natural Resource Conservation Commission, Austin, TX, Feb. 5, 2003. [Online]. Available: http://www.utexas.edu/research/ceer/texaqsarchive/pdfs/Emission%20Inventoryv3.pdf [accessed Apr. 3, 2003].

TNRCC (Texas Natural Resource Conservation Commission). 2003a. Targeted On-Road Testing (Remote Sensing). Vehicle Emissions Testing in Texas. [Online]. Available: http://www.tnrcc.state.tx.us/air/ms/vim.html#im3 [accessed Apr. 22, 2003].

TRB (Transportation Research Board). 1995. Land use and urban form. Pp. 174–209 in Expanding Metropolitan Highways: Implications for Air Quality and Energy Use. Special Report 245. Washington, DC: National Academy Press.

U.K. Ministry of Health. 1954. Mortality and Morbidity during the London Fog of December 1952. Reports on Public Health and Medical Subjects No. 95. London: HMSO.

USDA-ARS (U.S. Department of Agriculture-Agricultural Research Service). 1998. Effects of Ozone Air Pollution on Plants. Air Quality Program. NC State University, Agricultural Research Service, U.S. Department of Agriculture. [Online]. Available: http://www.ces.ncsu.edu/depts/pp/notes/Ozone/ozone.html [accessed Dec. 8, 2003].

USDA-ARS (U.S. Department of Agriculture-Agricultural Research Service). 2002. Air Quality-Plant Growth and Development Research Unit. Agricultural Research Service, U.S. Department of Agriculture. [Online]. Available: http://www.aqpgdr.saa.ars.usda.gov/ [accessed July 11, 2003].

VIEWS 2003. Visibility Information Exchange Web System. [Online]. Available: http://vista.cira.colostate.edu/views/ [accessed Jan. 30, 2004]

Volz, A., and D. Kley. 1988. Evaluation of the Montessori's series of ozone measurements made in the nineteenth century. Nature 332:240–242.

Wark, K., C.F. Warner, and W.T. Davis. 1998. Air pollution: Its Origin and Control, 3rd Ed. Menlo Park, CA: Addison-Wesley.

Watson, J.G. 2002. Visibility: Science and regulation. J. Air Waste Manage. Assoc. 52(6):628–713.

Watson, J.G., E. Fujita, J.C. Chow, B. Zielinska, L.W. Richards, W. Neff, and D. Dietrich. 1998. Northern Front Range Air Quality Study, Final Report. DRI 6580-685-8750.1F2. Prepared for The Office of the Vice President for Research and Information Technology, Colorado State University, Fort Collins, CO. June 30, 1998.

Watson, J.G., T. Zhu, J.C. Chow, J. Engelbrecht, E.M. Fujita, and W.E. Wilson. 2002. Receptor modeling application framework for particle source apportionment. Chemosphere 49(9):1093–1136.

Wayland, R.A., J.E. White, P.G. Dickerson, and T.S. Dye. 2002. Communicating real-time and forecasted air quality to the public. EM (Dec.):28–36.

Wayne, L.G., and Y. Horie. 1983. Evaluation of CARB's In-Use Vehicle Surveillance Program. CARB Contract No. A2-043-32. Prepared by Pacific Environmental Services, Inc., for California Air Resources Board, Sacramento, CA.

WDNR (Wisconsin Department of Natural Resources). 2000. Attainment Demonstration for Ozone for the Year 2007: The Phase 3 Attainment State Implementation Plan (SIP) for the Eastern Wisconsin Nonattainment Areas, Dec. 22, 2000. [Online]. Available: http://www.dnr.state.wi.us/org/aw/air/hot/1hrsip_p3.htm. [accessed Jan. 21, 2003].

Webb, A. 2001. In China, it's not easy being green: Beijing wants the nation's new light vehicles to meet strict clean-air standards. Enforcement: That's another matter. BW Online. May 29, 2001. Available: http://www.businessweek.com/bwdaily/dnflash/may2001/nf20010529_914.htm [accessed Jan. 15, 2003].

Whitall, D.R., and H.W. Paerl. 2001. Spatiotemporal variability of atmospheric nitrogen deposition to the Neuse River Estuary, North Carolina. J. Environ. Qual. 30(5):1508–1515.

Whitman, C.T. 2001. Administrator's Statement. Environmental Indicators Initiative, U.S. Environmental Protection Agency. [Online]. Available: http://www.epa.gov/indicate/statement.htm [accessed March, 11, 2003].

Wilhelm, M., and B. Ritz. 2003. Residential proximity to traffic and adverse birth outcomes in Los Angeles County, California, 1994–1996. Environ. Health Perspect. 111(2):207–216.

Wilkening, K., L.A. Barrie, and M. Engle. 2000. Atmospheric science: Trans-Pacific air pollution. Science 290(5489):65–66.

Williams, J. 2003. Airborne: Hazardous Air Pollutants Causing Human Health Risks and Ecological Damage Go Unregulated, A Report on EPA's Failure to Regulate Hazardous Air Pollutants. Sierra Club. [Online]. Available: http://www.sierraclub.org/toxics/Report.PDF [accessed June, 4, 2003].

Williams, R.C. 2002. Environmental tax interactions when pollution affects health or productivity. J. Environ. Econ. Manage. 44(2):261–270.

Winer, A.M., F.W. Lurmann, L.A. Coyner, S.D. Colome, and M.P. Poe. 1989. Characteriza-
tion of Air Pollutant Exposures in the South Coast Air Basin: Application of a New
Regional Human Exposure (REHEX) Model. Riverside, CA: Statewide Air Pollution
Research Center, University of California.

Wolff, G.T., A.M. Dunker, S.T. Rao, P.S. Porter, and I.G. Zurbe. 2001. Ozone air quality
over North America. Part I—A review of reported trends. J. Air Waste Manage. Assoc.
51(2):273–282.

WRAP (Western Regional Air Partnership). 2000. Voluntary Emissions Reduction Program
for Major Industrial Sources of Sulfur Dioxide in Nine Western States and a Backstop
Market Trading Program: An Annex to the Report of the Grand Canyon Visibility
Transport Commission. Prepared for U.S. Environmental Protection Agency, Washington, DC. Sept. 29, 2000.

WRAP. 2003. Western Regional Air Partnership. Western Governors' Association and the
National Tribal Environmental Council. [Online]. Available: http://www.wrapair.org/
[accessed Oct. 10, 2003].

Yuhnke, R.E. 1991. The amendments to reform transportation planning in the Clean Air
Act Amendments of 1990. Tulane Environ. Law J. 5(1):239–254.

Zimmerman, R. 1993. Social equity and environmental risk. Risk Anal. 13(6):649–666.

Zuckerman, B., and S.L. Weiner. 1998. Environmental Policymaking. A Workshop on
Scientific Credibility, Risk and Regulation, Sept. 24–25, 1998, Massachusetts Institute
of Technology Center for International Studies.

# Abbreviations

| | |
|---|---|
| ADOM | acid deposition modeling |
| AIRMoN | Atmospheric Integrated Research Monitoring Network |
| AIRS | Aerometric Information Retrieval System (EPA) |
| AQCR | air quality control region |
| AQM | air quality management |
| ARS | Agricultural Research Service (USDA) |
| ASPEN | Assessment System for Population Exposure Nationwide |
| ATS | Allowance Tracking System |
| BACT | best available control technology |
| BART | best available retrofit technology |
| BAT | best available technology |
| Btu | British thermal unit |
| CAA | Clean Air Act (1963) |
| CASAC | Clean Air Scientific Advisory Committee (of EPA) |
| CASTNet | Clean Air Status and Trends Network |
| CCRP | California Comparative Risk Project |
| CDC | Centers for Disease Control and Prevention |
| CEM | continuous emissions monitoring |
| CFCs | chlorofluorocarbons—compounds made up of chlorine, fluorine, and carbon |
| CFR | Code of Federal Regulations |
| CO | carbon monoxide |
| $CO_2$ | carbon dioxide |
| CPSC | Consumer Products Safety Commission |

| CTG | control technique guidelines |
| CTM | chemical transport model |
| 3D | three-dimensional—three in space |
| 4D | four-dimensional—three in space and one in time |
| DALYs | disability-adjusted life-years |
| DDT | dichlorodiphenyltrichloroethane |
| DOE | U.S. Department of Energy |
| ECO | employee commute options |
| EIIP | Emission Inventory Improvement Program |
| EKMA | empirical kinetic modeling approach |
| EPA | U.S. Environmental Protection Agency |
| ERC | emission-reduction credit |
| FACE | free air $CO_2$ enrichment |
| FEM | federal equivalent methods |
| FHWA | U.S. Federal Highway Administration |
| FIA/FHM | Forest Inventory and Analysis and Forest Health Monitoring Program |
| FIP | federal implementation plan |
| FRM | federal reference methods |
| FTIR | Fourier transform infrared spectroscopy |
| FTP | federal test procedure (for vehicle emissions) |
| G/bhp-hr | grams per brake horsepower hour |
| g/gal | grams per gallon |
| g/kWh | grams per kilowatt hour |
| GACT | generally available control technology |
| GAO | U.S. General Accounting Office |
| GC | gas chromatography |
| GPMN | Gaseous Pollutant Monitoring Network (NPS) |
| GPMP | Gaseous Pollutant Monitoring Program (NPS) |
| GPRA | Government Performance and Results Act (1993) |
| GVWR | gross vehicle weight rating (weight of vehicle plus rated cargo capacity) |
| $H_2S$ | hydrogen sulfide |
| HAP | hazardous air pollutant |
| HC | hydrocarbons |
| HCl | hydrogen chloride |
| HDV | heavy-duty vehicle |
| HEI | Health Effects Institute |
| HOV | high-occupancy vehicle |
| hp | horsepower |
| hr | hour |
| HUD | U.S. Department of Housing and Urban Development |
| I/M | vehicle inspection and maintenance |

| IM240 | vehicle emissions test using a dynamometer, which lasts for 240 seconds |
| IMPROVE | Interagency Monitoring of Protected Visual Environments |
| IRIS | Integrated Risk Information System |
| ISTEA | Intermodal Surface Transportation Efficiency Act (1991) |
| IWG | International Working Group on Environmental Justice |
| kW | kilowatts |
| LAER | lowest achievable emissions rate |
| lb | pounds |
| LDT | light-duty truck |
| LDV | light-duty vehicle |
| LEV | low-emissions vehicle |
| LPG | liquified petroleum gas |
| LTER | long-term ecological research |
| LTM | long-term monitoring |
| MACT | maximum achievable control technology |
| MATES | multiple air toxics exposure study |
| MDN | Mercury Deposition Network |
| $mg/m^3$ | milligrams per cubic meter |
| $\mu g/m^3$ | micrograms per cubic meter |
| MOBILE | EPA's computer program to estimate mobile-source emissions |
| MPO | Metropolitan Planning Organization |
| MSA | metropolitan statistical area |
| MTBE | methyl *tert*-butyl ether |
| NAA | nonattainment area |
| NAAQS | National Ambient Air Quality Standards |
| NADP | National Atmospheric Deposition Program |
| NAMS | national air monitoring stations |
| NAPAP | National Acid Precipitation Assessment Program |
| NATA | National Air Toxics Assessment |
| NCLAN | National Crop Loss Assessment Network |
| NCore | National Core Monitoring Network |
| NCS | National Children's Study |
| NEJAC | National Environmental Justice Advisory Council |
| NEP | National Estuaries Program |
| NEPA | National Environmental Policy Act |
| NERRS | National Estuarine Research Reserve System (NOAA) |
| NESHAPs | National Emission Standards for Hazardous Air Pollutants |
| NFS | National Forest System |
| NHANES | National Health and Nutrition Examination Survey |
| NHEXAS | National Human Exposure Assessment Survey |

| | |
|---|---|
| NLEV | national low-emission vehicle |
| NMHC | nonmethane hydrocarbons |
| NO | nitrogen oxide |
| $NO_2$ | nitrogen dioxide |
| $NO_x$ | oxides of nitrogen (NO and $NO_2$) |
| NOAA | U.S. National Oceanic and Atmospheric Administration |
| NONROAD | EPA's computer program to estimate emissions from mobile sources not used on roads (for example, aircraft, trains, farm equipment) |
| NPS | U.S. National Park Service |
| NRC | National Research Council |
| NSPS | New Source Performance Standards |
| NSR | new-source review |
| NTN | National Trends Network |
| $O_2$ | diatomic oxygen |
| $O_3$ | ozone |
| OBD | on-board diagnostics (indicates when vehicles' emission controls are not operating properly) |
| ODP | ozone-depleting potential |
| OH | hydroxyl radical |
| OSHA | U.S. Occupational Safety and Health Administration |
| OSTP | Office of Science and Technology Policy (White House) |
| OTAG | Ozone Transport Assessment Group |
| OTC | Ozone Transport Commission |
| OTR | Ozone Transport Region (the states from Maine to Virginia and Washington, DC) |
| PACE | Pollution Abatement Cost and Expenditures Survey |
| PAMS | photochemical assessment monitoring stations |
| Pb | lead |
| PBDEs | polybrominated diphenyl ethers |
| PCBs | polychlorinated biphenyls |
| PEM | parametric emissions monitoring |
| PHS | U.S. Public Health Service |
| PM | particulate matter |
| $PM_{2.5}$ | particulate matter with aerodynamic equivalent diameters of 2.5 micrometers ($\mu$m) or less |
| $PM_{10}$ | particulate matter with aerodynamic equivalent diameters of 10 $\mu$m or less |
| ppbv | parts per billion by volume |
| ppm | parts per million |
| PSD | prevention of significant deterioration |
| psi | pounds per square inch |
| QA/QC | quality assurance/quality control |

| | |
|---|---|
| RACT | reasonably available control technology |
| RADM | regional acid deposition model |
| RECLAIM | Regional Clean Air Management Program |
| REHEX | regional human exposure model |
| RFG | reformulated gasoline |
| RHC | reactive hydrocarbons |
| ROG | reactive organic gas |
| RVP | Reid vapor pressure (the vapor pressure of a petroleum product at 100°F) |
| SCAQMD | South Coast Air Quality Management District |
| SCR | selective catalytic reduction |
| SFTP | supplemental federal test procedure (for vehicle emissions) |
| SHED | sealed housing evaporative determination |
| SIP | state implementation plan |
| SLAMS | state and local air monitoring stations |
| $SO_2$ | sulfur dioxide |
| SOM | soil organic matter |
| STAPPA-ALAPCO | State and Territorial Air Pollution Program Administrators and Association of Local Air Pollution Control Officials |
| SUV | sport utility vehicle |
| TCM | transportation control measure |
| TCP | transportation control plan |
| TEA-21 | Transportation Equity Act for the 21st Century |
| TIME | temporally integrated monitoring of ecosystems |
| TIP | tribal implementation plan |
| TRI | toxic release inventory |
| UAM | urban airshed model |
| ULEV | ultra-low-emissions vehicle |
| USC | U.S. Code |
| USDA | U.S. Department of Agriculture |
| USGS | U.S. Geological Survey |
| VMT | vehicle miles traveled |
| VOC | volatile organic compound |
| vol | volume |
| WEPCO | Wisconsin Electric Power Company |
| WRAP | Western Regional Air Partnership |
| wt | weight |
| ZEV | zero-emission vehicle |

# Appendix A

# Committee Biosketches

**William Chameides** (Chair) is regents professor of earth and atmospheric studies at the Georgia Institute of Technology. His research interests include atmospheric chemistry; tropospheric gas-phase and aqueous-phase chemistry; air pollution; global chemical cycles; biospheric-atmospheric interaction; and global and regional environmental change. His NRC service includes being the chair of the Committee on Atmospheric Chemistry and the Committee on Ozone-Forming Potential of Reformulated Gasoline, and a member of the Committee on Tropospheric Ozone Formation and Measurement. He is a member of the National Academy of Sciences and a former member of the National Research Council (NRC) Board on Environmental Studies and Toxicology. Dr. Chameides has a B.A. degree from the State University of New York at Binghamton, and M.Ph. and Ph.D. degrees in geology and geophysics from Yale University.

**Daniel Greenbaum** (Vice Chair) is the president and chief executive officer of the Health Effects Institute, an independent research institute funded jointly by government and industry to provide research on the health effects of air pollution. At the Health Effects Institute, Mr. Greenbaum has overseen the development and implementation of a research plan that focuses the Institute's efforts on providing critical research and reanalysis on particulate matter, air toxics, and alternative fuels. In 1999, he served as chair of the U.S. Environmental Protection Agency (EPA) Blue Ribbon Panel on Oxygenates in Gasoline, which made recommendations on how to preserve the air pollution benefits of Reformulated Gasoline while preventing water contamination from methyl tertiary butyl ether (MTBE) and other additives. Prior to joining the Health Effects Institute, he served as commissioner of the Massachusetts Department of Environmental Protection. He currently serves as a member of the NRC Committee on Research Priorities for Airborne Particulate Matter and a member of the NRC Board on Environmental Studies and Toxicology. Mr. Greenbaum earned his Masters of City Planning from the Massachusetts Institute of Technology.

**Carmen Benkovitz** is a scientist at Brookhaven National Laboratory. Her research interests include mathematical modeling of transport and transformation of trace species in the atmosphere, and compilation and analyses of inventories of pollutant emissions to the atmosphere. She has previously served as principal investigator for a multitude of studies including Chemical and Microphysical Aerosol Model, Analysis of Air Pollution and Greenhouse Gases, Compilation and Analyses of Emissions Inventories for the National Oceanic and Atmospheric Administration's (NOAA's) Atmospheric Chemistry Project, and Global Emissions Inventories for Aerosol Research. Dr. Benkovitz received her Ph.D. in atmospheric sciences from New York University.

**Eula Bingham** is a professor of environmental health in the College of Medicine at the University of Cincinnati. Previously, she served as vice president and university dean for graduate studies at the University of Cincinnati and as Assistant Secretary of Labor for the Occupational Safety and Health Administration. She earned a Ph.D. in zoology at the University of Cincinnati. Dr. Bingham has served on several committees of the National Research Council including the Committee to Review the Structure and Performance of the Health Effects Institute and the Committee on Structure of Environmental Research in the United States. Her research interests include toxicology, chemical carcinogenesis, pulmonary defense mechanisms, regulatory toxicology, and occupational and environmental health. Dr. Bingham is a member of the Institute of Medicine.

**Michael Bradley** is president of M.J. Bradley & Associates. He formed MJB&A to provide private industry, nonprofit organizations, and government agencies with advice on air quality policy. Prior to founding MJB&A, Mr. Bradley was executive director of Northeast States for Coordinated Air Use Management (NESCAUM) for 12 years. As executive director, he played a lead role in the Ozone Transport Commission's development of the $NO_x$ budget program. Mr. Bradley also helped to shape the nonattainment, motor vehicle, and acid rain provisions in the 1990 Clean Air Act Amendments. In 1997, he founded the Clean Energy Group, which consists of electric generating companies committed to working with policy makers and other stakeholders to promote effective environmental policy options in the areas of air quality and climate change. He is a member of the EPA's Clean Air Act Advisory Committee. Mr. Bradley earned his M.S. degree in Environmental Management from the University of Washington.

**Richard Burnett** is a senior research scientist with the Healthy Environments and Consumer Safety Branch of Health Canada, where he has been working since 1983 on issues relating to the heath effects of outdoor air pollution. He is also an adjunct professor in the Department of Epidemiology and Community Medicine at the University of Ottawa. He received his Ph.D. from Queen's University in Mathematical Statistics. Dr. Burnett's work has focused on the use of administrative health and environmental information to determine the public health impacts of combustion related pollution using nonlinear random effects models, time series, and spatial analytical techniques.

**Dallas Burtraw** is a senior fellow in the Quality of the Environment Division at Resources for the Future. His research interests include restructuring of the electric utility market, the social costs of environmental pollution and benefit-cost and cost-effectiveness analysis of environmental regulation. Dr. Burtraw has investigated the effects on electric utilities of the emission-permit trading program legislated under the 1990 Amendments to the Clean Air Act. He has also evaluated the benefits of sulfur dioxide emission reductions as related to Title IV. He received his Ph.D. in economics from the University of Michigan.

**Laurence Caretto** is professor of mechanical engineering at California State University, Northridge (CSUN). He received his Ph.D. in engineering from the University of California, Los Angeles. Dr. Caretto served as chair of the Department of Mechanical and Chemical Engineering and Associate Dean of the College of Engineering and Computer Science. He was a partner at Sierra Research, a private firm that consults on federal and state regulations on air quality and emissions standards. Dr. Caretto served as a member and vice chair of the California Air Resources Board. His research interests are in combustion-generated air pollution and computational fluid dynamics.

**Costel Denson** is professor of chemical engineering at the University of Delaware. He received his Ph.D. from the University of Utah. His research has focused on the rheology and processing of polymeric materials and he is a fellow of the Society of Plastics Engineers. Dr. Denson has served as vice provost for Research at the University of Delaware where he was responsible for the administration of all aspects of the research enterprise. He has served as chair of EPA's Board of Scientific Counselors that conducted a management review of EPA's research laboratories and centers and also of its particulate matter research program. Dr. Denson has also served as a member of the National Science Foundation Advisory Committee for Environmental Research and Education. He is a past chair of the Materials Task Force at the Military Engineering Center of Excellence. Dr. Denson is currently a member of the NRC Board on Environmental Studies and Toxicology.

**Charles Driscoll** is a distinguished professor of civil and environmental engineering and university professor of environmental systems engineering at Syracuse University. His research interests include aquatic chemistry, biogeochemistry, soil chemistry, and water quality modeling. He is on the Board of Directors at the Upstate Freshwater Institute and the Hubbord Brock Research Foundation. He was a member of the NRC Panel on Processes of Lake Acidification. Dr. Driscoll received his Ph.D. in environmental engineering from Cornell University.

**Jane Hall** is a professor of economics at the College of Business and Economics, California State University, and co-director of the Institute for Economic and Environmental Studies. Dr. Hall has taught a variety of classes in the fields of economics and environmental studies, and her current research areas include assessing the value of environmental protection, economics of air pollution policy, environmental resource scarcity, and scarcity and conflict. She received her M.S. at the University of California, Berkeley, in agricultural and resource economics and her Ph.D. at the University of California, Berkeley, in energy and resources.

**Philip Hopke** is the Bayard D. Clarkson Distinguished Professor in the Departments of Chemical Engineering at Clarkson University and holds appointments in the Department of Civil and Environmental Engineering. Dr. Hopke received his Ph.D. in physical and environmental chemistry from Princeton University and was a research associate at Massachusetts Institute of Technology. After 4 years as an assistant professor of chemistry at the State University College at Fredonia, NY, he joined the University of Illiniois as a visiting assistant professor of chemistry. He then joined the Institute for Environmental Studies at the University of Illinois at Urbana-Champaign (UIUC) as an assistant professor of chemistry and eventually became professor of environmental chemistry with joint appointments in the Departments of Civil Engineering and Nuclear Engineering. He moved to Clarkson University in 1989. In 1991 he won the Principal Investigator Award in Air Quality Research from the Ontario Ministry of the Environment. Dr. Hopke has served on multiple NRC committees, including the Committee on Advances in Assessing Human Exposure to Airborne Pollutants, the Committee on Risk Assessment of Hazardous Air Pollutants, the Committee on Risk Assessment for Radon in Drinking Water, and the Committee on Research Priorities for Airborne Pollutants.

**Arnold Howitt** is executive director of the Taubman Center for State and Local Government at the Kennedy School of Government, Harvard University, where he is also adjunct lecturer in public policy. His recent research focuses on transportation, environmental regulation, and urban physical development issues. He also is director of a U.S. Justice Department-sponsored research program on domestic preparedness for terrorism. He is currently a member of an Institute of Medicine panel on evaluation of the Metropolitan Medical Response System program. Dr. Howitt has written widely on intergovernmental relations, including a report *Linking Transportation and Air Quality Planning: Implementation of the Transportation Conformity Regulations in 15 Nonattainment Areas*. He received his Ph.D. in political science from Harvard University.

**C. S. Kiang** is dean of the College of Environmental Sciences at Peking University. His research interests include atmospheric chemistry, numerical modeling, environmental science, and phase transition of global changes. Dr. Kiang was the founding director of the Southern Oxidant Studies, founding director of the Atmospheric Sciences program at Georgia Tech, and a member of the NRC Global Climate Change Study Panel in China and the Executive Committee for Global Atmospheric Measurement of Tropospheric Aerosol and Gases. Dr. Kiang earned his Ph.D. in physics at the Georgia Institute of Technology.

**Beverly Law** is research associate professor, Department of Forest Science, at Oregon State University, and science chair of the AmeriFlux network of approximately 45 research sites in the Americas. Her research interests include the influence of climate, age, and management on terrestrial ecosystem processes, ecophysiology, forest-atmosphere interaction, ecosystem process modeling, and remote sensing of vegetation characteristics. Previously, she participated in indicator development and design of EPA's Environmental Monitoring and Assessment Program—Forests (EMAP-Forests, now Forest Health Monitoring Program implemented by the U.S. Forest Service). She is on the editorial boards of the journals *Oecologia* and *Global Change Biology*.

**James Lents** is director of the Environmental Policy and Corporate Affiliates Program at the University of California at Riverside. Dr. Lents joined the University after a 27-year career in managing air quality improvement projects nationwide, including 11 years as executive officer of the South

Coast Air Quality Management District (SCAQMD) in Diamond Bar, California. His experience includes work in defining the emissions inventory development, modeling, and emissions control process for Chattanooga, Tennessee, and Denver, Colorado. His work in Colorado included oversight of the emissions inventory development and modeling and control efforts to evaluate oil production during the oil shale boom of the late 1970s and early 1980s. At the SCAQMD, Dr. Lents oversaw development of the first Air Quality Management Plan ever to be approved by EPA for the Los Angeles area. Dr. Lents' research experience includes work on combustion processes and measurement of air pollutants from jet and rocket engines and coal-fired electrical generation processes. Dr. Lents received his Ph.D. from Universty of Tennessee (Space Institute) in physics.

**Denise Mauzerall** is an assistant professor at the Woodrow Wilson School of Public and International Affairs at Princeton University. Her research examines transboundary air pollution from both the science and policy perspectives. She attempts to use the science of global change to contribute to the formation of farsighted environmental policy. Before coming to Princeton University, Dr. Mauzerall was a post-doc at the National Center for Atmospheric Research where she helped develop and used a global 3-dimensional photochemical model to examine the effect of fossil fuel combustion and biomass burning in Asia on global air pollution. She has also worked for EPA in the Global Change Division of the Office of Air and Radiation. Her current areas of research include a comparison of the relative contribution of different regions of the world to global air pollution and how those contributions will evolve as development progresses; the impact of ozone on agriculture in Asia; and an examination of the current nitric oxide emissions trading program in the United States. She is a contributing author to the 2001 Intergovernmental Panel on Climate Change (IPCC) WGI and WGIII assessments. Dr. Mauzerall received her Ph.D. in atmospheric chemistry from Harvard University.

**Thomas McGarity** is W. James Kronzer Chair in Trial and Appellate Advocacy at the University of Texas School of Law. He was articles editor of the *Texas Law Review*. Thomas McGrarity has studied both administrative law and environmental law. He also teaches torts. He is currently serving as co-reporter for rulemaking on the American Bar Association's restatement project of the Administrative Procedures Act and related statutes. He received his J.D. from the University of Texas.

**Jana Milford** is an associate professor of mechanical engineering at the University of Colorado at Boulder. Her research interests include photochemical air quality modeling, with applications to issues such as the effects on urban air quality of alternative fuels for motor vehicles and the effects on the upper troposphere of subsonic aircraft. Dr. Milford is a former member of the Colorado Air Quality Control Commission, which oversees state regulations for air quality. She has served on review boards for the Transportation Research Board of the National Academy of Sciences and is a consultant to the Science Advisory Board of EPA. Dr. Milford received her Ph.D. in engineering and public policy from Carnegie Mellon University. She is also a J.D. candidate at the University of Colorado School of Law.

**Michael Morris** is director of transportation for the North Central Texas Council of Governments, the metropolitan planning organization for Dallas-Fort Worth. He is responsible for travel demand forecasting and for conforming transportation planning with EPA's requirements. In addition, he administers the congestion mitigation/air quality program in four non-attainment counties. Mr. Morris holds a master's degree in civil engineering from the State University of New York at Buffalo. He is a licensed professional engineer and served on the National Academy Committee reviewing the EPA Mobile Source Emissions Factor (MOBILE) software.

**Spyros Pandis** is Elias Associate Professor of Chemical Engineering and Engineering and Public Policy in Carnegie Mellon University. His research interests include atmospheric chemistry, atmospheric pollution modeling, aerosol science, global change, and environmental policy analysis. He is the author of many articles and a book on these topics. He has served on several EPA, National Science Foundation (NSF), and North American Research Strategy for Tropospheric Ozone (NARSTO) review panels and committees. He is a former member of the NRC committee reviewing the Department of Energy (DOE) Office of Fossil Energy's research plan for fine particulates. Dr. Pandis received his Ph.D. in chemical engineering at the California Institute of Technology.

**P. Barry Ryan** is professor of exposure assessment and environmental chemistry in the Department of Environmental and Occupational Health at the Rollins School of Public Health of Emory University with a joint appointment as professor in the Department of Chemistry. Before joining the faculty at Emory in 1995, Dr. Ryan was associate professor of environmental health at the Harvard School of Public Health. He earned his Ph.D. in computational chemistry from Wesleyan University. Research conducted by Dr. Ryan focuses on multimedia, multipollutant human exposure assessment and nontraditional pathways of exposure.

**Adel Sarofim** is Presidential Professor in the College of Engineering at the University of Utah and senior technical advisor to Reaction Engineering

International in Salt Lake City. His research interests include radiative heat transfer, combustion, furnace design, applied chemical kinetics, and air pollution control and he is noted for his work with energy and the environment. Dr. Sarofim has served on several NRC committees including the Committee on Chemical Engineering Frontiers, the Committee on Chemicals in the Environment, and the Committee on Health Effects of Waste Incineration. He also served on EPA's Science Advisory Board and its Strategic Research Subcommittee. Dr. Sarofim received his Sc.D. in chemical engineering from Massachusetts Institute of Technology.

**Sverre Vedal** is a senior faculty member in the Division of Environmental and Occupational Health Sciences at the National Jewish Medical and Research Center in Denver, Colorado, and professor in the departments of preventive medicine and biometrics and of medicine, at the University of Colorado School of Medicine. His primary research interest is the health effects of environmental air pollution. He is currently a member of the Clean Air Scientific Advisory Committee (CASAC) of EPA and a member of the Review Committee of the Health Effects Institute. Dr. Vedal received his M.D. from the University of Colorado and an M.Sc. in epidemiology from the Harvard School of Public Health.

**Lauren Zeise** is chief of the Reproductive and Cancer Hazard Assessment Section of the California Environmental Protection Agency. Dr. Zeise's research focuses on modeling human interindividual variability in metabolism and risk. She has served on advisory boards of EPA, World Health Organization (WHO), U.S. Congressional Office of Technology Assessment (OTA), National Institute of Environmental Health Sciences (NIEHS), Institute of Medicine (IOM), and NRC committees on risk characterization, comparative toxicology of naturally occurring carcinogens, toxicology, and copper in drinking water. Dr. Zeise is a member of the NRC Board on Environmental Studies and Toxicology. She received her Ph.D. in applied sciences from Harvard University.

# Appendix B

# Statement of Task

Develop scientific and technical recommendations for strengthening the nation's air quality management system with respect to the way it identifies and incorporates important sources of exposure to humans and ecosystems and integrates new understandings of human and ecosystem risks. To this end, the committee will conduct a scientific and technical evaluation of the effectiveness of the major air quality provisions of the Clean Air Act and their implementation by federal, state, and local government agencies.

The committee's review will address scientific and technical aspects of the policies and programs that are intended to manage important air pollutants including, but not limited to, national ambient ("criteria") pollutants and air toxics. It will evaluate scientific and technical aspects of current approaches for health and environmental problem identification, regulatory standards development, air quality management plan development, plan implementation, and progress evaluation. Stratospheric ozone protection and greenhouse gas emissions will not be included in the scope of the study, except in regard to strategies to control emissions from sources in tropospheric air quality control programs.

The committee will address:

Scientific and technical bases for identifying and controlling air quality problems and for understanding the importance of various emissions sources;

Scientific and technical bases of current approaches used to set technology-based standards, emission standards, and ambient air quality standards;

Scientific and technical bases of current approaches for developing and implementing air quality management plans, including procedures for developing emissions inventories, models for evaluating management strategies and relating emissions to air quality; and the State Implementation Plans and other air quality management programs;

Measures of performance used to determine progress toward public health and environmental goals, and the use of these measures to modify management systems as needed;

Potential for new scientific concepts and methods, such as those related to human exposure assessment, intermedia transfer, and source-receptor modeling, to be utilized more effectively in the management of air quality;

Scientific and technical aspects of policies and tools (e.g., emissions trading) for air quality management;

Balance between the need for national consistency and the need for local flexibility in carrying out the major air quality provisions of the Clean Air Act;

The extent to which one can rely on anticipated technological advances for achieving emissions reductions in SIPs and other air quality management plans;

Adequacy of current and future expertise, resources, and infrastructure at federal, state, and local agencies to implement air quality management programs;

The effectiveness of federal research programs to enhance the nation's capacity to manage air pollution.

**Sponsor:** U.S. Environmental Protection Agency

# Appendix C

# 188 Hazardous Air Pollutants (HAPs)[a,b]

Acetaldehyde
Acetamide
Acetonitrile
Acetophenone
2-Acetylaminofluorene
Acrolein
Acrylamide
Acrylic acid
Acrylonitrile
Allyl chloride
4-Aminobiphenyl
Aniline
o-Anisidine
Asbestos
Benzene
Benzidine
Benzotrichloride
Benzyl chloride
Biphenyl
Bis(2-ethylhexyl) phthalate
Bis(chloromethyl) ether
1,1-Bis(4-chlorophenyl)
  ethane (DDE)
Bromoform

1,3-Butadiene
Calcium cyanamide
Captan
Carbaryl
Carbon disulfide
Carbon tetrachloride
Carbonyl sulfide
Catechol
Chloramben
Chlordane
Chlorine
Chloroacetic acid
2-Chloroacetophenone
Chlorobenzene
Chlorobenzilate
Chloroform
Chloromethyl methyl ether
Chloroprene
Cresols/cresylic acid
  o-Cresol
  m-Cresol
  p-Cresol
Cumene

2,4-Dichlorophenoxyacetic acid
  (including salts and esters)
  (2,4-D)
Diazomethane
Dibenzofurans
1,2-Dibromo-3-chloropropane
Dibutylphthalate
1,4-Dichlorobenzene(p)
3,3'-Dichlorobenzidine
Dichloroethyl ether
1,3-Dichloropropene
Dichlorvos
Diethanolamine
Diethyl sulfate
3,3'-Dimethoxy benzidine
*p*-Dimethyl amino azobenzene
N,N-Dimethyl aniline
3,3'-Dimethyl benzidine
Dimethyl carbamoyl chloride
N,N-Dimethyl formamide
1,1-Dimethyl hydrazine
Dimethyl phthalate
Dimethyl sulfate
4,6-Dinitro-*o*-cresol
  (including salts)
2,4-Dinitrophenol
2,4-Dinitrotoluene
1,4-Dioxane
1,2-Diphenylhydrazine
Epichlorohydrin
1,2-Epoxybutane
Ethyl acrylate
Ethyl benzene
Ethyl carbamate
Ethyl chloride
Ethylene dibromide
Ethylene dichloride
Ethylene glycol
Ethylene imine
Ethylene oxide
Ethylene thiourea
Ethylidene dichloride
Formaldehyde
Heptachlor

Hexachlorobenzene
Hexachlorobutadiene
Hexachlorocyclopentadiene
Hexachloroethane
Hexamethylene-1,6-diisocyanate
Hexamethylphosphoramide
Hexane
Hydrazine
Hydrochloric acid
Hydrogen fluoride
Hydroquinone
Isophorone
Lindane (all isomers)
Maleic anhydride
Methanol
Methoxychlor
Methyl bromide
Methyl chloride
Methyl chloroform
Methyl ethyl ketone
Methyl hydrazine
Methyl iodide
Methyl isobutyl ketone
Methyl isocyanate
Methyl methacrylate
Methyl *tert*-butyl ether
4,4'-Methylene-bis(2-chloroaniline)
Methylene chloride
4,4'-Methylene diphenyl
  diisocyanate
4,4'-Methylene dianiline
Naphthalene
Nitrobenzene
4-Nitrobiphenyl
4-Nitrophenol
2-Nitropropane
N-Nitroso-N-methylurea
N-Nitrosodimethylamine
N-Nitrosomorpholine
Parathion
Pentachloronitrobenzene
Pentachlorophenol
Phenol
*p*-Phenylenediamine

Phosgene
Phosphine
Phosphorus
Phthalic anhydride
Polychlorinated biphenyls
1,3-Propane sultone
ß-Propiolactone
Propionaldehyde
Propoxur (Baygon)
Propylene dichloride
Propylene oxide
1,2-Propylenimine
Quinoline
Quinone (*p*-benzoquinone)
Styrene
Styrene oxide
2,3,7,8-Tetrachloro-
dibenzo-*p*-dioxin
1,1,2,2-Tetrachloroethane
Tetrachloroethylene
Titanium tetrachloride
Toluene
2,4-Toluene diamine
2,4-Toluene diisocyanate
*o*-Toluidine
Toxaphene
1,2,4-Trichlorobenzene
1,1,2-Trichloroethane
Trichloroethylene
2,4,5-Trichlorophenol

2,4,6-Trichlorophenol
Triethylamine
Trifluralin
2,2,4-Trimethylpentane
Vinyl acetate
Vinyl bromide
Vinyl chloride
Vinylidene chloride
Xylenes (mixed isomers)
  *o*-Xylenes
  *m*-Xylenes
  *p*-Xylenes
Antimony compounds
Arsenic compounds (inorganic)
Beryllium compounds
Cadmium compounds
Chromium compounds
Cobalt compounds
Coke oven emissions
Cyanide compounds[c]
Glycol ethers[d]
Lead compounds
Manganese compounds
Mercury compounds
Fine mineral fibers[e]
Nickel compounds
Polycyclic organic matter[f]
Radionuclides (including radon)
Selenium compounds

---

Note: For all listings above that contain the word "compounds" and for glycol ethers, the following applies: Unless otherwise specified, these listings are defined as including any unique chemical substance that contains the named chemical (antimony, arsenic, etc.).

[a]The original list established under section 112(b)(1) of the Clean Air Act contained 189 HAPs. Caprolactam was removed June 18, 1996 (see 61 Fed. Reg. 30816 [1996]).

[b]On May 30, 2003, EPA proposed to remove the compound methyl ethyl ketone (MEK) from the HAPs list (see 68 Fed. Reg. 32605 [2003]), and on November 21, 2003, it proposed to remove ethylene glycol monobutyl ether from the list (see 68 Fed. Reg. 65648 [2003]).

[c]X'CN where X = H' or any other group where a formal dissociation may occur. For example, KCN or Ca(CN)$_2$.

[d]The definition of glycol ethers has been modified to exclude surfactant alcohol ethoxylates and their derivatives (SAEDs) (65 Fed. Reg. 47342 [2000]).

[e]Includes mineral fiber emissions from facilities manufacturing or processing glass, rock, or slag fibers (or other mineral derived fibers) of average diameter of 1 micrometer or less.

[f]Includes organic compounds with more than one benzene ring and that have a boiling point greater than or equal to 100°C.

# Appendix D

## Recommendations for Continuous Development and Implementation of Measurements to Determine Status and Trends in Ecosystem Exposure and Condition

### RESEARCH AND DEVELOPMENT RECOMMENDATIONS

1. *Develop a comprehensive understanding of chronic effects of multiple air pollutants on ecosystems at ecologically relevant temporal and spatial scales.*

   This includes feedbacks and interactions among the atmosphere, plants, soil, and animals in terrestrial ecosystems, and interactions of organisms, habitat, and water quality in interconnected aquatic ecosystems. Because different species and different developmental stages of these species respond differently to air pollutants, integrated field studies of air pollutant and climatic effects should cover a range of vegetation types and a range of age classes. Integrated *in situ* experiments should link co-located ambient air pollution measurements with observations of ecosystem response.

Specific Improvements Needed:

- Field observations and experiments on interactive effects of ozone, reactive nitrogen deposition, atmospheric carbon dioxide, and climatic stress (for example, water availability) on responses of different terrestrial ecosystems (including forests at different developmental stages)
- Studies on ecological consequences of depletion of nutrient cations from soil and enrichment of nitrogen and sulfur in soil
- Long-term mechanistic studies to evaluate the consequences for plant physiological processes, of changes in soil chemistry that occur as a result of

369

air pollution, taking into account other factors such as reduced resistance to disease and climate extremes

• Develop methods to combine data from various sampling designs (for example, data assimilation and meta-analysis)

• Technological improvements in ozone measurement for biologically relevant ozone data (for example, appropriate temporal and spatial resolution)

• Modeling of the spatial distribution of dry deposition and ozone concentrations as a function of source, terrain, and meteorology. Mesoscale meteorological models require improved treatments of surface-atmosphere interactions, simulations in complex terrain, and simulation of flows under stagnating conditions often associated with ozone episodes.

• Develop coupled ecological process modeling and air quality modeling for evaluating responses spatially and temporally, and for scenario testing. The atmospheric modeling can provide spatial information on air quality to the ecosystem process models, which provide estimates of response to air pollution.

• Monitoring data are needed for verification of model predictions of exposure, particularly in rural areas.

• Continued studies on bioaccumulation of pollutants in aquatic and terrestrial ecosystems, and terrestrial/aquatic interactions.

• Exposure, transfer, and bioaccumulation of mercury in terrestrial and aquatic ecosystems. Studies on the effects of mercury on soil microbial processes and subsequent effects on plant processes in terrestrial ecosystems, including deposition in forests, terrestrial accumulation, transport to aquatic ecosystems and conversion of inorganic mercury to methyl mercury.

2. *Improve process-based models of ecosystem response to pollutants for regional assessments.*

Development and testing of spatially-explicit ecosystem models need to be accelerated for determining response at scales of air pollutant exposure. In terrestrial ecosystems, field studies should be linked with improved modeling of ecosystem exposure and responses to multiple controlling factors, cumulative stress, and plant community dynamics. This includes interaction of multiple cumulative effects of air pollutants and climatic factors on physiology of receptor plants (e.g., photosynthate allocation) and belowground processes, differential sensitivies of species, and linking shifts in biogeochemical processes to changes in community dynamics and species composition.

3. *Develop and test tools for assessing impacts of pollutants on biological species, populations, and ecosystems.*

As new technologies—such as stable isotope analysis and remote sensing from aircraft and satellites—develop, continued studies are needed to identify biological indicators for detecting responses to pollutants at various levels of biological organization. This understanding is particularly important for terrestrial ecosystems, as relatively little progress has been made in this area compared to that in aquatic ecosystems.

4. *Improve methods for monitoring ambient air quality in ecosystems, and ecosystem response.*
Continue to conduct studies to determine appropriate suites of measurements, sample design, and sampling intensity to detect changes in ecosystem condition in response to pollutants. This is particularly important for terrestrial ecosystems, where little progress has been made. Biogeochemistry, habitat and biodiversity, and the linkage between diversity and productivity, are important factors for which a comprehensive suite of indicators should be developed. Indicators should include intermediate variables (for example, leaf area index and the foliar chemistry used to model productivity) as well as final variables (for example, mortality). Examine the possibility of using critical loads to quantify impacts on terrestrial ecosystems.

5. *Conduct risk assessment research.*
Develop methods for quantifying susceptibility of ecosystems to multiple stressors at multiple scales. See for example Linthurst et al. (2000).

## IMPLEMENTATION RECOMMENDATIONS

There has been a lack of coordination in research on measuring the responses of ecosystems to pollutants, and in implementation of monitoring programs that use the new knowledge gained. Likewise, responsibility for monitoring of ecosystem conditions has been divided among agencies, and offices within agencies, so that a cohesive long-term program on monitoring terrestrial and aquatic ecosystem conditions does not exist. Analysis and reporting of results have been spotty, and very much delayed (for example, U.S. Department of Agriculture Forest Service Forest Health Monitoring), and data and reports are not easily located.

Key elements for implementation of measurements to determine status and trends in air pollution exposure and ecosystem condition are:

1. *The institutional framework for monitoring exposure and ecosystem response.*
True coordination between federal and state agencies is necessary to implement a unified cohesive program for monitoring air quality and ter-

restrial and aquatic ecosystems, including planning, implementation, analysis, and reporting of results.

2. *Transfer of knowledge gained in research and development to monitoring programs.*

There should be a process for moving models and measurements from research and development to implementation in monitoring—requiring coordination between responsible agencies.

### 3. Establish baselines of ecosystem condition.

Baseline ecosystem condition can be identified as the initial condition for establishing trends and detecting changes. Reliable baseline conditions have not been identified and reported for terrestrial, aquatic, and coastal ecosystems across the United States. Reports to date on conditions have not been based on representative surveys of ecosystem condition, nor have they incorporated meaningful biological measures of ecosystem condition.

4. *A comprehensive suite of indicators should be measured consistently in terrestrial and aquatic ecosystems with appropriate coverage across the United States, and baseline conditions of ecosystems should be established in one widely distributed consolidated report.*

An ecosystem perspective is essential in monitoring at all sites, including ecological structure and function, such as soil condition in terrestrial ecosystems and habitat condition in aquatic ecosystems. Ecosystem characteristics that are important process model input variables (for example, foliar and soil carbon and nitrogen, leaf area index, and tree heights), or are important for testing or constraining models (for example, productivity) should be monitored. Soil samples from terrestrial ecosystems and sediment or tissue samples from aquatic systems should be archived for analyses in the future, when new approaches or techniques become available.

5. *Co-locate long-term measurements of air quality, meteorology, and ecosystem responses (e.g., along pollution gradients).*

The density and distribution of air quality monitoring stations in rural, agricultural, and remote forest areas should be increased, aided by a statistical design that will improve spatial and temporal estimates of exposure. Measurement locations should maximize coverage along gradients from urban to remote areas, and encompass a range of topographic and microclimate conditions. For ozone effects on terrestrial ecosystems, sampling should be conducted on plants with multiple years of foliage to determine cumulative effects over several years; on deciduous species, sampling should be conducted at the end of the summer, to allow summer-long cumulative effects to be determined (U.S. Forest Service Forest Inventory and Analysis [FIA] sampling actually occurs all summer) (Campbell et al. 2000). The sampling intensity for ozone injury should be increased, and foliar sampling should be co-located with active ozone monitors.

6. *Probability sample designs for monitoring will ensure coverage within a domain of existing populations.*

For example, sampling can be proportional to the area coverage of different forest types or the number of first order streams (Olsen et al. 1999). The plot design and the survey design need to be sufficient to determine differences between sites.

7. *Conduct intensive ecosystem studies at a subset of representative plots.*

Such studies will increase the understanding of mechanisms of response to multiple factors, including air quality and climate.

8. *Improve National Weather Service meteorological data by adding measurement of incident total solar radiation.*

Incident radiation strongly influences ozone formation. It is a key variable used in environmental analysis, and is critical for spatial modeling of meteorology, ozone distribution, and ecosystem effects.

9. *Release for regional analysis the exact locations of forest survey plots on public lands.*

An amendment to the Food Security Act (2001) prohibits release of exact locations of Forest Inventory and Analysis and Forest Health Monitoring (FIA/FHM) plots, including those on public lands. The release of exact plot locations will allow scientists outside FIA/FHM to investigate air pollution effects on forest ecosystems in a spatial context. Currently, investigators who are collaborating with FIA personnel may make complicated arrangements with some regions to allow FIA personnel to run specific analyses for them, but the practice is inconsistent among regions, it is usually impractical because of the complexity of the analyses, and because there are extremely long delays in data processing (processing often delayed for years or never accomplished).

10. *Expand the EPA Temporally Integrated Monitoring of Ecosystems/Long-Term Monitoring (TIME/LTM) program for monitoring surface waters to include sites from regions sensitive to atmospheric deposition (for example, Southeast, Upper Midwest, West).*

Measurements should be long-term to ensure continuity for trends detection. The TIME/LTM program should be integrated with national networks to quantify atmospheric deposition and watershed information on soil, forest vegetation, and other attributes. Measurements should be linked with biogeochemistry modeling to predict both effects of acidic deposition and response to proposed air quality controls. EPA should also consider expanding TIME/LTM to include monitoring of acid-sensitive biological indicators at a subset of sites. This would allow managers to assess linkages between changes in the acid-base status of surface waters and biological responses to changing chemical conditions.

11. *Improve methods and facilities on regional scales to evaluate the status, trends and response to controls on atmospheric sources of nitrogen (see, for example, ESA [1997b]).*

There is a need for basic estimates of organic nitrogen deposition rates and loadings, and for linking delivered atmospheric loads to sources. For example, the effects of forest management and agricultural practices on delivery of nitrogen to coastal waters should be examined. Assessments of atmospheric deposition profiles are needed for all key estuarine/coastal systems along the Atlantic and Gulf coasts. These profiles should be developed using a consistent set of methodologies or tailored to specific regions using a common base of information.

12. *Expand the existing estuarine monitoring programs (National Estuaries Program and National Estuarine Standards Reserve System) to address existing gaps in knowledge of effects of atmospherically deposited chemicals in the coastal zone.*

Where atmospheric deposition data are lacking in some systems, expansion would have to include wet and dry deposition measurements (requiring significant funding and technical expertise). Monitoring at these sites should be expanded to quantify mass inflow of reactive nitrogen and other pollutants (for example, mercury) that are derived from atmospheric sources. The monitoring program should be coordinated with measurements of estuarine conditions, including nitrogen species, dissolved oxygen, chlorophyll, and sea grass biomass. A second option is to foster the formation of integrated estuarine studies (for example, MODMON, the Neuse River Estuary Modeling and Monitoring program at the University of North Carolina (at MODMON 2001). Using either approach, the monitoring program should be linked to modeling efforts to simulate emissions and atmospheric deposition and the transport and fate of atmospherically derived pollutants within watersheds and estuarine ecosystems.

# Index

Low-emissions vehicle (LEV), 121, 141
Lowest achievable emissions rate (LAER),
94, 178, 180–183
LPG. *See* Liquefied petroleum gas
LRTAP. *See* Long-Range Transboundary
Air Pollution
LTER. *See* Long-term ecological research
LTM. *See* Long-term monitoring

# M

MACT. *See* Maximum achievable control
technology
Main components of an attainment-
demonstration SIP, 96
Major stationary sources, permits and
standards for new or modified, 177–
186
Maleic anhydride, 366
MANE-VU. *See* Mid-Atlantic/Northeast
Visibility Union
Manganese compounds, 367
Manmade total, and mobile-source total,
143
Market-based approaches, using whenever
practical and effective, 18–19, 294
MATES. *See* Multiple air toxics exposure
study
Maximum achievable control technology
(MACT), 56–57, 67, 174, 186–190,
213–214, 272, 298, 304, 310–311
MDN. *See* Mercury Deposition Network
MDPVs. *See* Medium-duty passenger
vehicles
Measurements of air quality, meteorology,
and ecosystem responses, co-locating
long-term, 372–373
Measuring the progress and assessing the
benefits of AQM, 216–267
assessing ecosystem benefits from
improved air quality, 252–261
assessing the economic benefits of air
quality improvements, 261–265
limitations of techniques for tracking
progress in AQM, 266–267
monitoring air quality, 220–241
monitoring pollutant emissions, 216–
220
State of the Environment report as
indicating a new paradigm emerging
at the EPA, 267

strengths of techniques for tracking
progress in AQM, 265–266
Medium- and heavy-duty trucks
average $NO_x$ emissions by vehicle model
years for, 153
average PM23 emissions by vehicle
model years for, 152
Medium-duty passenger vehicles (MDPVs),
134, 141
Mercury compounds, 55, 367, 370
Mercury Deposition Network (MDN), 230–
231, 257
Methane ($CH_4$), 205
Methanol, 78, 366
Methoxychlor, 366
Methyl bromide, 366
Methyl chloride, 366
Methyl chloroform, 366
Methyl ethyl ketone, 55n, 366
Methyl hydrazine, 366
Methyl iodide, 366
Methyl isobutyl ketone, 366
Methyl isocyanate, 366
Methyl methacrylate, 366
Methyl *tert*-butyl ether (MTBE), 366
4,4'-Methylene-bis(2-chloroaniline), 366
Methylene chloride, 366
4,4'-Methylene dianiline, 366
4,4'-Methylene diphenyl diisocyanate, 366
Metropolitan planning organizations
(MPOs), 165–166
Metropolitan statistical area (MSA), 90
Mid-Atlantic/Northeast Visibility Union
(MANE-VU), 123
Midwest Regional Planning Organization
(Midwest RPO), 123
Mineral fibers, fine, 367
MOBILE. *See* U.S. Environmental
Protection Agency (EPA)
Mobile-source emission-control programs,
167–172
conformity, 170–172
controls on motorists' behaviors, 170
high-emitting gasoline vehicles, 167–169
promotion of new technologies using
vehicle emission standards, 167–168
reducing emissions from older and
nonroad diesel engines, 169
regulating the content of gasoline and
diesel fuels, 169–170
Mobile-source emission inventories, 101–103

# O

O$_2$. *See* Diatomic oxygen
O$_3$. *See* Ozone
OAQPS. *See* Office of Air Quality Planning and Standards (of EPA)
OBD. *See* On-board diagnostics
OBDII, 149
Observation-based model for O$_3$, 111–112
ODPs. *See* Ozone-depleting potentials
Off-normal emissions, 195–196
Office of Air Quality Planning and Standards (of EPA) (OAQPS), 50
Office of Environmental Justice (of EPA), 50
Office of Management and Budget (of the White House) (OMB), 39, 51, 262
Office of Pollution Prevention and Toxics (of EPA), 78n, 272n
Office of Research and Development (of EPA) (ORD), 48
Office of Science and Technology Policy (of the White House) (OSTP), 39, 261
Older facilities remaining in operation, and the application of NSR and PSD, 183–184
On-board diagnostics (OBD), 149
    OBDII, 149
One atmosphere approach, for assessing and controlling air pollutants, 278–279
Open-market
    and other forms of trading, 210–212
    and other noncapped forms of trading, 211
Operation of the NSR and PSD requirements, 180–181
ORD. *See* Office of Research and Development (of EPA)
OSHA. *See* U.S. Occupational Safety and Health Administration (OSHA)
OSTP. *See* Office of Science and Technology Policy (of the White House)
OTAG. *See* Ozone Transport Assessment Group
OTC. *See* Ozone Transport Commission
OTR. *See* Ozone Transport Region
Over-reliance on models for O$_3$ SIPs, 113–114
Oxides of nitrogen (NO, NO$_2$, or N$_x$). *See* Nitrogen dioxide; Nitrogen oxides

Ozone-depleting potentials (ODPs), 198
Ozone Monitoring Sites in the United States, 234
Ozone (O$_3$), 4, 24, 28–29, 33, 37, 41–43, 58, 73–77, 90–96, 107–117, 121, 129, 132, 175, 181, 238, 268
    emissions-based models for, 105
    foliar injury to cotton induced by chronic exposure to, 54
    health impact on the human respiratory system, 70
    observation-based model for, 111–112
Ozone Transport Assessment Group (OTAG), 103n, 121–122, 131, 270, 275
Ozone Transport Commission (OTC), 120–121, 131, 137, 214, 255, 275
Ozone Transport Region (the states from Maine to Virginia and Washington, DC) (OTR), 121–122, 203
Ozonesonde sites, in North America, 224

# P

PACE. *See* Pollution Abatement Cost and Expenditures Survey
PAMS. *See* Photochemical assessment monitoring stations
Parametric emissions monitoring (PEM), 192–194
Parathion, 366
Particulate matter (PM), 4, 26, 37–41, 59, 72, 77, 84, 105, 130–132, 151, 169, 217, 226–227, 235–236
    with aerodynamic equivalent diameters of 2.5 micrometers (mm) or less (PM$_{2.5}$), 27, 41, 48, 52, 91, 93, 111, 115, 168, 173, 227, 235, 275
    with aerodynamic equivalent diameters of 10 (mm) or less (PM$_{10}$), 14, 27, 48, 76, 93, 126, 142, 235, 262
Partnership, between technical and regulatory communities, 112
Passenger car exhaust emissions, evolution of California and federal tailpipe standards on, 138–139
Pb. *See* Lead
PBDEs. *See* Polybrominated diphenyl ethers
PCBs. *See* Polychlorinated biphenyls
PEM. *See* Parametric emissions monitoring